Universitext

Springer
Berlin
Heidelberg
New York
Hong Kong
London
Milan
Paris
Tokyo

Michèle Audin

Geometry

Michèle Audin
Université Louis Pasteur
IRMA
7 rue René Descartes
67084 Strasbourg, France
e-mail: maudin@math.u-strasbg.fr

Cataloging-in-Publication Data applied for

Die Deutsche Bibliothek - CIP-Einheitsaufnahme

Audin, Michèle:
Geometry / Michèle Audin. - Berlin ; Heidelberg ; New York ; Barcelona ;
Hong Kong ; London ; Milan ; Paris ; Tokyo : Springer, 2002
(Universitext)
Einheitssacht.: Geometrie <engl.>
ISBN 3-540-43498-4

The original French edition was published in 1998 under the title „Géométrie" by Éditions BELIN, Paris, et Éditions ESPACES 34, Montpellier

ISBN 3-540-43498-4 Springer-Verlag Berlin Heidelberg New York

Mathematics Subject Classification (2000): 51XX, 53XX

This work is subject to copyright. All rights are reserved, whether the whole or part of the material is concerned, specifically the rights of translation, reprinting, reuse of illustrations, recitation, broadcasting, reproduction on microfilm or in any other way, and storage in data banks. Duplication of this publication or parts thereof is permitted only under the provisions of the German Copyright Law of September 9, 1965, in its current version, and permission for use must always be obtained from Springer-Verlag. Violations are liable for prosecution under the German Copyright Law.

Springer-Verlag Berlin Heidelberg New York
a member of BertelsmannSpringer Science+Business Media GmbH

http://www.springer.de

© Springer-Verlag Berlin Heidelberg 2003
Printed in Italy

The use of general descriptive names, registered names, trademarks etc. in this publication does not imply, even in the absence of a specific statement, that such names are exempt from the relevant protective laws and regulations and therefore free for general use.

Cover design: *design & production*, Heidelberg
Typesetting by the author using a TEX macro package
Printed on acid-free paper SPIN 10751784 41/3142db - 5 4 3 2 1 0

Contents

Introduction

> I remember that I tried several times to use a slide rule, and that, several times also, I began modern maths textbooks, saying to myself that if I were going slowly, if I read all the lessons in order, doing the exercises and all, there was no reason why I should stall.
>
> Georges Perec, *in* [**Per78**].

1. This is a book...

This is a book written for students who have been taught a small amount of geometry at secondary school and some linear algebra at university. It comes from several courses I have taught in Strasbourg.

Two directing ideas. The first idea is to give a rigorous exposition, based on the definition of an affine space *via* linear algebra, but not hesitating to be elementary and down-to-earth. I have tried both to explain that linear algebra is useful for elementary geometry (after all, this is where it comes from) and to show "genuine" geometry: triangles, spheres, polyhedra, angles at the circumference, inversions, parabolas...

It is indeed very satisfying for a mathematician to define an affine space as being a set acted on by a vector space (and this is what I do here) but this formal approach, although elegant, must not hide the "phenomenological" aspect of elementary geometry, its own aesthetics: yes, Thales' theorem expresses the fact that a projection is an affine mapping, no, you do not need to orient the plane before defining oriented angles... but this will prevent neither the Euler circle from being tangent to the incircle and excircles, nor the Simson lines from enveloping a three-cusped hypocycloid!

This makes you repeat yourself or, more accurately, go back to look at certain topics in a different light. For instance, plane inversions, considered in a naïve way in Chapter III, make a more abstract comeback in the chapter on projective geometry and in that on quadrics. Similarly, the study of projective conics in Chapter VI comes after that of affine conics... although it would have been simpler—at least for the author!—to deduce everything from the projective treatment.

The second idea is to have an as-open-as-possible text: textbooks are often limited to the program of the course and do not give the impression that mathematics is a science in motion (nor in feast, actually!). Although the program treated here is rather limited, I hope to interest *also* more advanced readers.

Finally, mathematics is a human activity and a large part of the contents of this book belongs to our most classical cultural heritage, since are evoked the rainbow according to Newton, the conic sections according to Apollonius, the difficulty of drawing maps of the Earth, the geometry of Euclid and the parallel axiom, the measure of latitudes and longitudes, the perspective problems of the painters of the Renaissance(1) and the Platonic polyhedra. I have tried to show this in the way of writing the book(2) and in the bibliographical references.

2. How to use this book

Prerequisites. They consist of the basics of linear algebra and quadratic forms(3), a small amount of abstract algebra (groups, subgroups, group actions...)(4) and of topology of $\mathbf{R}^n$ and the definition of a differentiable mapping and—for the last chapter only—the usual various avatars of the implicit function theorem, and for one or two advanced exercises, a drop of complex analysis.

Exercises. All the chapters end with exercises. It goes without saying (?) that you *must* study and solve the exercises. They are of three kinds:

- There are firstly proofs or complements to notions that appear in the main text. These exercises are not difficult and it is necessary to solve them in order to check that you have understood the text. They are a complement to the reading of the main text; they are often referenced there and should be done as you go along reading the book.

(1) The geometry book of Dürer [**Dür95**] was written for art amateurs, not for mathematicians.

(2) The way to write mathematics is also part of the culture. Compare the "eleven properties of the sphere" in [**HCV52**] with the "fourteen ways to describe the rain" of [**Eis41**].

(3) There is a section reminding the readers of the properties of quadratic forms in the chapter on conics and quadrics.

(4) Transformation groups are the essence of geometry. I hope that this ideology is transparent in this text. To avoid hiding this essence, I have chosen *not* to write a section of general nonsense on group actions. The reader can look at [**Per96**], [**Art90**] or [**Ber94**].

– There are also "just-exercises", often quite nice: they contain most of the phenomena (of plane geometry, for instance) evoked above.
– There are also more theoretical exercises. They are not always more difficult to solve but they use more abstract notions (or the same notions, but considered from a more abstract viewpoint). They are especially meant for the more advanced students.

Hints of solutions to many of these exercises are grouped at the end of the book.

About the references. The main reason to have written this book is of course the fact that I was not completely satisfied with other books: there are numerous geometry books, the good ones being often too hard or too big for students (I am thinking especially of [**Art57**], [**Fre73**], [**Ber77**], [**Ber94**]). But there are many good geometry books... and I hope that this one will entice the reader to read, in addition to the three books I have just mentioned [**CG67**], [**Cox69**], [**Sam88**], [**Sid93**] and the more recent [**Sil01**].

To write this book and more precisely the exercises, I have also raided (shamelessly, I must confess) quite a few French secondary school books of the last fifty years, that might not be available to the English-speaking readers but deserve to be mentioned: [**DC51**], [**LH61**], [**LP67**] and [**Sau86**].

3. About the English edition

This is essentially a translation of the French *Géométrie* published in 1998 by Belin and Espaces34. However, I have also corrected some of the errors of the French edition and added a few figures together with better explanations (in general due to discussions with my students in Strasbourg) in a few places, especially in the chapter on quadrics, either in the main text or in the solutions to the exercises.

I must confess that I have had a hard time with the terminology. Although I am almost bilingual in differential or algebraic geometry, I was quite amazed to realize that I did not know a single English word dealing with elementary geometry. I have learnt from [**Cox69**], from (the English translation [**Ber94**] of) [**Ber77**] and from [**Sil01**].

4. Acknowledgements

I wish to thank first all the teachers, colleagues and students, who have contributed, for such a long time, to my love of the mathematics I present in this book.

It was is Daniel Guin who made me write it. Then Nicole Bopp carefully read an early draft of the first three chapters. Both are responsible for the existence of this book. I thank them for this.

A preliminary version was tested by the Strasbourg students[5] during the academic year 1997–98. Many colleagues looked at it and made remarks, suggestions and criticisms, I am thinking mainly of Olivier Debarre, Paul Girault and Vilmos Komornik[6]. The very latest corrections to the English edition were suggested by Ana Cannas da Silva and Mihai Damian. I thank them, together with all those with whom I have had the opportunity to discuss the contents of this book and its style, especially Myriam Audin and Juliette Sabbah[7] for their help with the writing of the exercises about caustics.

Laure Blasco carefully read the preliminary version and criticized in great detail the chapter on quadrics. She has helped me to look for a better balance between the algebraic presentation and the geometric properties. For her remarks and her discrete way of insisting, I thank her.

Pierre Baumann was friendly enough to spend much of his precious time reading this text. He explained to me his disagreements with tenacity and kindness in pleasant discussions. In addition to thousands of typographic and grammatical corrections, innumerable ameliorations are due to him, all converging to more rigor but also to a better appropriateness to the expected audience. For the time he spent, for his humor and his spidery scrawl, I thank him.

I was very pleased that Daniel Perrin read the preliminary version with a lot of care and his sharp and expert eye, he explained me his many disagreements and has (almost always) convinced me that I was wrong. This book owes to him better presentation of the relation linear algebra/geometry, a few arrows, a great principle, numerous insertions of "we have", and a lot of (minor or not) corrections together with several statements and exercises (and probably even an original result, in Exercise V.38). Is it necessary to add that I am grateful to him?

Finally, I am grateful to all the students who have suffered the lectures this book comes from and all those who have worked hard because of the errors and clumsiness of the preliminary version and even of the French edition. I cannot name them all, but among them, I want to mention especially Nadine Baldensperger, Régine Barthelmé, Martine Bourst, Sophie Gérardy, Catherine Goetz, Mathieu Hibou, Étienne Mann, Nicolas Meyer, Myriam

---

(5) To be quite honest, I should say that I have used these students as guinea pigs.

(6) I was also very pleased to include his elegant short proof of the Erdős–Mordell theorem (Exercise III.25) in this edition.

(7) Who has also drawn some of the pictures.

Oyono-Oyono, Magali Pointeaux, Sandrine Zitt and all those who have asked questions, raised criticisms and even made suggestions that were very useful, but, more importantly, have given some sense and life to the final writing of this version.

◊

For this
book, I have
used the LaTeX 2_ε
packages of the Société
mathématique de France. I cannot thank myself for writing, translat-
ing and typing this text, or for solving most of
the three hundred and fifty-five exercises and
"drawing" most of the one hundred and
seventy-two pictures it contains,
but I can thank Claude
Sabbah for his sin-
gular, stylistic,
etc. help.

◊

Chapter I

Affine Geometry

An affine space is a set of points; it contains lines, etc. and affine geometry[1] deals, for instance, with the relations between these points and these lines (collinear points, parallel or concurrent lines...). To define these objects and describe their relations, one can:

- Either state a list of axioms, describing incidence properties, like "through two points passes a unique line". This is the way followed by Euclid (and more recently by Hilbert). Even if the process and *a fortiori* the axioms themselves are not explicitly stated, this is the way used in secondary schools.
- Or decide that the essential thing is that two points define a *vector* and define everything starting from linear algebra, namely from the axioms defining the vector spaces.

I have chosen here to use the *second* method, because it is more abstract and neater, of course, but also, mainly, because I think that it is time, at this level, to prove to students that the linear algebra they were taught is "useful" for something!

1. Affine spaces

Definition 1.1. A set $\mathcal{E}$ is endowed with the structure of an *affine space* by the data of a vector space[2] E and a mapping Θ that associates a vector of

[1] "Pure" affine geometry, in the sense that there are no distances, angles, perpendiculars, these belonging to Euclidean geometry, which is the subject of the following chapters.

[2] This is a vector space over a commutative field $\mathbf{K}$ of characteristic 0 that it is not useful to put explicitly in the definition. The readers can imagine that this is $\mathbf{R}$ or $\mathbf{C}$.

E with any ordered pair of points in $\mathcal{E}$,

$$\begin{aligned} \mathcal{E} \times \mathcal{E} &\longrightarrow E \\ (A, B) &\longmapsto \overrightarrow{AB} \end{aligned}$$

(Figure 1) such that:

- for any point A of $\mathcal{E}$, the partial map $\Theta_A : B \mapsto \overrightarrow{AB}$ is a bijection from $\mathcal{E}$ to E;
- for all points A, B and C in $\mathcal{E}$, we have $\overrightarrow{AB} = \overrightarrow{AC} + \overrightarrow{CB}$ (Chasles' relation, Figure 2).

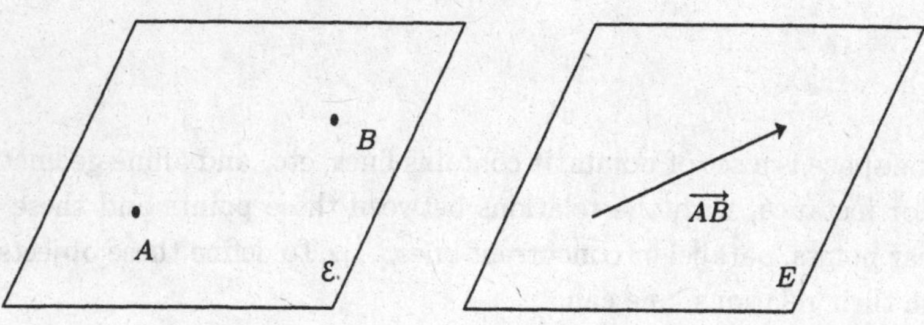

Fig. 1

The vector space E is the *direction* of $\mathcal{E}$, or its *underlying vector space*, the elements of $\mathcal{E}$ are called *points* and the dimension of the vector space E is called the *dimension* of $\mathcal{E}$.

Examples 1.2

(1) With this definition, the empty set is an affine space (directed by any vector space) of which it is wise to agree that it has no dimension.

(2) Any vector space has the natural structure[(3)] of an affine space: the mapping $\Theta : E \times E \to E$ is simply the mapping that associates with the ordered pair (u, v) the vector $v - u$.

(3) If $\mathcal{E}_1$ and $\mathcal{E}_2$ are two affine spaces directed respectively by E_1 and E_2, the Cartesian product $\mathcal{E}_1 \times \mathcal{E}_2$ is an affine space directed by $E_1 \times E_2$: the mapping

$$\Theta : (\mathcal{E}_1 \times \mathcal{E}_2) \times (\mathcal{E}_1 \times \mathcal{E}_2) \to E_1 \times E_2$$

is the one that associates, with the ordered pair $((A_1, A_2), (B_1, B_2))$, the ordered pair of vectors $(\overrightarrow{A_1B_1}, \overrightarrow{A_2B_2})$.

Properties. Chasles' relation gives directly $\overrightarrow{AA} = 0$ and $\overrightarrow{AB} = -\overrightarrow{BA}$.

---

(3) It is *natural* because it is defined by the bare vector space structure (with no additional choice). It would be more exact, but less natural (!) to say that it is "canonical".

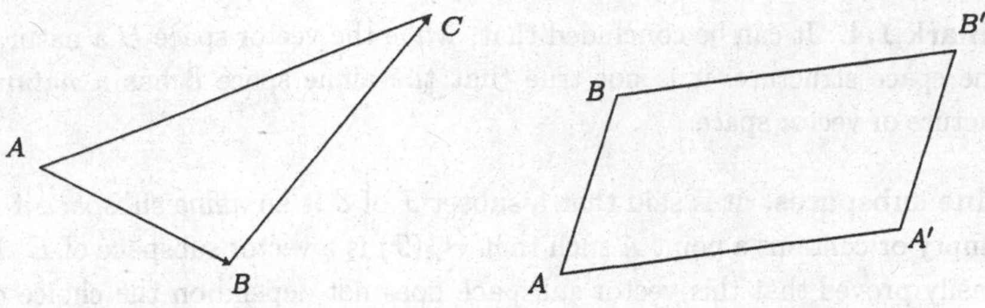

Fig. 2. Chasles' relation **Fig. 3.** The parallelogram rule

The parallelogram rule. This says that the two equalities $\overrightarrow{AB} = \overrightarrow{A'B'}$ and $\overrightarrow{AA'} = \overrightarrow{BB'}$ are equivalent. It is proved using Chasles' relation:

$$\overrightarrow{AB} = \overrightarrow{AA'} + \overrightarrow{A'B'} + \overrightarrow{B'B},$$

which is also written

$$\overrightarrow{AB} - \overrightarrow{A'B'} = \overrightarrow{AA'} - \overrightarrow{BB'}.$$

When one of the relations is satisfied, it is said that $AA'B'B$ is a *parallelogram* (Figure 3).

Remark 1.3. If A is a point of the affine space $\mathcal{E}$ and if u is a vector of the vector space E underlying it, the unique point B of $\mathcal{E}$ such that $\overrightarrow{AB} = u$ is sometimes denoted

$$B = A + u.$$

This notation is consistent since we have

$$(A + u) + v = A + (u + v)$$

(this is a translation of Chasles' relation). A justification will be found in Exercise I.47.

Vectorializations of an affine space. Once a point A has been chosen in an affine space $\mathcal{E}$, it is possible to give $\mathcal{E}$ the structure of a vector space. This vector space is denoted $\mathcal{E}_A$. The mapping

$$\begin{aligned} \Theta_A : \mathcal{E} &\longrightarrow E \\ M &\longmapsto \overrightarrow{AM} \end{aligned}$$

is a bijection, and it allows us to carry the vector space structure of E to $\mathcal{E}$: it is said that $M + N = Q$ if $\overrightarrow{AM} + \overrightarrow{AN} = \overrightarrow{AQ}$. Notice that the vector space structure defined this way depends very heavily on the point A, which becomes the zero vector of the vector space $\mathcal{E}_A$ due to the relation $\overrightarrow{AA} = 0$.

Remark 1.4. It can be concluded that, while the vector space E a natural affine space structure, it is not true that the affine space $\mathcal{E}$ has a *natural* structure of vector space.

Affine subspaces. It is said that a subset $\mathcal{F}$ of $\mathcal{E}$ is an *affine subspace* if it is empty or contains a point A such that $\Theta_A(\mathcal{F})$ is a vector subspace of E. It is easily proved that this vector subspace does not depend on the choice of the point A. More precisely:

Proposition 1.5. *Let $\mathcal{F}$ be an affine subspace of $\mathcal{E}$. There exists a vector subspace F of E such that, for any point B of $\mathcal{F}$, $\Theta_B(\mathcal{F}) = F$. The subspace $\mathcal{F}$ is an affine space directed by F.*

The proof is an exercise (Exercise I.2). □

Remark 1.6. If M and N are points of $\mathcal{F}$, the vector $\overrightarrow{MN}$ belongs to F.

Conversely, we have, almost by definition:

Proposition 1.7. *Let F be a vector subspace of E and let A be a point of $\mathcal{E}$. There exists a unique affine subspace directed by F and passing through A.*

Proof. If $\mathcal{F}$ is an affine subspace through A directed by F, then $\Theta_A(\mathcal{F}) = F$ and

$$\mathcal{F} = \left\{ M \in \mathcal{E} \mid \overrightarrow{AM} \in F \right\}.$$

Conversely, this equality defines an affine subspace directed by F and passing through A. □

Examples 1.8

(1) An affine space of dimension 0 consists of a unique point (why?). All the points of an affine space $\mathcal{E}$ are affine subspaces. The subspaces of dimension 1 (*resp.* 2) are called *lines* (*resp. planes*).

(2) Let E and F be two vector spaces and let $f : E \to F$ be a linear mapping. For any v in the image of f in F, the inverse image $f^{-1}(v)$ is an affine subspace of E (here we consider, of course, that E is endowed with its natural affine structure) directed by $\operatorname{Ker} f$ (Figure 4).

Proof. Let $v \in \operatorname{Im} f$. We want to prove that, given u in $\mathcal{F} = f^{-1}(v)$, we have

$$\Theta_u(f^{-1}(v)) = \operatorname{Ker} f.$$

But $\Theta_u(x) = x - u$ by definition. If $y \in \operatorname{Ker} f$, for $x = y + u$, $f(x) = f(u) = v$, thus $x \in \mathcal{F}$ and we have indeed $y = x - u = \Theta_u(x)$ for some x in $\mathcal{F}$. Hence $\operatorname{Ker} f \subset \Theta_u(\mathcal{F})$.

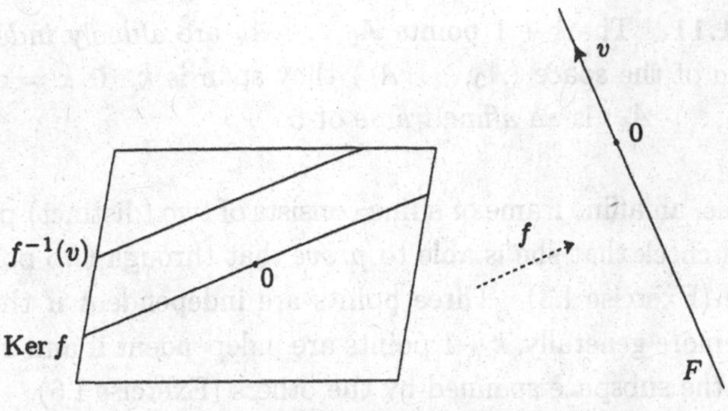

Fig. 4

Conversely, if $y \in \Theta_u(\mathcal{F})$, $y = x - u$ for some x of $\mathcal{F}$ and thus $f(y) = 0$. □

For instance, the set of solutions of a linear system, if nonempty, is an affine subspace directed by the set of solutions of the associated homogeneous system. The equation $\sum_{i=1}^{n} a_i x_i = b$ defines an affine subspace of the vector space $\mathbf{R}^n$ (or $\mathbf{C}^n$, or $\mathbf{K}^n$).

(3) More generally, the affine subspaces of a vector space E are the subspaces of the form $F + u_0$, where F is a vector subspace and u_0 is a vector of E. The vector subspaces are thus the affine subspaces that contain 0.

Intersection of affine subspaces, subspace spanned by a subset of $\mathcal{E}$

Proposition 1.9. *Any intersection of affine subspaces is an affine subspace.*

Proof. Let $(\mathcal{F}_i)_{i\in I}$ be a family of affine subspaces of $\mathcal{E}$. Let $\mathcal{F}$ be their intersection. If it is empty, this is an affine subspace. Otherwise, choose a point A in it. Any $\Theta_A(\mathcal{F}_i)$ is a vector subspace F_i of the direction E of $\mathcal{E}$.

Let F be the intersection of the subspaces F_i in E. This is a vector subspace (is this clear?) and $\mathcal{F}$ is the affine subspace passing through A and directed by F: a point M of $\mathcal{E}$ is in $\mathcal{F}$ if and only if it belongs to each $\mathcal{F}_i$, namely if and only if $\overrightarrow{AM}$ is in all the F_i's, hence in F. □

Proposition 1.10. *Let S be a subset of $\mathcal{E}$. The intersection of all the affine subspaces of $\mathcal{E}$ containing S is the smallest affine subspace containing S.* □

This subspace is the *subspace spanned by* (or *generated by*) S. It is denoted $\langle S\rangle$. For instance, if $S = \{A_0, \ldots, A_k\}$ is a finite set, $\langle A_0, \ldots, A_k\rangle$ is the affine subspace containing A_0 and directed by the vector space spanned by the vectors $\overrightarrow{A_0A_1}, \ldots, \overrightarrow{A_0A_k}$. In particular, its dimension is at most k.

Definition 1.11. The $k+1$ points $A_0, \ldots, A_k$ are *affinely independent* if the dimension of the space $\langle A_0, \ldots, A_k \rangle$ they span is k. If $k = \dim \mathcal{E}$, it is said that $(A_0, \ldots, A_k)$ is an *affine frame* of $\mathcal{E}$.

For instance, an affine frame of a line consists of two (distinct) points. The reader should check that she is able to *prove* that through two points passes a unique line (Exercise I.3). Three points are independent if they are not collinear and more generally, $k+1$ points are independent if and only if none of them is in the subspace spanned by the others (Exercise I.6).

Remark 1.12. An affine frame $(A_0, \ldots, A_k)$ of a space $\mathcal{E}$ can be considered as the datum of an *origin* A_0 and a basis $(\overrightarrow{A_0A_1}, \ldots, \overrightarrow{A_0A_k})$ of its direction. This allows us to give to any point M of $\mathcal{E}$ *coordinates*, the components of the vector $\overrightarrow{A_0M}$ in the base under consideration. See § 6.

Notation. The symbol $\langle A, B \rangle$ denotes, if A and B are distinct, the line through A and B. It will also be denoted AB, of course. Let us take the opportunity to give a notation for the segments, in the case of real affine spaces, of course(4). If A and B are two points, the set of points M of the line AB such that $\overrightarrow{AM} = \lambda \overrightarrow{AB}$ with $0 \leqslant \lambda \leqslant 1$, in natural language the segment AB, will be denoted, in case there is a risk of ambiguity, $[AB]$. I will always write "the line AB", "the segment AB"...

Relative position of two affine subspaces, parallelism

Definition 1.13. It is said that two affine subspaces $\mathcal{F}$ and $\mathcal{G}$ of $\mathcal{E}$ are *parallel* (notation $\mathcal{F}//\mathcal{G}$, which is read "$\mathcal{F}$ is parallel to $\mathcal{G}$") if they have the same direction.

Remark 1.14. With this definition, two subspaces can be disjoint (that is, $\mathcal{F} \cap \mathcal{G} = \varnothing$) without being parallel. For instance, a line is never parallel to a plane (Figure 5). On the contrary, in a plane, two lines that do not intersect are parallel. Some authors use *weakly parallel* to describe two subspaces $\mathcal{F}$ and $\mathcal{G}$ whose directions satisfy $F \subset G$ (as in Figure 5). I do not like this terminology very much, mainly because the "weak parallelism" is not an equivalence relation (is this clear?).

Example 1.15. If $f : E \to F$ is a linear mapping, all the subspaces $f^{-1}(v)$ (for v in the image of f) are parallel since they are all directed by $\operatorname{Ker} f$.

Proposition 1.16. *If $\mathcal{F}$ is parallel to $\mathcal{G}$, then $\mathcal{F}$ and $\mathcal{G}$ are equal or disjoint.*

---

(4) Why, by the way?

Proof. Assume that $\mathcal{F} \cap \mathcal{G}$ is not empty. Let A be a point in the intersection. The direction F of $\mathcal{F}$ and the point A define a *unique* affine subspace, according to Proposition 1.7. Thus, if $\mathcal{F} \cap \mathcal{G}$ is not empty, we have the equality $\mathcal{F} = \mathcal{G}$. □

Notice that the "parallel axiom[5]" is true in the affine spaces (Figure 6):

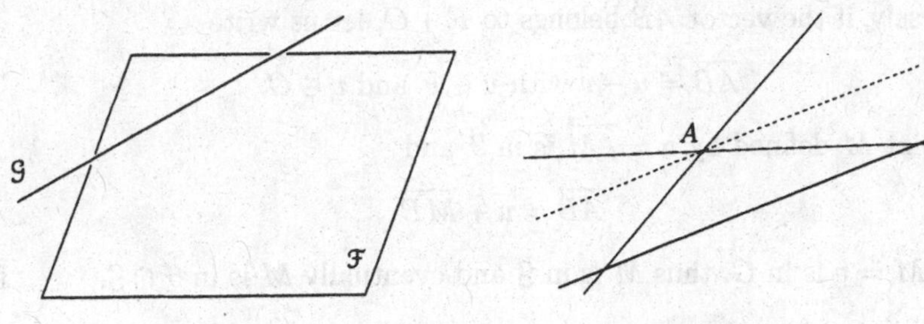

Fig. 5 **Fig. 6.** The parallel axiom

Proposition 1.17 ("Parallel axiom"). *Through any point of an affine space passes a unique line parallel to a given line.*

Proof. The point A and the direction D of the line $\mathcal{D}$ determine the expected parallel:

$$\mathcal{D}' = \left\{ M \in \mathcal{E} \mid \overrightarrow{AM} \in D \right\}.$$

□

Proposition 1.18. *Let $\mathcal{F}$ and $\mathcal{G}$ be two affine subspaces of an affine space $\mathcal{E}$, directed respectively by F and G. Assuming that F and G generate E (in symbols, $F + G = E$), any subspace parallel to $\mathcal{G}$ intersects $\mathcal{F}$ somewhere.*

The proof is based on the following lemma.

Lemma 1.19. *Let $\mathcal{F}$ and $\mathcal{G}$ be two affine subspaces of an affine space $\mathcal{E}$, directed respectively by F and G. Let A be a point of $\mathcal{F}$, B a point of $\mathcal{G}$. The subspace $\mathcal{F} \cap \mathcal{G}$ is not empty if and only if the vector $\overrightarrow{AB}$ belongs to $F + G$.*

Proof of the proposition. Let A be a point of $\mathcal{F}$ and B be a point of $\mathcal{E}$. We want to prove that the affine subspace $\mathcal{G}'$ passing through B and parallel to $\mathcal{G}$ intersects $\mathcal{F}$. We write

$$\overrightarrow{AB} \in E = F + G$$

and apply the lemma to $\mathcal{F}$ and $\mathcal{G}'$ to conclude. □

---

(5) It is contained in the definition of parallelism and thus, eventually, in the definition of a vector space. See also Corollary III-1.15, Corollary IV-3.2 and Exercise V.50.

Proof of the lemma. If the intersection $\mathcal{F} \cap \mathcal{G}$ is not empty, we can choose there a point M. We then have

$$\overrightarrow{AM} = u \in F, \quad \overrightarrow{BM} = v \in G,$$

hence

$$\overrightarrow{AB} = \overrightarrow{AM} - \overrightarrow{BM} \in F + G.$$

Conversely, if the vector $\overrightarrow{AB}$ belongs to $F + G$, let us write

$$\overrightarrow{AB} = u - v \text{ with } u \in F \text{ and } v \in G.$$

The point M defined by $u = \overrightarrow{AM}$ is in $\mathcal{F}$ and

$$\overrightarrow{AB} = u + \overrightarrow{MB}$$

thus $\overrightarrow{BM} = v$ is in G, thus M is in $\mathcal{G}$ and eventually M is in $\mathcal{F} \cap \mathcal{G}$. □

2. Affine mappings

Affine mappings are to affine geometry what linear mappings are to linear algebra.

Definition 2.1. Let $\mathcal{E}$ and $\mathcal{F}$ be two affine spaces directed respectively by E and F. A mapping $\varphi : \mathcal{E} \to \mathcal{F}$ is said to be *affine* if there exists a point O in $\mathcal{E}$ and a linear mapping $f : E \to F$ such that

$$\forall M \in \mathcal{E}, \quad f(\overrightarrow{OM}) = \overrightarrow{\varphi(O)\varphi(M)}.$$

Remark 2.2. The linear mapping f does not depend on the choice of the point O. Indeed, if O' is another point, we have

$$\begin{aligned}\overrightarrow{\varphi(O')\varphi(M)} &= \overrightarrow{\varphi(O')\varphi(O)} + \overrightarrow{\varphi(O)\varphi(M)}\\ &= -\overrightarrow{\varphi(O)\varphi(O')} + \overrightarrow{\varphi(O)\varphi(M)}\\ &= -f(\overrightarrow{OO'}) + f(\overrightarrow{OM})\\ &= f(\overrightarrow{OM} - \overrightarrow{OO'}) \text{ since } f \text{ is linear}\\ &= f(\overrightarrow{O'M}).\end{aligned}$$

□

Notation. As the linear mapping f depends only on φ, we have a mapping from the set of affine mappings to that of linear mappings. I will denote by $\overrightarrow{\varphi}$ the image of φ by this mapping (hence, here $f = \overrightarrow{\varphi}$).

What we have just proved can be written, in this notation:

$$\overrightarrow{\varphi(A)\varphi(B)} = \overrightarrow{\varphi}(\overrightarrow{AB}) \text{ for all points } A, B \text{ of } \mathcal{E}.$$

I am going to use affine mappings and linear mappings. To help the reader, I use Latin letters f, g *etc.* for linear mappings and Greek letters φ, ψ *etc.* for affine mappings. I will of course specify which mappings are used in each case.

Remark 2.3. Let φ be a mapping from the affine space $\mathcal{E}$ to the affine space $\mathcal{F}$. Let O be a point of $\mathcal{E}$. Let us vectorialize $\mathcal{E}$ at O and $\mathcal{F}$ at $\varphi(O)$. We have linear isomorphisms

$$(\Theta_O)^{-1} : E \longrightarrow \mathcal{E}_O \quad \text{and} \quad \Theta_{\varphi(O)} : \mathcal{F}_{\varphi(O)} \longrightarrow F.$$

Consider the composed mapping $\overrightarrow{\varphi}$:

$$\begin{array}{ccccccc} E & \longrightarrow & \mathcal{E}_O & \xrightarrow{\varphi} & \mathcal{F}_{\varphi(O)} & \longrightarrow & F \\ u & \longmapsto & M & \longmapsto & \varphi(M) & \longmapsto & v \end{array}$$

where M and v are defined by $\overrightarrow{OM} = u$ and $v = \overrightarrow{\varphi(O)\varphi(M)}$. The mapping $\overrightarrow{\varphi}$ satisfies

$$\overrightarrow{\varphi}(\overrightarrow{OM}) = \overrightarrow{\varphi(O)\varphi(M)}.$$

To say that the mapping φ is affine is to say that $\overrightarrow{\varphi}$ is linear, and this is equivalent to saying that φ is itself *linear* as a mapping from the vector space $\mathcal{E}_O$ to the vector space $\mathcal{F}_{\varphi(O)}$.

Examples 2.4

(1) The "constant mapping" that maps $\mathcal{E}$ to a point is affine; the associated linear mapping is the zero mapping.

(2) If $\mathcal{E} = \mathcal{F} = \mathbf{R}$, the affine mappings are the mappings of the form $x \mapsto ax + b$ (the associated linear mapping is $x \mapsto ax$).

(3) More generally, if E and F are two vector spaces endowed with their natural affine structures, a mapping

$$\varphi : E \longrightarrow F$$

is affine if and only if there exists a vector v_0 in F and a linear mapping

$$f : E \longrightarrow F$$

such that $\varphi(u) = f(u) + v_0$ for all u in E.

Proof. We choose $O = 0$, the relation

$$\overrightarrow{\varphi}(\overrightarrow{OM}) = \overrightarrow{\varphi(O)\varphi(M)}$$

is written, by definition of the affine structure of the vector space F,

$$\overrightarrow{\varphi}(u) = \varphi(u) - \varphi(0),$$

this being the expected relation, with $v_0 = \varphi(0)$ and $f = \overrightarrow{\varphi}$. □

The linear mappings from E to F are thus the affine mappings that maps 0 to 0.

(4) Assume that $\mathcal{E} = \mathcal{F}$. The affine mappings whose associated linear mapping is Id_E are the mappings

$$\varphi : \mathcal{E} \longrightarrow \mathcal{E}$$

such that $\overrightarrow{\varphi(A)\varphi(B)} = \overrightarrow{AB}$ for all A and B in $\mathcal{E}$. The parallelogram rule then gives $\overrightarrow{A\varphi(A)} = \overrightarrow{B\varphi(B)}$ for all A and B. In other words, the vector $\overrightarrow{M\varphi(M)}$ is a constant vector u. It is said that φ is the *translation* of vector u. It is denoted by t_u (Figure 7).

(5) Let O be a point, λ be a scalar and φ be the mapping defined by $\overrightarrow{O\varphi(M)} = \lambda\overrightarrow{OM}$. This is an affine mapping. The associated linear mapping is the (linear) dilatation of ratio λ. The point O is fixed, φ is called the *(central) dilatation* of center O and ratio λ and it is denoted by $h(O, \lambda)$ (Figure 8).

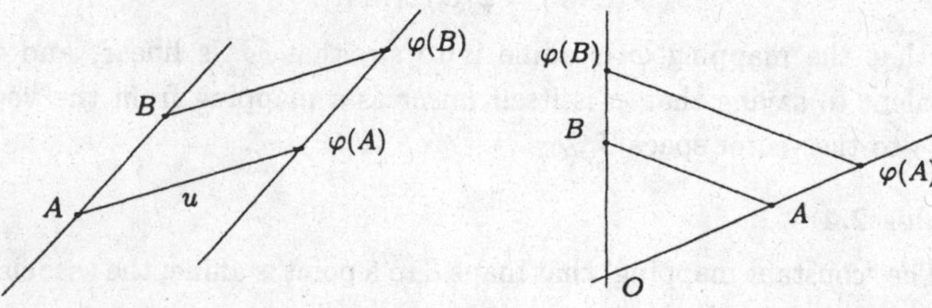

Fig. 7. A translation **Fig. 8.** A central dilatation

Effect on the subspaces. In the same way that (and because) the image of a linear subspace by a linear mapping is a linear subspace, we have:

Proposition 2.5. *The image of an affine subspace by an affine mapping is an affine subspace.*

Proof. Let $\varphi : \mathcal{E} \to \mathcal{E}'$ be an affine mapping. Let $\mathcal{F} \subset \mathcal{E}$ be an affine subspace of direction F. If $\mathcal{F}$ is empty, its image by φ is empty and, in particular, this is an affine subspace. Otherwise, let A be a point of $\mathcal{F}$. It is clear that $\varphi(\mathcal{F})$ is the affine subspace of $\mathcal{E}'$ directed by $\overrightarrow{\varphi}(F)$ and passing through $\varphi(A)$. □

Corollary 2.6. *Any affine mapping maps three collinear points to three collinear points.* □

The same argument, namely the use of the analogous linear algebra result, allows us to prove:

Proposition 2.7. *The inverse image of an affine subspace by an affine mapping is an affine subspace.*

The proof is left as a exercise for the reader. □

Effect on barycenters[6]**.** The translation into affine terms of linearity is often phrased "affine mappings preserve barycenters" and more precisely:

Proposition 2.8. *The image of the barycenter of* $((A_1,\alpha_1),\ldots,(A_k,\alpha_k))$ *by the affine mapping* φ *is the barycenter of* $((\varphi(A_1),\alpha_1),\ldots,(\varphi(A_k),\alpha_k))$. *Conversely, if a mapping* φ *transforms the barycenter of any system of punctual masses* $((A,\alpha),(B,1-\alpha))$ *into that of the system* $((\varphi(A),\alpha),(\varphi(B),1-\alpha))$, *then this is an affine mapping.*

Proof. Let us assume first that φ is an affine mapping. Denoting systematically by M' the image $\varphi(M)$, we have, for any point P of $\mathcal{E}$,

$$\begin{aligned}\overrightarrow{\varphi}(\alpha_1\overrightarrow{PA_1}+\cdots+\alpha_k\overrightarrow{PA_k}) &= \alpha_1\overrightarrow{\varphi}(\overrightarrow{PA_1})+\cdots+\alpha_k\overrightarrow{\varphi}(\overrightarrow{PA_k})\\ &= \alpha_1\overrightarrow{P'A'_1}+\cdots+\alpha_k\overrightarrow{P'A'_k}.\end{aligned}$$

If P is the barycenter of the system (A_i,α_i), the relation gives

$$\alpha_1\overrightarrow{P'A'_1}+\cdots+\alpha_k\overrightarrow{P'A'_k}=0$$

and P' is indeed the barycenter of the image system.

Conversely, let G be the barycenter of the system $((A,\alpha),(B,1-\alpha))$. For any point O of $\mathcal{E}$, we have

$$\overrightarrow{OG}=\alpha\overrightarrow{OA}+(1-\alpha)\overrightarrow{OB}.$$

As G' is the barycenter of $((A',\alpha),(B',1-\alpha))$, it satisfies a relation of the same type for any point and in particular for the image O' of O:

$$\overrightarrow{O'G'}=\alpha\overrightarrow{O'A'}+(1-\alpha)\overrightarrow{O'B'}.$$

A point O and its image O' by φ being given, we define a mapping f by $f(\overrightarrow{OM})=\overrightarrow{O'M'}$... and we want to prove that f is linear. Notice first that by "putting $M=O$", we find that $f(0)=0$. We have then

$$f(\alpha\overrightarrow{OA}+(1-\alpha)\overrightarrow{OB})=f(\overrightarrow{OG})=\overrightarrow{O'G'}=\alpha\overrightarrow{O'A'}+(1-\alpha)\overrightarrow{O'B'}$$

and thus, for all α, A and B,

$$f(\alpha\overrightarrow{OA}+(1-\alpha)\overrightarrow{OB})=\alpha f(\overrightarrow{OA})+(1-\alpha)f(\overrightarrow{OB}).$$

In particular, if $B=O$, putting $u=\overrightarrow{OA}$,

$$f(\alpha u)=\alpha f(u)+(1-\alpha)f(0)=\alpha f(u).$$

(6) See § 4 for a summary of the definition and first properties of the barycenter.

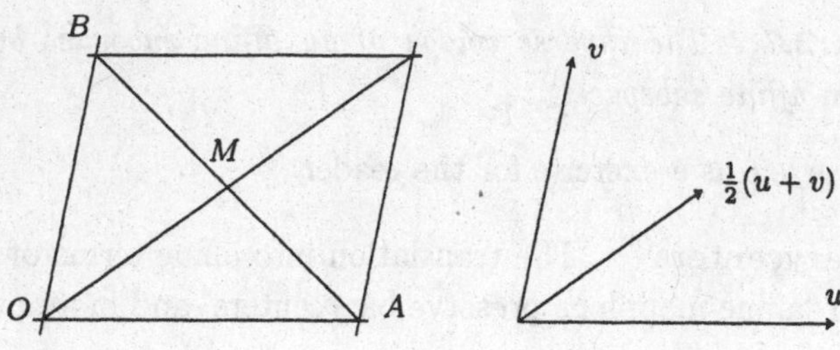

Fig. 9

Finally, if u and v are two vectors of E, we define A and B so that $u = \overrightarrow{OA}$ and $v = \overrightarrow{OB}$. For $\alpha = 1/2$, we have[(7)]

$$\frac{1}{2}f(u+v) = f\left(\frac{u+v}{2}\right)$$

because of what we have just done. But $(u+v)/2 = \overrightarrow{OM}$ where M is the midpoint of AB (Figure 9), thus

$$f\left(\frac{u+v}{2}\right) = \frac{1}{2}\overrightarrow{O'A'} + \frac{1}{2}\overrightarrow{O'B'} = \frac{1}{2}\left(f(u) + f(v)\right).$$

We deduce (multiplying by 2) that

$$f(u+v) = f(u) + f(v).$$

Therefore f is indeed linear and thus φ is affine, with $\overrightarrow{\varphi} = f$. □

From the direct (and easy) statement of Proposition 2.8, we deduce for instance, considering the barycenters of two points endowed with positive masses (in the case of a real affine space):

Corollary 2.9. *The image of a segment by an affine mapping is a segment.* □

Affine mappings and frames. In the same way as a linear mapping is determined by the images of the vectors of a basis, an affine mapping is determined by the image of an affine frame (this is straightforward, Exercise I.16). The next consequence of this property is very useful:

Proposition 2.10. *The only affine mapping of the n-dimensional affine space $\mathcal{E}$ that fixes $n+1$ independent points is the identity mapping.* □

For instance, an affine mapping from the plane to itself that has three noncollinear fixed points is the identity; an affine mapping that has two (distinct) fixed points fixes the line they span.

---

(7) We use here the fact that the characteristic of the field is different from 2.

The affine group. Let us begin by looking at the composition of two affine mappings.

Proposition 2.11. *The composition $\psi \circ \varphi$ of two affine mappings $\varphi : \mathcal{E} \to \mathcal{F}$ and $\psi : \mathcal{F} \to \mathcal{G}$ is an affine mapping. The associated linear mapping is the composition of the associated linear mappings (in formulas, $\overrightarrow{\psi \circ \varphi} = \overrightarrow{\psi} \circ \overrightarrow{\varphi}$).*

An affine mapping φ is bijective if and only if the associated linear mapping $\overrightarrow{\varphi}$ is. Then φ^{-1} is affine and the linear mapping associated with it is the inverse mapping of $\overrightarrow{\varphi}$ (in formulas $\overrightarrow{\varphi^{-1}} = \overrightarrow{\varphi}^{-1}$).

Proof. The assertion on the composition is clear. With obvious notation, we have:

$$\overrightarrow{\psi(P')\psi(M')} = \overrightarrow{\psi}(\overrightarrow{P'M'}) = \overrightarrow{\psi} \circ \overrightarrow{\varphi}(\overrightarrow{PM}).$$

Assume now that $\overrightarrow{\varphi}$ is a bijective linear mapping. For M in $\mathcal{F}$, we look for the points P of $\mathcal{E}$ such that $\varphi(P) = M$, but this is equivalent to $\overrightarrow{O'P'} = \overrightarrow{O'M}$, that is, to $\overrightarrow{\varphi}(\overrightarrow{OP}) = \overrightarrow{O'M}$ and finally to $\overrightarrow{OP} = \overrightarrow{\varphi}^{-1}(\overrightarrow{O'M})$, whence the existence and uniqueness of P. The affine mapping φ is thus indeed bijective.

Conversely, if φ is bijective, let us consider a vector u of F and let us look for the vectors v such that $\overrightarrow{\varphi}(v) = u$. Let us fix a point O and its image O', together with the point M such that $\overrightarrow{O'M} = u$. As φ is bijective, there is a unique point P in $\mathcal{E}$ such that $M = \varphi(P)$. We then have

$$\overrightarrow{\varphi}(\overrightarrow{OP}) = \overrightarrow{\varphi(O)\varphi(P)} = \overrightarrow{O'M} = u$$

and $\overrightarrow{OP}$ is the unique solution, therefore $\overrightarrow{\varphi}$ is bijective. Moreover, $\overrightarrow{OP} = (\overrightarrow{\varphi})^{-1}(u)$ so that $(\overrightarrow{\varphi})^{-1} = \overrightarrow{\varphi^{-1}}$. □

Corollary 2.12. *The affine bijections from $\mathcal{E}$ to itself form a group, the affine group $GA(\mathcal{E})$.* □

Proposition 2.13. *The mapping from the affine group to the linear group,*

$$\begin{aligned} GA(\mathcal{E}) &\longrightarrow GL(E) \\ \varphi &\longmapsto \overrightarrow{\varphi} \end{aligned}$$

which maps an affine mapping to the associated linear mapping is a surjective group homomorphism, the kernel of which is the group of translations of $\mathcal{E}$, isomorphic with the vector space E.

Proof. This is a direct consequence of what precedes. The kernel consists of the affine mappings whose associated linear mapping is Id_E, namely the translations, as we have already said. It is clear that the group of translations is isomorphic to the additive group of E: this is only a pedantic way to say that $t_u \circ t_v = t_{u+v}$.

The only thing which is left to prove is the surjectivity of our homomorphism. We prove a slightly more precise result, which deserves to be stated separately:

Lemma 2.14. *Let O be a point of $\mathcal{E}$. Let f be a linear isomorphism of E. There exists a unique affine mapping φ whose linear map is f and that fixes O (in formulas $f = \overrightarrow{\varphi}$ and $\varphi(O) = O$).*

This lemma ends the proof of the proposition. □

Its proof is straightforward: $\varphi(M)$ is the point defined by

$$\overrightarrow{O\varphi(M)} = f(\overrightarrow{OM}).$$

This argument gives also a more general result:

Lemma 2.15. *Let $f : E \to F$ be a linear mapping. Let $\mathcal{E}$ and $\mathcal{F}$ be two affine spaces directed by E and F respectively. For all points O of $\mathcal{E}$, O' of $\mathcal{F}$, there exists a unique affine mapping*

$$\varphi : \mathcal{E} \longrightarrow \mathcal{F}$$

that maps O to O' and whose linear mapping is f. □

Notice that these statements give a lot of examples of affine mappings. They also say:

Corollary 2.16. *Given a point O in $\mathcal{E}$, any affine mapping φ from $\mathcal{E}$ to itself can be written in a unique way in the form*

$$\varphi = t_u \circ \psi$$

where ψ fixes O. □

Remark 2.17. We have already noticed an equivalent property by vectorializing $\mathcal{E}$ at O (see Examples 2.4).

Conjugation of translations. We could have replaced, in the statement of Corollary 2.16

$$\varphi = t_u \circ \psi \text{ by } \varphi = \psi \circ t_v.$$

The affine mapping fixing O and having the same associated linear mapping as φ is indeed the same in the two writings. The vectors of the translations, however, are different: we have

$$t_u = \psi \circ t_v \circ \psi^{-1}.$$

The two translations are conjugated.

Proposition 2.18. *The conjugated mapping $\varphi \circ t_v \circ \varphi^{-1}$ of a translation by an element φ of the affine group $GA(\mathcal{E})$ is the translation of vector $\overrightarrow{\varphi}(v)$.*

Proof. Let M be a point of $\mathcal{E}$ and let $M' = t_v(M)$ be its image by the translation. We have

$$\overrightarrow{\varphi(M)\varphi(M')} = \overrightarrow{\varphi}(\overrightarrow{MM'}) = \overrightarrow{\varphi}(v)$$

so that $\varphi(M') = t_{\overrightarrow{\varphi}(v)}(\varphi(M))$. We thus have, for every point M' of $\mathcal{E}$,

$$\varphi \circ t_v \circ \varphi^{-1}(M') = \varphi \circ t_v(M) = \varphi(M') = t_{\overrightarrow{\varphi}(v)}(M'),$$

and this says indeed that

$$\varphi \circ t_v \circ \varphi^{-1} = t_{\overrightarrow{\varphi}(v)}.$$

□

A digression: conjugation. The previous statement can be read in the following way: when a translation is conjugated (by an affine transformation[8]), it remains a translation. Moreover, the vector of the new translation is given, in terms of the original vector and the transformation used to conjugate, by the only formula that makes sense.

This is an illustration of a general "principle" in geometry that we will have the opportunity to meet quite often in this book and whose (very vague) statement is the following.

"Principle" 2.19 (Conjugation principle). *If τ is an element of a group of transformations G,*

- *its conjugate $\varphi \circ \tau \circ \varphi^{-1}$ by an element of G is an element "of the same geometrical nature" as τ,*
- *the elements defining this "nature" are, for the conjugate $\varphi \circ \tau \circ \varphi^{-1}$, the images of those of τ by φ.*

This is not a theorem but rather a general framework that may be transformed, in each particular situation, in a precise statement by replacing the words between quotation marks by precisely defined objects, for instance:

- translations, vectors of translations,
- central symmetries, centers, or more generally
- dilatations, centers (see Exercise I.18),
- (in the symmetric group) transpositions, elements exchanged by the transposition.

[8] A *transformation* of $\mathcal{E}$ is a bijective mapping from $\mathcal{E}$ to itself.

Fixed points. The reader will have understood (see if necessary Remark 2.3) that an affine mapping form $\mathcal{E}$ to itself that has (at least) a fixed point can be considered as a linear mapping from the vectorialized of $\mathcal{E}$ at this point to itself. It is thus interesting to know when an affine mapping has a fixed point.

Proposition 2.20. *Let φ be an affine transformation of $\mathcal{E}$. It has a unique fixed point in $\mathcal{E}$ if and only if the isomorphism $\overrightarrow{\varphi}$ has no other fixed vector than* 0.

Remark 2.21. The condition means that 1 is not an eigenvalue of $\overrightarrow{\varphi}$.

Proof. If φ has a fixed point O, we vectorialize $\mathcal{E}$ at O to see that the other fixed points of φ correspond to the nonzero fixed vectors of $\overrightarrow{\varphi}$. Thus O is the *unique* fixed point if and only if the only fixed vector of $\overrightarrow{\varphi}$ is the zero vector.

If $\overrightarrow{\varphi}$ has no nonzero fixed vector (in other words, if this endomorphism does not have the eigenvalue 1), we look for the fixed points of φ. Let O be a point and O' its image. A fixed point M satisfies

$$\overrightarrow{\varphi}(\overrightarrow{OM}) = \overrightarrow{O'\varphi(M)} = \overrightarrow{O'M} = \overrightarrow{O'O} + \overrightarrow{OM},$$

that is

$$\overrightarrow{\varphi}(\overrightarrow{OM}) - \overrightarrow{OM} = \overrightarrow{O'O}.$$

By hypothesis, $\overrightarrow{\varphi} - \mathrm{Id}$ is injective. As this is an endomorphism of a finite-dimensional space, it is also onto and the equation $\overrightarrow{\varphi}(u) - u = \overrightarrow{O'O}$ has a unique solution $\overrightarrow{OM}$. Thus φ has a fixed point.

Eventually, if M and N are two fixed points of φ, $\overrightarrow{\varphi}(\overrightarrow{MN}) = \overrightarrow{MN}$, so that $\overrightarrow{MN} = 0$ and $M = N$. Therefore φ has a unique fixed point. The proposition follows. □

Under certain assumptions, it is possible to specify what happens when the linear mapping has the eigenvalue 1:

Proposition 2.22. *Let φ be an affine transformation of $\mathcal{E}$. Assume that there is a direct sum decomposition*

$$E = \operatorname{Ker}(\overrightarrow{\varphi} - \mathrm{Id}) \oplus \operatorname{Im}(\overrightarrow{\varphi} - \mathrm{Id}).$$

Then there exists a unique vector v and a unique affine mapping ψ having a fixed point such that

- $\overrightarrow{\varphi}(v) = v$,
- $\varphi = t_v \circ \psi$.

Moreover, t_v and ψ commute. The affine mapping φ has a fixed point if and only if $v = 0$, in which case the set of fixed points of φ is an affine subspace directed by the eigenspace of fixed vectors of $\overrightarrow{\varphi}$.

Proof. We start with any point O of $\mathcal{E}$ and use the decomposition of E as a direct sum to write

$$\overrightarrow{O\varphi(O)} = v + \overrightarrow{\varphi}(z) - z, \text{ where } v \text{ satisfies } \overrightarrow{\varphi}(v) = v.$$

Let us consider the point A defined by $z = \overrightarrow{AO}$. We have

$$\overrightarrow{A\varphi(A)} = \overrightarrow{AO} + \overrightarrow{O\varphi(O)} + \overrightarrow{\varphi(O)\varphi(A)} = z + v + \overrightarrow{\varphi}(z) - z + \overrightarrow{\varphi}(\overrightarrow{OA}) = v.$$

We thus put $\psi = t_{-v} \circ \varphi$. This new affine mapping satisfies

$$\psi(A) = t_{-v}(\varphi(A)) = A$$

and thus it has fixed points. We also have

$$\psi \circ t_v \circ \psi^{-1} = t_{\overrightarrow{\psi}(v)} = t_{\overrightarrow{\varphi}(v)} = t_v$$

(using Proposition 2.18) so that ψ and t_v commute. Let us prove now the uniqueness of the pair (v, ψ). We assume that

$$\varphi = t_v \circ \psi = t_{v'} \circ \psi'$$

where the two mappings ψ and ψ' have fixed points, *e.g.*, $\psi(A) = A$ and $\psi'(A') = A'$, and v, v' are two vectors fixed by $\overrightarrow{\varphi}$. Then $\overrightarrow{A\varphi(A)} = v$ and $\overrightarrow{A'\varphi(A')} = v'$, thus

$$\overrightarrow{AA'} = \overrightarrow{A\varphi(A)} + \overrightarrow{\varphi(A)\varphi(A')} + \overrightarrow{\varphi(A')A'} = v + \overrightarrow{\varphi}(\overrightarrow{AA'}) - v'$$

and

$$\overrightarrow{AA'} - \overrightarrow{\varphi}(\overrightarrow{AA'}) = v - v'.$$

This vector belongs to $\operatorname{Ker}(\overrightarrow{\varphi} - \operatorname{Id}_E) \cap \operatorname{Im}(\overrightarrow{\varphi} - \operatorname{Id}_E) = 0$, therefore $v = v'$ and $\psi = \psi'$. If $v = 0$, we have $\varphi = \psi$ so that φ has a fixed point. Conversely, if φ has a fixed point, we use it as the point O, hence $v = 0$. Vectorializing the affine space $\mathcal{E}$ at O, it is seen that the fixed points of φ correspond to the fixed vectors of $\overrightarrow{\varphi}$. The set of fixed points is thus the affine subspace through O directed by the vector subspace of fixed vectors $\overrightarrow{\varphi}$. □

3. Using affine mappings: three theorems in plane geometry

We are now in an affine plane.

Thales' theorem. This theorem expresses simply the fact that projections are affine mappings (Exercise I.12).

Theorem 3.1. *Let d, d' and d'' be three distinct parallel lines, $\mathcal{D}_1$ and $\mathcal{D}_2$ two lines that are not parallel to d. Let $A_i = \mathcal{D}_i \cap d$, $A'_i = \mathcal{D}_i \cap d'$, $A''_i = \mathcal{D}_i \cap d''$. Then the equality*

$$\frac{\overrightarrow{A_1A''_1}}{\overrightarrow{A_1A'_1}} = \frac{\overrightarrow{A_2A''_2}}{\overrightarrow{A_2A'_2}}$$

holds. Conversely, if a point B of $\mathcal{D}_1$ satisfies the equality

$$\frac{\overrightarrow{A_1B}}{\overrightarrow{A_1A'_1}} = \frac{\overrightarrow{A_2A''_2}}{\overrightarrow{A_2A'_2}},$$

then it belongs to d'' (and $B = A''_1$).

Remark 3.2. If A, B, C and D are four collinear points (with $C \neq D$), the ratio $\overrightarrow{AB}/\overrightarrow{CD}$ is well defined; this is the scalar λ such that $\overrightarrow{AB} = \lambda\overrightarrow{CD}$.

Proof. See Figure 10. Let π be the projection on $\mathcal{D}_2$ in the direction of d and let p be the associated linear projection. Then π maps A_1 to A_2 *etc.* Moreover, if $\overrightarrow{A_1A''_1} = \lambda\overrightarrow{A_1A'_1}$, we have $p(\overrightarrow{A_1A''_1}) = \lambda p(\overrightarrow{A_1A'_1})$ because p is linear. This is to say that $\overrightarrow{A_2A''_2} = \lambda\overrightarrow{A_2A'_2}$, an equality from which the direct sense of the theorem follows.

The converse statement is a consequence of the direct one: we have

$$\overrightarrow{A_1A''_1} = \frac{\overrightarrow{A_2A''_2}}{\overrightarrow{A_2A'_2}}\overrightarrow{A_1A'_1}$$

so that $\overrightarrow{A_1B} = \overrightarrow{A_1A''_1}$ and thus that $B = A''_1$. □

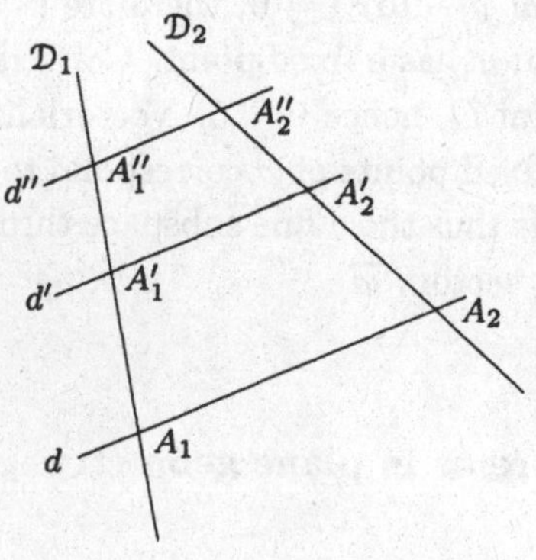

Fig. 10. Thales' theorem

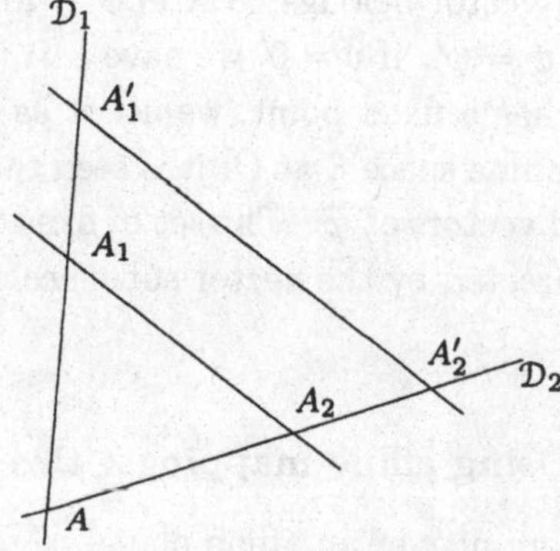

Fig. 11. and one of its corollaries

Corollary 3.3. *Let $\mathcal{D}_1$ and $\mathcal{D}_2$ be two lines intersecting at A, and d and d' two parallel lines intersecting $\mathcal{D}_i$ at A_i, A'_i distinct from A. Then*

$$\frac{\overrightarrow{AA_1}}{\overrightarrow{AA'_1}} = \frac{\overrightarrow{AA_2}}{\overrightarrow{AA'_2}} = \frac{\overrightarrow{A_1A_2}}{\overrightarrow{A'_1A'_2}}.$$

Proof. See Figure 11. Draw a line parallel to d and d' through A and apply Thales' theorem, which gives the first equality but also shows that the dilatation of center A mapping A_1 to A'_1 maps A_2 to A'_2. The second equality follows. □

Pappus' theorem. We now use dilatations and more precisely the fact that two dilatations of the same center commute.

Theorem 3.4. *Let A, B, C be three points of a line $\mathcal{D}$ and A', B', C' be three points of a line $\mathcal{D}'$ distinct from $\mathcal{D}$. If AB' is parallel to BA' and BC' is parallel to CB', then AC' is parallel to CA'.*

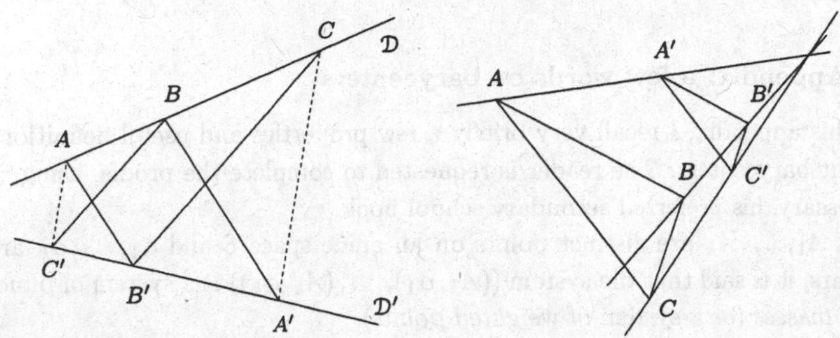

Fig. 12. Pappus' theorem **Fig. 13.** Desargues' theorem

Proof. See Figure 12. Assume first that $\mathcal{D}$ and $\mathcal{D}'$ are not parallel. Let O be their intersection point. Let φ be the dilatation of center O that maps A to B and ψ the one that maps B to C.

Then, thanks to Thales' theorem[(9)], φ maps B' to A' and ψ maps C' to B'. Hence $\psi \circ \varphi$ maps A to C and $\varphi \circ \psi$ maps C' to A'. As φ and ψ are dilatations of the same center, they commute. There is thus a dilatation of center O that maps A to C and C' to A', but then AC' and CA' are parallel, as is asserted by (the converse of) Thales' theorem.

If $\mathcal{D}$ and $\mathcal{D}'$ are parallel, one can replace central dilatations by translations in the proof. □

(9) Or simply because a linear dilatation maps any vector to a collinear vector.

Desargues' theorem

Theorem 3.5. *Let ABC and $A'B'C'$ be two triangles without common vertex and whose sides are respectively parallel. Then the lines AA', BB' and CC' are concurrent or parallel.*

Proof. See Figure 13. If AA' and BB' intersect at O, the dilatation φ of center O that maps A to A' also maps B to B' (Thales again). Let λ be its ratio and let $C'' = \varphi(C)$. Thus $\overrightarrow{OC''} = \lambda\overrightarrow{OC}$. But $\overrightarrow{OA'} = \lambda\overrightarrow{OA}$ and thus $A'C''$ and AC are parallel. Then C'' lies on the parallel to AC through A', namely $A'C'$, but it also lies on the parallel to BC through B', namely $B'C'$. Hence $C'' = C'$. But of course, O, C and C'' are collinear, so that CC' passes through O.

If AA' and BB' are parallel, we argue the same way using translations. □

Other versions of the theorems of Pappus and Desargues can be found in Exercises I.39 and I.40. They will show their deep nature and unity in Chapter V.

4. Appendix: a few words on barycenters

In this appendix, I recall very briefly a few properties and useful definitions about barycenters. The reader is requested to complete the proofs, using, if necessary, his preferred secondary school book.

If $A_1, \ldots, A_k$ are distinct points on an affine space $\mathcal{E}$ and $\alpha_1, \ldots, \alpha_k$ are scalars, it is said that the system $((A_1, \alpha_1), \ldots, (A_k, \alpha_k))$ is a *system of punctual masses* (or a *system of weighted points*).

Proposition 4.1. *If $((A_1, \alpha_1), \ldots, (A_k, \alpha_k))$ is a system of punctual masses such that $\sum \alpha_i \neq 0$, there exists a unique point G of $\mathcal{E}$ satisfying the equality*

$$\sum \alpha_i \overrightarrow{GA_i} = 0.$$

Moreover, for any point O of $\mathcal{E}$, we have

$$\left(\sum \alpha_i\right) \overrightarrow{OG} = \sum \alpha_i \overrightarrow{OA_i}.$$

The unique point G defined by this proposition is called the *barycenter* of the system. The reader is requested to consider what happens when the sum of the masses is zero.

Notice that the barycenter of the system $((A_1, \lambda\alpha_1), \ldots, (A_k, \lambda\alpha_k))$ is, by definition and for any nonzero scalar λ, the same as that of the system $((A_1, \alpha_1), \ldots, (A_k, \alpha_k))$.

When all the masses α_i are equal, the barycenter is called the *equibarycenter* of the points $A_1, \ldots, A_n$. The equibarycenter of two points A and B

is the *midpoint* of the segment AB; that of three noncollinear points A, B and C is called the *centroid* or *center of gravity* of the triangle ABC.

The barycenter (or rather the "operation of barycentration") satisfies an associativity property that is tedious to express but very easy to check:

Proposition 4.2 (Associativity of the barycenter). *Given scalars*

$$\alpha_{1,1},\ldots,\alpha_{1,k_1},\ldots,\alpha_{r,1},\ldots,\alpha_{r,k_r}$$

such that none of the sums $\sum_j \alpha_{i,j}$, $\sum_{i,j}\alpha_{i,j}$ *is zero, let*

$$((A_{1,1},\alpha_{1,1}),\ldots,(A_{1,k_1},\alpha_{1,k_1}),\ldots,(A_{r,1},\alpha_{r,1}),\ldots,(A_{r,k_r},\alpha_{r,k_r}))$$

be a system of punctual masses and let G *be its barycenter. Let* B_i *be the barycenter of the system* $((A_{i,1},\alpha_{i,1}),\ldots,(A_{i,k_i},\alpha_{i,k_i}))$. *The barycenter of the system*

$$\Big((B_1,\textstyle\sum_j\alpha_{1,j}),\ldots,(B_r,\sum_j\alpha_{r,j})\Big)$$

is G. □

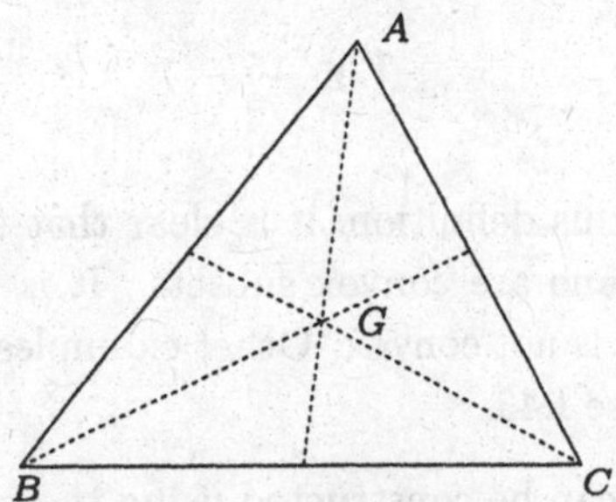

Fig. 14. The centroid of a triangle

In other words, to find the barycenter of a large system, we can first group the terms and consider their barycenters, then take the barycenter of the system consisting of these new points, weighted by suitable coefficients (sums of the masses of the points used in the grouping). A simple and useful special case gives the following corollary:

Corollary 4.3. *Let* G *be the barycenter of* $((A,\alpha),(B,\beta),(C,\gamma))$, *where* $\alpha+\beta+\gamma\neq 0$ *and* $\beta+\gamma\neq 0$. *The intersection point* A' *of* AG *and* BC *is the barycenter of* $((B,\beta),(C,\gamma))$. □

As a consequence, in a triangle, the three medians (lines from the vertices to the midpoints of the opposite sides) are concurrent at the centroid of the triangle (Figure 14).

5. Appendix: the notion of convexity

Convexity is an important and useful notion of affine geometry (useful not only in geometry!). It will be used in Chapter IV. We give here a few rudiments, referring the reader to [**Ber94**] for more information.

We consider now a real affine space (to be able to speak of "segment"). A subset $\mathcal{C} \subset \mathcal{E}$ is said to be *convex* if for all points A and B of $\mathcal{C}$, the segment AB is contained in $\mathcal{C}$ (Figure 15).

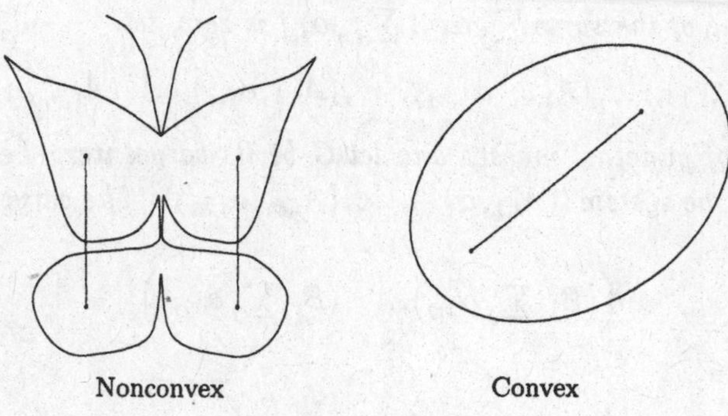

Fig. 15

Example 5.1. With this definition, it is clear that the empty set, a point, a segment, a line, a plane are convex subsets. It is also clear that the set consisting of two points is not convex. Other examples and counter-examples will be found in Exercise I.43.

New convex subsets can be constructed using the next proposition, whose (easy) proof is left to the reader.

Proposition 5.2. *Any intersection of convex subsets is convex.* □

The union of two convex subsets is not convex in general (think of a subset consisting of two points).

Convexity is an "affine notion"; this can be expressed by the following statement.

Proposition 5.3. *The image of any convex subset by an affine mapping is convex.*

Proof. Let $\mathcal{C}$ be a convex subset of the affine space $\mathcal{E}$ and let φ be an affine mapping from $\mathcal{E}$ to $\mathcal{E}'$. Let A' and B' be any two points of $\varphi(\mathcal{C})$. We want to prove that the segment $A'B'$ is contained in $\mathcal{C}$. We can find two points A and B of $\mathcal{C}$ of whom A' and B' are the images. As $\mathcal{C}$ is convex, it contains the segment AB. The image of this segment by φ is the segment $A'B'$

(this is what Corollary 2.9 says), and thus this segment is indeed contained in $\varphi(\mathcal{C})$. □

In the same way, one proves (and the reader will check) that

Proposition 5.4. *The inverse image of a convex subset by an affine mapping is convex.* □

Definition 5.5. The intersection of all the convex subsets containing the subset S of $\mathcal{E}$ is a convex subset of $\mathcal{E}$, called the *convex hull* of S (see Figure 16).

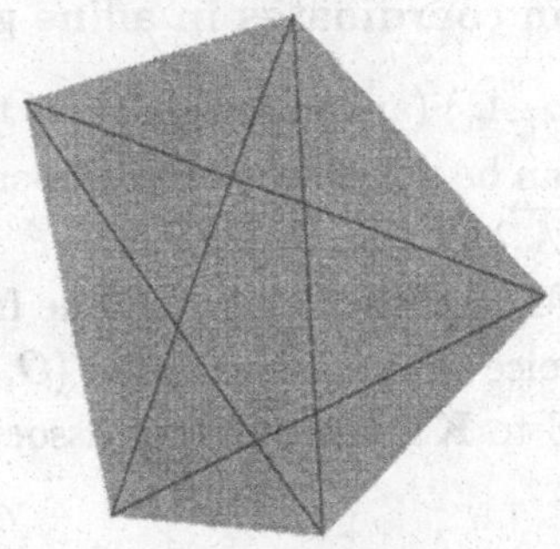

Fig. 16. The convex hull of a star-shaped polygon

Proposition 5.6. *The convex hull of S is the set of barycenters of the points of S endowed with positive masses.*

Proof. Denote by $\mathcal{C}(S)$ the convex hull of S and by S_+ the set of barycenters of the points of S endowed with positive masses.

Notice first that S_+ is convex: if M and N are two barycenters of points of S endowed with positive masses, the points of the segment MN are the barycenters of the systems $((M,\alpha),(N,1-\alpha))$ with $0 \leqslant \alpha \leqslant 1$. Thanks to the associativity of the barycentration, these points belong to S_+.

As any point M of S is the barycenter of the system $((M,1))$, we have the inclusion $S \subset S_+$ and thus also, S_+ being convex,

$$\mathcal{C}(S) \subset S_+.$$

Lot us piove conversely that S_+ is contained in $\mathcal{C}(S)$. We prove, by induction on k, that, if $\alpha_1,\ldots,\alpha_k$ are nonnegative real numbers and $A_1,\ldots,A_k$ points of $\mathcal{C}(S)$, then the barycenter of $((A_1,\alpha_1),\ldots,(A_k,\alpha_k))$ belongs to the convex hull $\mathcal{C}(S)$.

For $k = 1$, the barycenter is A_1 and the assertion is true. For $k = 2$, the barycenter is a point of the segment A_1A_2; it belongs to $\mathcal{C}(S)$ by convexity.

We explain now how to pass from a system of $k-1$ points to a system of k points.

If one of the α_i's is nonzero, we can remove (A_i, α_i) from the list. Let us thus assume that none of the α_i's is zero. In particular, α_k is not zero and neither is $\alpha_1 + \cdots + \alpha_{k-1}$. Let G' be the barycenter of the system $((A_1, \alpha_1), \ldots, (A_{k-1}, \alpha_{k-1}))$. The point G we are interested in is the barycenter of $((G', \alpha_1 + \cdots + \alpha_{k-1}), (A_k, \alpha_k))$. This is a point of the segment $G'A_k$. Now G' is in the convex hull $\mathcal{C}(S)$ by induction hypothesis, thus, using the convexity of $\mathcal{C}(S)$, G belongs to $\mathcal{C}(S)$. □

6. Appendix: Cartesian coordinates in affine geometry

An affine frame $(O, A_1, \ldots, A_n)$ (and an origin O) of the affine space $\mathcal{E}$ being given, any point M of $\mathcal{E}$ can be defined by the components of the vector $\overrightarrow{OM}$ in the basis $(\overrightarrow{OA_1}, \ldots, \overrightarrow{OA_n})$ of the vector space E directing $\mathcal{E}$, called the *Cartesian coordinates* of M in the considered affine frame.

In other words, the choice of the affine frame $(O, A_1, \ldots, A_n)$ defines an affine isomorphism from $\mathcal{E}$ to $\mathbf{K}^n$, the one that associates with the point M its Cartesian coordinates.

Affine subspaces. An affine subspace can be described by a point A and a direction, the latter being defined by a basis $(u_1, \ldots, u_k)$, like this:

$$\mathcal{F} = \left\{ M \in \mathcal{E} \mid \overrightarrow{AM} = \sum \lambda_i u_i \right\}.$$

And this is translated, in terms of the coordinates $(x_1, \ldots, x_n)$ of M, by

$$\begin{cases} x_1 = a_1 + \lambda_1 u_1^1 + \cdots + \lambda_k u_k^1 \\ \vdots \\ x_n = a_n + \lambda_1 u_1^n + \cdots + \lambda_k u_k^n \end{cases}$$

where $(a_1, \ldots, a_n)$ are the coordinates of the point A, $(u_i^1, \ldots, u_i^n)$ are the components of the vector u_i in the basis $(\overrightarrow{OA_1}, \ldots, \overrightarrow{OA_n})$ of E. These equations are called a *system of parametric equations* of $\mathcal{F}$.

For instance, if $(b_1, \ldots, b_n)$ are the coordinates of a point B distinct from A, the equations

$$\begin{cases} x_1 = a_1 + \lambda(b_1 - a_1) \\ \vdots \\ x_n = a_n + \lambda(b_n - a_n) \end{cases}$$

describe the line AB (see more generally Exercise I.8).

An affine subspace can also be described by *Cartesian equations.* A basis of E being given, the vector subspace F can be described by a system of Cartesian equations

$$\begin{cases} \alpha_{1,1}x_1 + \cdots + \alpha_{1,n}x_n = 0 \\ \vdots \\ \alpha_{m,1}x_1 + \cdots + \alpha_{m,n}x_n = 0. \end{cases}$$

The points M of the affine subspace $\mathcal{F}$ are characterized by the equation $\overrightarrow{AM} \in \mathcal{F}$, namely by

$$\begin{cases} \alpha_{1,1}(x_1 - a_1) + \cdots + \alpha_{1,n}(x_n - a_n) = 0 \\ \vdots \\ \alpha_{m,1}(x_1 - a_1) + \cdots + \alpha_{m,n}(x_n - a_n) = 0 \end{cases}$$

or also by

$$\begin{cases} \alpha_{1,1}x_1 + \cdots + \alpha_{1,n}x_n = b_1 \\ \vdots \\ \alpha_{m,1}x_1 + \cdots + \alpha_{m,n}x_n = b_m. \end{cases}$$

These are Cartesian equations of $\mathcal{F}$. For instance the system

$$\begin{cases} 2x + 3y - 5z = 1 \\ 2x - 2y + 3z = 0 \end{cases}$$

describes an affine line in $\mathbf{R}^3$ (why?).

The Cartesian equations describe the affine subspace as being

$$\mathcal{F} = f^{-1}((b_1, \ldots, b_m))$$

where $f : E \to \mathbf{K}^m$ is the linear mapping whose matrix in the bases

- $(\overrightarrow{OA_1}, \ldots, \overrightarrow{OA_n})$, basis of E, on the one hand,
- canonical basis of $\mathbf{K}^m$, on the other,

is

$$\begin{pmatrix} \alpha_{1,1} & \cdots & \alpha_{1,n} \\ \vdots & & \vdots \\ \alpha_{m,1} & \cdots & \alpha_{m,n} \end{pmatrix}.$$

The reader is requested to check that she has understood by solving Exercise I.9.

Affine mappings. Let us consider now an affine mapping $\varphi : \mathcal{E} \to \mathcal{E}'$, assuming that the affine space $\mathcal{E}'$ is also endowed with an affine frame, denoted $(O', A'_1, \ldots, A'_m)$. The points of $\mathcal{E}'$ are represented by their coordinates $(x'_1, \ldots, x'_m)$...; in other words we use, in addition to the isomorphism $\mathcal{E} \to \mathbf{K}^n$, an isomorphism $\mathcal{E}' \to \mathbf{K}^m$.

Through these isomorphisms[10], φ becomes an affine mapping

$$\mathbf{K}^n \longrightarrow \mathbf{K}^m$$

which we know (see Examples 2.4) has the form "linear mapping plus constant". This is to say that the coordinates $(x'_1, \ldots, x'_m)$ of the image $\varphi(M)$ of the point M of coordinates $(x_1, \ldots, x_n)$ are given by

$$\begin{cases} x'_1 &= a_{1,1}x_1 + \cdots + a_{1,n}x_n + b_1 \\ \vdots & \\ x'_m &= a_{m,1}x_1 + \cdots + a_{m,n}x_n + b_m. \end{cases}$$

The reader is invited to check that he has understood this writing by trying to change the frames (Exercise I.22).

Exercises and problems

Some of the exercises that follow use well-known Euclidean notions (orthocenter, area...) that it would be a pity not to use under the pretext that they have not yet been defined in this book.

Following the example of [**CG67**], I have often omitted the words "Prove that" to write these statements.

Italic letters (E, F...) always denote the directions of the affine spaces denoted by the corresponding roundhand letter ($\mathcal{E}$, $\mathcal{F}$...).

Direct applications and complements

Exercise I.1. The diagonals of a parallelogram intersect at their midpoints (if $AA'B'B$ is a parallelogram and if M satisfies $\overrightarrow{AB'} = 2\overrightarrow{AM}$, then it also satisfies $\overrightarrow{A'B} = 2\overrightarrow{A'M}$).

Exercise I.2. Let $\mathcal{F}$ be a subset of an affine space $\mathcal{E}$, directed by the vector space E and let A be a point of $\mathcal{F}$ such that $\Theta_A(\mathcal{F})$ is a linear subspace F of E. Prove that, for any point B of $\mathcal{F}$, one has $\Theta_B(\mathcal{F}) = F$ (this is Proposition 1.5).

Exercise I.3. Through two (distinct) points of an affine space passes a unique line.

Exercise I.4. Let $\mathcal{F}_1$ and $\mathcal{F}_2$ be two affine subspaces of an affine space $\mathcal{E}$. Under which condition is $\mathcal{F}_1 \cup \mathcal{F}_2$ an affine subspace?

Exercise I.5. A subset $\mathcal{F}$ of an affine space is an affine subspace if and only if for all points A and B of $\mathcal{F}$, the inclusion $\langle A, B\rangle \subset \mathcal{F}$ holds.

---

[10] If $\eta : \mathcal{E} \to \mathbf{K}^n$ and $\eta' : \mathcal{E}' \to \mathbf{K}^m$ are the names of the isomorphisms defined by the frames, this is $\eta' \circ \varphi \circ \eta^{-1}$ that we are describing here.

Exercise I.6. The points $A_0, \dots, A_k$ are affinely independent if and only if

$$\forall i \in \{0, \dots, k\}, \quad A_i \notin \langle A_0, \dots, A_{i-1}, A_{i+1}, \dots, A_k \rangle.$$

Exercise I.7. The points $A_0, \dots, A_k$ are affinely independent if and only if

$$\forall i \in \{1, \dots, k\}, \quad A_i \notin \langle A_0, \dots, A_{i-1} \rangle.$$

Exercise I.8. The affine space is endowed with an affine frame. Describe the affine subspace spanned by the points $B_0, \dots, B_k$ by a system of parametric equations.

Exercise I.9. Let A be a matrix of m lines and n columns with entries in $\mathbf{K}$ and let B be a (column) vector in $\mathbf{K}^m$. Let $\mathcal{F}$ be the subset of $\mathbf{K}^n$ defined by

$$\mathcal{F} = \{X \in \mathbf{K}^n \mid AX = B\}.$$

Explain why $\mathcal{F}$ is an affine subspace. What is its direction? When is it empty? Express its dimension in terms of the rank r of the matrix A.

Exercise I.10. Under which conditions do the two equations

$$a_1x_1 + \cdots + a_nx_n = b \quad \text{and} \quad a'_1x_1 + \cdots + a'_nx_n = b'$$

describe parallel hyperplanes in $\mathbf{K}^n$?

Exercise I.11. Under which conditions do the two systems of equations

$$\begin{cases} a_1x + b_1y + c_1z = 0 \\ a_2x + b_2y + c_2z = 0 \end{cases} \quad \text{and} \quad \begin{cases} a'_1x + b'_1y + c'_1z = 0 \\ a'_2x + b'_2y + c'_2z = 0 \end{cases}$$

describe (vector) lines of $\mathbf{K}^3$? the same line of $\mathbf{K}^3$?

Under which conditions do the two systems of equations

$$\begin{cases} a_1x + b_1y + c_1z = d_1 \\ a_2x + b_2y + c_2z = d_2 \end{cases} \quad \text{and} \quad \begin{cases} a'_1x + b'_1y + c'_1z = d'_1 \\ a'_2x + b'_2y + c'_2z = d'_2 \end{cases}$$

describe affine lines of $\mathbf{K}^3$? parallel affine lines of $\mathbf{K}^3$?

Exercise I.12 (Projections). Let $\mathcal{D}$ be a line in an affine plane $\mathcal{P}$ and let d' be a line direction, not parallel to $\mathcal{D}$. The mapping $\pi : \mathcal{P} \to \mathcal{P}$ such that $M' = \pi(M) \in \mathcal{D}$ and $\overrightarrow{MM'} \in d'$ is called *projection* on $\mathcal{D}$ in the direction d'. Prove that π is an affine mapping. What is the associated linear mapping?

Let $\mathcal{F}$ be an affine subspace of a space $\mathcal{E}$, and let G be a linear subspace of E such that $F \oplus G = E$. Define an (affine) projection on $\mathcal{F}$ in the direction G.

Let $\pi : \mathcal{E} \to \mathcal{E}$ be an affine mapping satisfying $\pi \circ \pi = \pi$. Prove that π is a projection.

Exercise I.13 (Thales' theorem, in any dimension). In an affine space, consider three parallel hyperplanes H, H' and H'' together with two lines $\mathcal{D}_1$ and $\mathcal{D}_2$ none of which is parallel to H. Let $A_i = \mathcal{D}_i \cap H$, $A'_i = \mathcal{D}_i \cap H'$, $A''_i = \mathcal{D}_i \cap H''$. Prove that

$$\frac{\overrightarrow{A_1A''_1}}{\overrightarrow{A_1A'_1}} = \frac{\overrightarrow{A_2A''_2}}{\overrightarrow{A_2A'_2}}.$$

Prove moreover that, if a point B of $\mathcal{D}_1$ satisfies

$$\frac{\overrightarrow{A_1B}}{\overrightarrow{A_1A'_1}} = \frac{\overrightarrow{A_2A''_2}}{\overrightarrow{A_2A'_2}},$$

then it is H'' (and $B = A''_1$).

Let $\mathcal{D}_1$ and $\mathcal{D}_2$ be two lines intersecting at A, and H and H' two parallel hyperplanes intersecting $\mathcal{D}_i$ at A_i, A'_i distinct from A. Prove that

$$\frac{\overrightarrow{AA_1}}{\overrightarrow{AA'_1}} = \frac{\overrightarrow{AA_2}}{\overrightarrow{AA'_2}} = \frac{\overrightarrow{A_1A_2}}{\overrightarrow{A'_1A'_2}}.$$

Exercise I.14 (Symmetries). If F and G are two complementary linear subspaces in a vector space E, one defines the symmetry s_F about F in the direction G by

$$s_F(u+v) = u - v \text{ if } u \in F \text{ and } v \in G.$$

Check that this is a linear mapping and an involution.

Let $\mathcal{F}$ be an affine subspace of an affine space $\mathcal{E}$ and G be a direction of affine subspaces such that $F \oplus G = E$. One chooses a point $O \in \mathcal{F}$ and one defines $\sigma : \mathcal{E} \to \mathcal{E}$ by

$$\overrightarrow{O\sigma(M)} = s_F(\overrightarrow{OM}).$$

If $O' \in \mathcal{F}$, check that

$$\overrightarrow{OM'} = s_F(\overrightarrow{OM}) \iff \overrightarrow{O'M'} = s_F(\overrightarrow{O'M}).$$

Prove that σ is an affine mapping (symmetry) and that it does not depend on the choice of O.

Exercise I.15 (Glide symmetries). Let φ be an affine transformation of the affine space $\mathcal{E}$. Assume that the associated linear mapping $\overrightarrow{\varphi}$ is a symmetry. Prove that φ can be written in a unique way as the composition of an affine symmetry σ and a translation of vector v that is fixed by $\overrightarrow{\varphi}$.

Let φ be an affine transformation of $\mathcal{E}$. Assume that the associated linear mapping is involutional (namely, satisfies $\overrightarrow{\varphi}^2 = \mathrm{Id}_E$). Does it follow that φ is involutional (namely, satisfies $\varphi^2 = \mathrm{Id}_{\mathcal{E}}$)?

Exercise I.16. An affine mapping is determined by the image of an affine frame.

Exercise I.17. A linear mapping $f : E \to F$ is given. Describe all the affine mappings $\varphi : \mathcal{E} \to \mathcal{F}$ of which f is the associated linear mapping.

Exercise I.18. What is the conjugate $\varphi \circ h(O, \lambda) \circ \varphi^{-1}$ of the dilatation $h(O, \lambda)$ by the affine transformation φ?

Exercise I.19. What is the composition $h(B, \lambda') \circ h(A, \lambda)$ of two dilatations? Is the set of affine central dilatations a group? What is the subgroup[11] that it generates in the affine group?

Exercise I.20. If $h(A, \lambda) \circ h(B, \mu) = h(C, \nu)$, the three points A, B and C are collinear.

Exercise I.21. The affine space $\mathcal{E}$ is endowed with an affine frame. Describe in Cartesian coordinates the following affine mappings:

- translation of vector v,
- dilatation of center A and ratio λ.

Exercise I.22. The affine spaces $\mathcal{E}$ and $\mathcal{E}'$ are endowed with affine frames. An affine mapping from $\mathcal{E}$ to $\mathcal{E}'$ is then given in matrix form by

$$X' = AX + B$$

where A is a matrix of m lines and m columns and B is a (column) vector of $\mathbf{K}^m$. How is this writing transformed under a change of affine frames in $\mathcal{E}$ and $\mathcal{E}'$?

Exercise I.23 (A very useful "trick"). Let f be a linear mapping from E to itself. Assume that the image by f of any vector is a collinear vector. Write this assumption with mathematical symbols and quantifiers. Write in similar terms the definition of a linear dilatation. Compare the two writings and prove that f is, nevertheless, a linear dilatation.

Exercise I.24. Let $\mathcal{E}$ be an affine space of dimension at least 2 and φ be an affine mapping $\mathcal{E} \to \mathcal{E}$ such that the image of any line is a line that is parallel to it. Prove that φ is a translation or a dilatation.

Exercise I.25 (Affine group of the line). Recall what the linear group of the (vector) line $\mathbf{K}$ is. Describe the affine group of this same line.

Exercise I.26. In a real affine plane, describe the interior of a triangle in terms of barycenters.

---

(11) This subgroup is often called the *group of dilatations*.

Exercise I.27. Let $ABCD$ be a (scalene) tetrahedron. Prove that its center of gravity G is the midpoint of the segments joining the midpoints of two opposite edges.

If A' is the centroid of the triangle BCD, prove that G belongs to the segment AA', and that it is located at the three-quarters of this segment, starting from A.

Exercise I.28 (Affine frame and barycentric coordinates). Let $(A_0, \ldots, A_n)$ be an affine frame in the space $\mathcal{E}$. Prove that any point M of $\mathcal{E}$ is a barycenter of $(A_0, \ldots, A_n)$ for some masses $(\lambda_0, \ldots, \lambda_n)$. Are these masses unique? The system $(\lambda_0, \ldots, \lambda_n)$ is called a *system of barycentric coordinates* of the point M in the frame $(A_0, \ldots, A_n)$.

Exercise I.29 (Affinities). A real number $\alpha \neq 1$ is given. Find the affine mappings such that, if M' denotes the image of M and M'' that of M',

$$\overrightarrow{M'M''} = \alpha \overrightarrow{MM'}.$$

Exercise I.30. Prove that the complement of a line in a real affine plane has two connected components. What can be said of the complement of a line in a real affine space of dimension 3 or more? of the complement of a line in a complex affine plane?

Exercises

Exercise I.31. Two points A and B are given in an affine plane $\mathcal{P}$. With any point M of the plane, the mapping φ associates the centroid of the triangle AMB. Is this an affine mapping? Same question with the orthocenter, $\mathcal{P}$

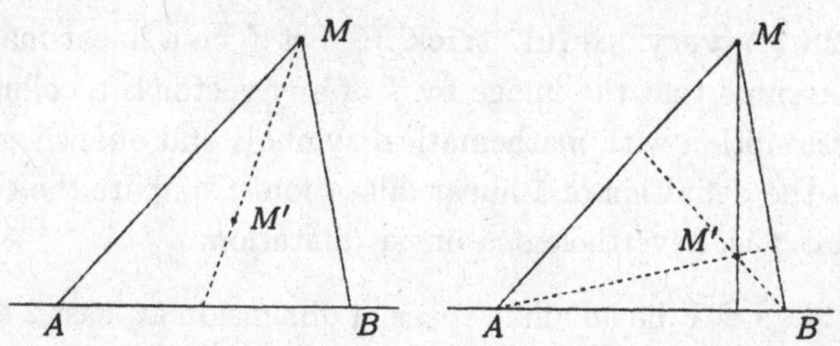

Fig. 17

being Euclidean (see Figure 17).

Exercise I.32. Let ABC be a triangle. Let M_0 be a point of the side AB. The parallel to BC through M_0 intersects AC at M_1. The parallel to AB through M_1 intersects BC at M_2 *etc.* (Figure 18). This defines points M_i (for $i \geqslant 0$). Prove that $M_6 = M_0$.

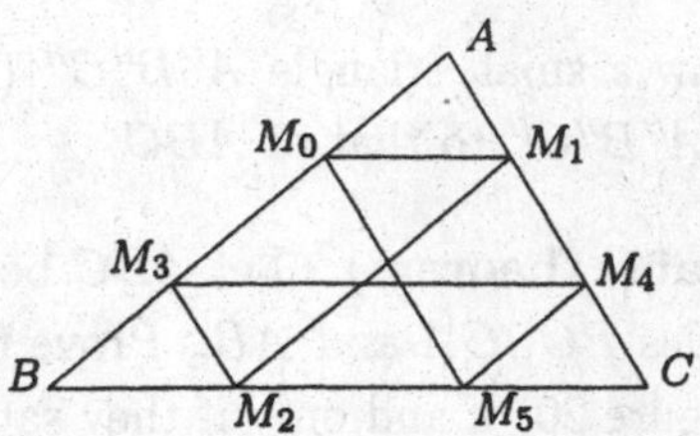

Fig. 18

Exercise I.33. A bounded subset of an affine space cannot have more than one symmetry center.

Exercise I.34. Given n points $A_1, \ldots, A_n$ in an affine plane $\mathcal{P}$, is it possible to find n points $B_1, \ldots, B_n$ such that $A_1, A_2, \ldots, A_n$ are the midpoints, respectively, of $B_1B_2, B_2B_3, \ldots, B_nB_1$ (Figure 19)? Consider in particular the cases $n = 3$ and $n = 4$.

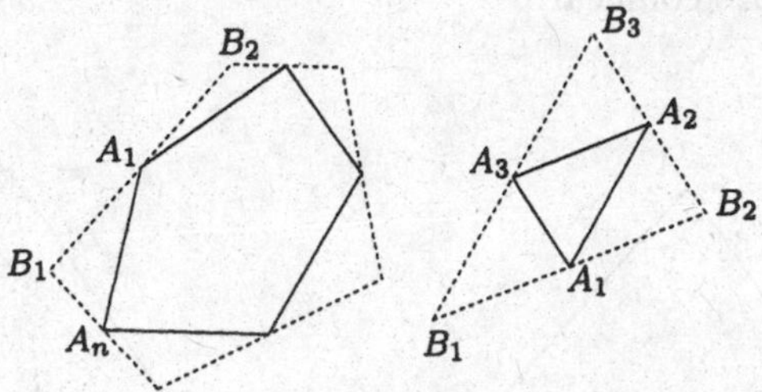

Fig. 19

Exercise I.35. Given a triangle ABC, construct three points A', B' and C' such that B' is the midpoint of AC', C' that of BA' and A' that of CB'.

Exercise I.36. On the three sides of a triangle ABC, put three points A', B', C' in such a way that $\overrightarrow{AC'} = \frac{2}{3}\overrightarrow{AB}$, $\overrightarrow{BA'} = \frac{2}{3}\overrightarrow{BC}$, $\overrightarrow{CB'} = \frac{2}{3}\overrightarrow{CA}$. The lines

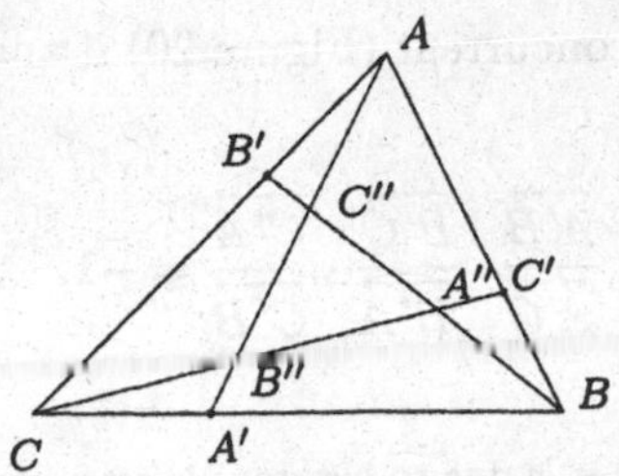

Fig. 20

AA', BB' and CC' draw a small triangle $A''B''C''$ (Figure 20). Compute the ratio of the area of $A''B''C''$ to that of ABC.

Exercise I.37 (Menelaüs' theorem). Let ABC be a triangle, and A', B' and C' points of the sides BC, CA and AB. Prove that the points A', B' and C' are collinear (Figure 20) if and only if they satisfy the equality:

$$\frac{\overrightarrow{A'B}}{\overrightarrow{A'C}} \cdot \frac{\overrightarrow{B'C}}{\overrightarrow{B'A}} \cdot \frac{\overrightarrow{C'A}}{\overrightarrow{C'B}} = 1.$$

Applications. Assume that A', B' and C' are collinear.

Let A'', B'' and C'' be the symmetric points of A', B' and C' about the midpoints of the corresponding sides. Prove that A'', B'' and C'' are collinear.

Let I, J and K be the midpoints of AA', BB' and CC'. We want to prove that I, J and K are collinear(12). Let E, F and G be the midpoints of $B'C'$, AC' and AB' respectively. Prove that $I \in FG$, $J \in GE$ and $K \in EF$, then that I, J and K are collinear.

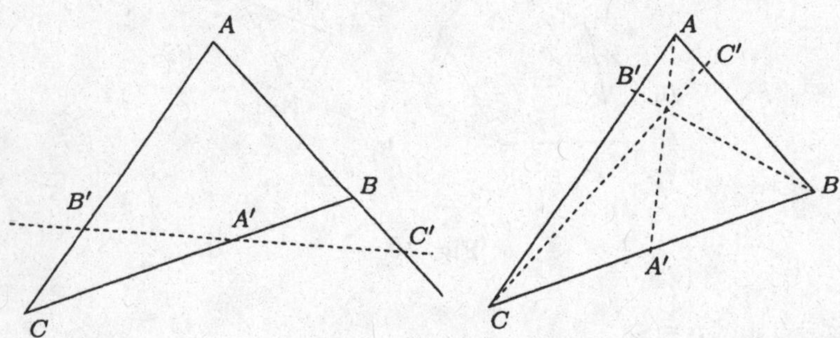

Fig. 21. Menelaüs and Ceva (hand in hand)

Exercise I.38 (Ceva's theorem). Let A', B' and C' be three points on the sides of a triangle as in Exercise I.37. Prove that the lines AA', BB' and CC' are parallel or concurrent (Figure 20) if and only if they satisfy the equality

$$\frac{\overrightarrow{A'B}}{\overrightarrow{A'C}} \cdot \frac{\overrightarrow{B'C}}{\overrightarrow{B'A}} \cdot \frac{\overrightarrow{C'A}}{\overrightarrow{C'B}} = -1.$$

(12) This result (it seems that it is due to Newton) is often stated this way: in a complete quadrilateral, here the one defined by the four lines AB, AC, BC and $B'C'$, the midpoints of the diagonals are collinear.

Exercise I.39 (Pappus, new version, as a corollary of Menelaüs). Let $\mathcal{D}$ and $\mathcal{D}'$ be two lines intersecting at O. Let A, B and C be three points of $\mathcal{D}$, and A', B' and C' be three points of $\mathcal{D}'$. Assume that $B'C$ and $C'B$ intersect at α, $C'A$ and $A'C$ at β, and $A'B$ and $B'A$ at γ. Prove that α, β and γ are collinear (see Figure 6 in Chapter V).

Exercise I.40 (Desargues, new version, as a corollary of Menelaüs). Let ABC and $A'B'C'$ be two triangles. Assume that BC and $B'C'$ intersect at α, CA and $C'A'$ at β, and AB and $A'B'$ at γ. Prove that α, β and γ are collinear if and only if AA', BB' and CC' are concurrent or parallel (see Figure 7 of Chapter V)[(13)].

Exercise I.41. Let ABC be a triangle, P a point of the line BC, M a point of the line AP. The parallels to CM through P and to AP through B intersect at B'. The parallels to BM through P and to AP through C intersect at C'. Let I, J and K be the midpoints of PM, BB' and CC'. Prove that I, J and K are collinear, then that M, B' and C' are collinear.

Exercise I.42. On a sheet of paper, two nonparallel lines $\mathcal{D}$ and $\mathcal{D}'$ have been drawn, but their intersection point is not on the sheet. On the same sheet, a point M is taken, away from $\mathcal{D}$ and $\mathcal{D}'$. Construct the line joining M to the intersection point of $\mathcal{D}$ and $\mathcal{D}'$.

Exercise I.43 (Convexity). Among the following subsets of an affine space, which ones are convex: a circle, a disk, a half-line, a half-plane, a sphere?

Let ABC be a triangle and φ be an affine transformation. Prove that the image by φ of the interior of ABC is the interior of the triangle $\varphi(A)\varphi(B)\varphi(C)$.

Exercise I.44. Let S be a finite set of points in an affine space. Prove that its convex hull $\mathcal{C}(S)$ is compact.

Exercise I.45. Let S be a subset of an affine space. Assume that S is not contained in a hyperplane. Prove that the interior of the convex hull $\mathcal{C}(S)$ is not empty.

Exercise I.46 (Half-planes). Let $\mathcal{D}$ be a line in an affine plane $\mathcal{P}$ and A be a point of $\mathcal{P}$ away from $\mathcal{D}$. Let

$$\mathcal{P}_A = \{M \in \mathcal{P} \mid [AM] \cap \mathcal{D} = \varnothing\}.$$

(13) All these applications of Menelaüs' theorem are classical, in fact Desargues used Menelaüs' theorem to prove his own theorem. See [CG67].

Let Δ be a line through A meeting $\mathcal{D}$ at a point I. Prove that M belongs to $\mathcal{P}_A$ if and only if the projection of M on Δ in the direction $\mathcal{D}$ belongs to the open half-line of origin I containing A.

Prove that:

- if B is a point of $\mathcal{P}_A$, then $\mathcal{P}_B = \mathcal{P}_A$,
- if $B \notin \mathcal{P}_A \cup \mathcal{D}$, then $\mathcal{P}$ is the disjoint union $\mathcal{P}_A \cup \mathcal{P}_B \cup \mathcal{D}$,
- $\mathcal{P}_A$ is convex.

It is said that $\mathcal{P}_A$ is the *open half-plane defined by $\mathcal{D}$ and containing* A. The (closed) half-plane is $\mathcal{P}_A \cup \mathcal{D}$.

Define more generally the half-spaces of an affine space defined by a hyperplane and prove that they have analogous properties.

More theoretical exercises

Exercise I.47. Let $\mathcal{E}$ be an affine space directed by E. One defines a mapping

$$E \times \mathcal{E} \longrightarrow \mathcal{E}$$

associating, with the vector u and the point M, the unique point M' of $\mathcal{E}$, which can be denoted

$$M' = M + u$$

(see Remark 1.3) and such that $\overrightarrow{MM'} = u$. Check that this is an action of the additive group of the vector space E on $\mathcal{E}$, and that it is transitive (there is a unique orbit) and free (all the stabilizers are trivial).

Conversely, prove that if the additive group of a vector space E acts freely and transitively on a set $\mathcal{E}$, the latter is endowed, by this action, with the structure of an affine space.

Exercise I.48. Let E be a vector space, F be a linear subspace and X be an equivalence class modulo F. Prove that X is an affine space directed by F.

Exercise I.49 (Center of the affine group). The center of the affine group $GA(\mathcal{E})$ (set of elements of the group that commute with everything) consists just of the identity transformation.

Exercise I.50 (Affine group (continuation)). Prove that there exists a bijection

$$GA(\mathcal{E}) \longrightarrow GL(E) \times E.$$

Does there exist a group isomorphism?

Exercise I.51 (The (so-called) "fundamental" theorem of affine geometry). Let $\mathcal{E}$ and $\mathcal{E}'$ be two (real) affine spaces of the same dimension $n \geqslant 2$. Let $\varphi : \mathcal{E} \to \mathcal{E}'$ be a *bijection*. We assume that

$$(*) \qquad A, B, C \text{ collinear} \Longrightarrow \varphi(A), \varphi(B), \varphi(C) \text{ collinear}$$

and we want to prove that then, φ is an affine mapping[14]. The images are denoted $A' = \varphi(A)$, *etc.*

(1) (a) Prove that the assumption "φ is bijective" is necessary.
(b) Prove that the assumption $n \geqslant 2$ is necessary.
In both cases, one can (if the assumption is not satisfied) look for a *non* affine mapping satisfying $(*)$.

(2) (a) Prove that, if $\mathcal{F}$ is an affine subspace of $\mathcal{E}'$, $\varphi^{-1}(\mathcal{F})$ is an affine subspace of $\mathcal{E}$ (one can use Exercise I.5).
(b) Prove that, if the $n+1$ points $A_0, \dots, A_n$ constitute an affine frame of $\mathcal{E}$, then their images $\varphi(A_0), \dots, \varphi(A_n)$ are affinely independent points of $\mathcal{E}'$.
(c) Deduce that, if $A_0, \dots, A_k$ are affinely independent points of $\mathcal{E}$, then their images $A'_0, \dots, A'_k$ are affinely independent points in $\mathcal{E}'$.
(d) Deduce that the image of any line $\mathcal{D}$ is a line $\mathcal{D}'$ and that the image of any plane is a plane, and
(e) that if $\mathcal{D}_1$ and $\mathcal{D}_2$ are two parallel lines, their images $\mathcal{D}'_1$ and $\mathcal{D}'_2$ are two parallel lines.

Let a point O be given in $\mathcal{E}$ and let O' be its image in $\mathcal{E}'$.

(3) Let A and B be two points of $\mathcal{E}$ and let C be such that $\overrightarrow{OC} = \overrightarrow{OA} + \overrightarrow{OB}$. Assume that A, B and C are not collinear. Using 2(e), prove that $\overrightarrow{O'C'} = \overrightarrow{O'A'} + \overrightarrow{O'B'}$.

(4) Let $\mathcal{D}$ be a line through O and let $\mathcal{D}'$ be its image. Fix a point $A \in \mathcal{D}$, distinct from O, and its image A'. For every $\lambda \in \mathbf{R}$, consider the point M such that $\overrightarrow{OM} = \lambda\overrightarrow{OA}$. Check that its image M' satisfies $\overrightarrow{O'M'} = \mu\overrightarrow{O'A'}$ for a unique $\mu \in \mathbf{R}$. This way, a mapping

$$\sigma_{\mathcal{D}} : \mathbf{R} \longrightarrow \mathbf{R}$$
$$\lambda \longmapsto \mu$$

has been defined.
(a) Let M and N be such that $\overrightarrow{OM} = \lambda\overrightarrow{OA}$, $\overrightarrow{ON} = \lambda'\overrightarrow{OA}$. Using a point B away from the line $\mathcal{D}$ and parallel lines, construct the points P and Q of $\mathcal{D}$ such that $\overrightarrow{OP} = (\lambda+\lambda')\overrightarrow{OA}$, $\overrightarrow{OQ} = \lambda\lambda'\overrightarrow{OA}$.

---

[14] Beyond the interest of this theorem, notice, in its proof, the geometric "reconstruction" of the algebraic operations of the field, in question (4). Speaking of that, the interested reader should see **[Art57]**.

(b) Prove that $\sigma_{\mathcal{D}}(\lambda + \lambda') = \sigma_{\mathcal{D}}(\lambda) + \sigma_{\mathcal{D}}(\lambda')$ and that $\sigma_{\mathcal{D}}(\lambda\lambda') = \sigma_{\mathcal{D}}(\lambda)\sigma_{\mathcal{D}}(\lambda')$. Deduce that $\sigma_{\mathcal{D}}$ is a field automorphism of the field **R**.

(c) Let σ be an automorphism of the field **R**. Check that $\sigma(1) = 1$, that $\sigma|_{\mathbf{Q}} = \mathrm{Id}_{\mathbf{Q}}$ and that σ is an increasing mapping. Deduce that $\sigma = \mathrm{Id}$.

(d) Prove the φ is an affine mapping.

(5) Where did you use the fact that φ is bijective? that $n \geqslant 2$?

(6) What happens in the complex case?

Exercise I.52. Redo Exercise I.24 without assuming *a priori* that φ is affine.

Chapter II

Euclidean Geometry, Generalities

In this chapter, all the vector spaces are defined over the field $\mathbf{R}$ of real numbers. The spaces under consideration all have finite dimension.

We now introduce tools allowing us to measure distances (and even angles).

1. Euclidean vector spaces, Euclidean affine spaces

Recall that a *scalar product* (or *dot product*) on a vector space E is a "positive-definite-symmetric-bilinear-form", namely a mapping

$$\begin{aligned} \Phi : E \times E &\longrightarrow \mathbf{R} \\ (u, v) &\longmapsto \Phi(u, v) \end{aligned}$$

- linear in both variables,
- symmetric (namely such that $\Phi(v, u) = \Phi(u, v)$ for all vectors u and v of E)
- and such that $\Phi(u, u) \geqslant 0$ for all u, the equality $\Phi(u, u) = 0$ holding if and only if $u = 0$.

Notations. We use here the most standard notation possible, that is, we write $u \cdot v$ for $\Phi(u, v)$ and $\|u\|^2$ for $u \cdot u$ (understanding that $\|u\|$ is a non-negative number). We also write $u \perp v$ when $u \cdot v = 0$, this defining a relation among subspaces, orthogonality. We denote by $F^\perp$ the orthogonal of subspace F, in symbols

$$F^\perp = \{x \in E \mid x \cdot y = 0 \text{ for all } y \text{ in } F\}.$$

The space decomposes as $E = F \oplus F^\perp$. More generally, if S is a subset of E, we denote by $S^\perp$ the set of vectors of E that are orthogonal to all the vectors of S. This is a vector subspace, the orthogonal of the subspace spanned by S.

Definition 1.1. A vector space endowed with a scalar product is said to be a *Euclidean vector space.* A *Euclidean affine space* is an affine space directed by a Euclidean vector space. The *distance* of two points A and B is defined by $d(A,B) = \|\overrightarrow{AB}\|$.

We will of course abbreviate $d(A,B)$ to AB.

Triangle inequality. The Cauchy–Schwarz inequality implies that $\|\cdot\|$ is a norm (a Euclidean space is a normed vector space) and that d is a distance (a Euclidean affine space is a metric space). That is, in particular, d satisfies the triangle inequality

$$d(A,B) \leqslant d(A,C) + d(C,B)$$

with the precision that the equality holds only if the points A, C and B are collinear, in that order. See Exercise II.1.

Isometries. A *linear isometry* is a linear mapping that preserves the norm, that is, a linear mapping

$$f : E \longrightarrow F$$

(where E and F are Euclidean vector spaces) such that $\|f(u)\| = \|u\|$ for any vector u of E. Since the scalar product can be expressed in terms of the norm, as we have

$$u \cdot v = \frac{1}{4}\left(\|u+v\|^2 - \|u-v\|^2\right),$$

isometries preserve the scalar product and, in particular, the orthogonality.

Similarly, an affine mapping

$$\varphi : \mathcal{E} \longrightarrow \mathcal{F}$$

(where $\mathcal{E}$ and $\mathcal{F}$ are Euclidean affine spaces) is an *affine isometry* if

$$d(\varphi(A), \varphi(B)) = d(A,B)$$

for all A and B in $\mathcal{E}$... this being of course equivalent to saying that the associated linear mapping is a linear isometry.

It is clear that the composition of two isometries is an isometry. Let $O(E)$, *resp.* $\operatorname{Isom}(\mathcal{E})$, denote the set of (linear) isometries from E to E, *resp.* (affine) isometries from $\mathcal{E}$ to $\mathcal{E}$.

Theorem 1.2. *The sets* $O(E)$, $\operatorname{Isom}(\mathcal{E})$, *endowed with the composition of mappings, are groups.*

Proof. We must prove that the isometries are bijections. We know that an affine mapping is bijective if and only if the associated linear mapping is (see Proposition I-2.11). Therefore, it suffices to prove that the linear isometries are bijections.

Now, an endomorphism of a finite-dimensional vector space is bijective if and only if it is injective[(1)]. It is thus enough to prove that any linear mapping $f : E \to E$ that preserves the norm is injective, and this is quite easy:

$$f(u) = 0 \quad \Longrightarrow \quad 0 = \|f(u)\| = \|u\| \quad \Longrightarrow \quad u = 0.$$

Moreover, the inverse bijection of an isometry is again an isometry. Let us write the proof of this assertion in the affine case, for a change. Let φ be an affine isometry. The inverse bijection ψ is again an affine mapping according to the proposition mentioned above and

$$d(\varphi(A), \varphi(B)) = d(A, B) \quad \Longrightarrow \quad d(A', B') = d(\psi(A'), \psi(B'))$$

so that ψ is indeed an isometry.

Eventually, the reader will certainly notice that the identity mapping Id_E is a linear isometry (and $\mathrm{Id}_{\mathcal{E}}$ an affine isometry). □

Examples 1.3

(1) Translations are isometries (the linear mapping associated with a translation is the identity).

(2) In general, a dilatation of ratio λ multiplies the lengths by $|\lambda|$. A dilatation is thus an isometry only if its ratio is ± 1... that is, if it is the identity mapping (case where $\lambda = 1$) or a central symmetry ($\lambda = -1$). This does not prevent dilatations from being very useful in Euclidean geometry (see, *e.g.*, Exercises II.13, II.17, II.19 and II.20 to convince yourself).

(3) The orthogonal symmetries are isometries. If $F \subset E$ is a vector subspace, the orthogonal symmetry s_F is the linear mapping equal to Id_F on F and to $-\mathrm{Id}_G$ on $G = F^\perp$, that is, $s_F(y + z) = y - z$ if $y \in F$ and $z \in G$ (Figure 1).

    Orthogonal *affine* symmetries are defined similarly: if $\mathcal{F}$ is an affine subspace of $\mathcal{E}$, $M' = \sigma_{\mathcal{F}}(M)$ is defined by $\overrightarrow{OM'} = s_F(\overrightarrow{OM})$ where F is of course the direction of $\mathcal{F}$ and O is any point of $\mathcal{F}$. It is easy but necessary to check that the result does not depend of the choice of O in $\mathcal{F}$ (Exercise I.14). It has probably already been noticed that symmetries are involutions (in symbols $\sigma_{\mathcal{F}}^2 = \mathrm{Id}_{\mathcal{E}}$).

[(1)] One of the miracles of linearity in finite dimension.

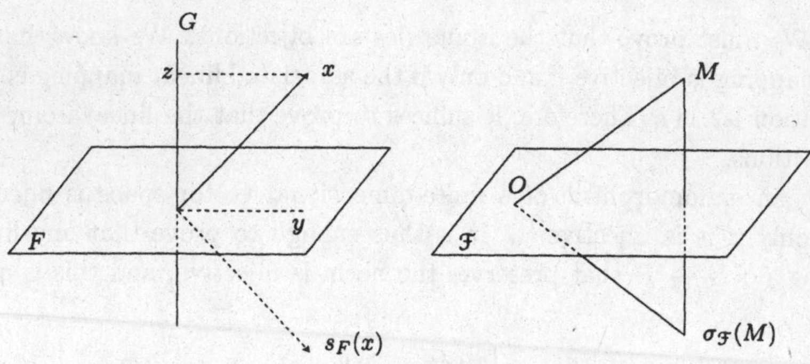

Fig. 1

(4) In particular, the *reflections*, which are the orthogonal symmetries about hyperplanes, are isometries. See Exercise II.7 for an explicit formula.

2. The structure of isometries

I prove now that any isometry from $\mathcal{E}$ to $\mathcal{E}$ can be written as the composition of a certain number of reflections. In a more pedantic way: "The reflections generate the group of isometries". I prove this first for *plane* affine isometries; I then do it again in all dimensions (beginning with the vectorial case).

Plane isometries. Let us consider first an isometry φ of an affine Euclidean plane $\mathcal{P}$. If φ has three noncollinear fixed points, we know (Proposition I-2.10) that φ is the identity.

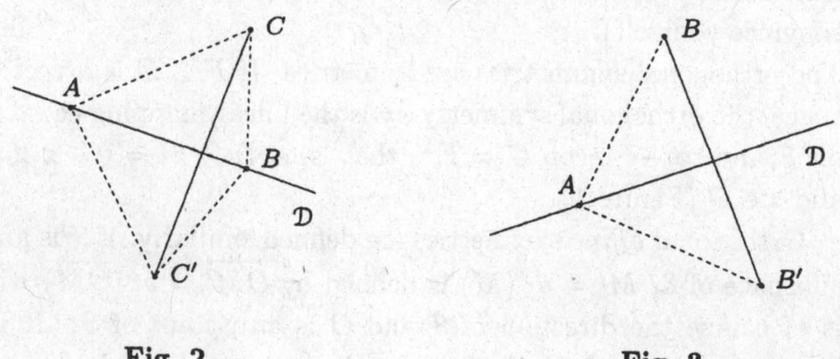

Fig. 2 **Fig. 3**

If φ has two distinct fixed points A and B (but not three noncollinear fixed points), we consider the line $\mathcal{D} = \langle A, B \rangle$ and a point C not on $\mathcal{D}$ and its image C'. Since φ is an isometry, $AC = AC'$ and $BC = BC'$, thus $\mathcal{D}$ is the perpendicular bisector[2] of CC' (Figure 2). Then $\sigma_{\mathcal{D}} \circ \varphi$ fixes A, B and C, thus $\sigma_{\mathcal{D}} \circ \varphi$ is the identity and $\varphi = \sigma_{\mathcal{D}}$, φ is a reflection.

If φ has a unique fixed point A, let B be another point and B' its image. As $AB = AB'$, the perpendicular bisector $\mathcal{D}$ of BB' passes through A (Figure 3) and $\sigma_{\mathcal{D}} \circ \varphi$ has two fixed points A and B, hence this is a reflection and φ is the composition of two reflections.

If φ has no fixed point, let A and A' be a point and its image. If $\mathcal{D}$ is the perpendicular bisector of AA', $\sigma_{\mathcal{D}} \circ \varphi$ fixes A hence this is the composition of one or two reflections, thus φ is the composition of two or three reflections.

This way, we have proved, in dimension 2, a quite precise version of the announced result, Theorem 2.2 below.

Let us now prove this result in any dimension, using the same idea, namely using fixed points to reduce to simpler or already understood cases. Let us begin with the linear case, for which the precise statement is:

Theorem 2.1. *Let E be a Euclidean vector space of dimension n. Any isometry of E can be written at the composition of p reflections for some integer $p \leqslant n$.*

We will then prove the corresponding affine result (Theorem 2.2 below), as a consequence both of the linear result and of its proof.

Proof. It is done by induction on the dimension n of E.

The case where $n = 1$ is quite simple because we know the list of all the isometries of a Euclidean line: the linear mappings have all the form $x \mapsto \lambda x$, and such a mapping is an isometry if and only if $|\lambda| = 1$, hence the isometries are the identity (composition of zero reflection) and the central symmetry $-\mathrm{Id}$, that is a reflection. The theorem is true (by direct inspection) for $n = 1$.

Assume now (this is the induction hypothesis) that the theorem is true in all Euclidean vector spaces of dimension $\leqslant n-1$ and consider a Euclidean vector space E of dimension n. Let f be an isometry.

To use the induction hypothesis, we need a way to reduce to a space of smaller dimension. Let us use a vector $x_0 \neq 0$. Then, either it is fixed by f (namely $f(x_0) = x_0$), or it is not ($f(x_0) \neq x_0$).

In the first case, we consider the hyperplane $S = x_0^{\perp}$. Since f preserves the scalar product and fixes x_0, f preserves S:

$$y \in S \implies f(y) \cdot x_0 = f(y) \cdot f(x_0) = y \cdot x_0 = 0 \implies f(y) \in S.$$

[2] The line which is the set of points equidistant from C and C' (see Exercise II.7).

Let us thus consider the restriction $f' = f|_S$ of f to S; this is an isometry of S, that is of dimension $n-1$. We can apply the induction hypothesis to it, so that there exist q hyperplanes $H'_1, \ldots, H'_q$ of S (with $q \leqslant n-1$) such that

$$f' = s_{H'_1} \circ \cdots \circ s_{H'_q}.$$

Starting from these hyperplanes of S, we construct hyperplanes of E. Let

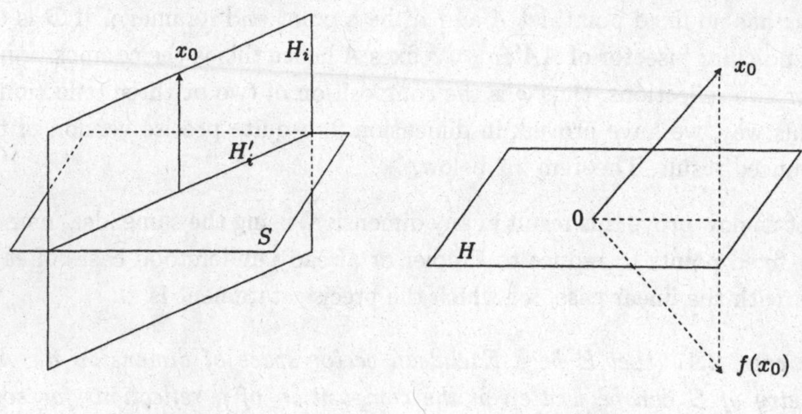

Fig. 4 Fig. 5

H_i the hyperplane of E spanned by H'_i and x_0 (Figure 4). We have

$$s_{H_1} \circ \cdots \circ s_{H_q} = f.$$

To be convinced that this is true, it suffices to decompose E as $E = S \oplus \langle x_0 \rangle$ and to notice that we have

$$\begin{cases} s_{H_1} \circ \cdots \circ s_{H_q}(\lambda x_0) = \lambda x_0 = f(\lambda x_0) \\ s_{H_1} \circ \cdots \circ s_{H_q}(y) = s_{H'_1} \circ \cdots \circ s_{H'_q}(y) = f'(y) = f(y) \quad \text{for } y \in S. \end{cases}$$

This finishes the proof (with a number $q \leqslant n-1$ of reflections) in the case where $f(x_0) = x_0$.

Let us consider now the case where $f(x_0) \neq x_0$. The idea is to reduce to the previous case by "making x_0 fixed". Let H be the perpendicular bisector of x_0 and $f(x_0)$. It is *a priori* an affine hyperplane of E. Since $\|f(x_0)\| = \|x_0\|$, the vector 0 is in H, which is thus a vector subspace. We have actually

$$H = (x_0 - f(x_0))^{\perp}$$

(Figure 5), so that $s_H(x_0) = f(x_0)$. Then

$$s_H \circ f(x_0) = s_H(f(x_0)) = x_0.$$

Hence $s_H \circ f$ has a fixed point and the result obtained in the first case can be applied:

$$s_H \circ f = s_{H_1} \circ \cdots \circ s_{H_q} \quad q \leqslant n-1,$$

thus

$$f = s_H \circ s_{H_1} \circ \cdots \circ s_{H_q}$$

is indeed the composition of p reflections (with $p \leqslant n$). □

Theorem 2.2. *Let $\mathcal{E}$ be a Euclidean affine space of dimension n. Any isometry of $\mathcal{E}$ can be written as the composition of p reflections for some integer $p \leqslant n+1$.*

Proof. We take our inspiration from the last part of the previous proof to reduce to the linear case.

If the affine isometry φ has a fixed point $A \in \mathcal{E}$, we only have to vectorialize $\mathcal{E}$ at A to reduce to the linear case. Hence, applying Theorem 2.1, we obtain that φ is the composition of p reflections about hyperplanes through A (for a number $p \leqslant n$).

Otherwise, let A be any point and $A' \neq A$ be its image. Let $\mathcal{H}$ denote the perpendicular bisector of AA' (see if necessary Exercise II.7), so that $\sigma_{\mathcal{H}} \circ \varphi$ is an affine isometry that fixes A and to which what precedes can be applied. Thus

$$\sigma_{\mathcal{H}} \circ \varphi = \sigma_{\mathcal{H}_1} \circ \cdots \circ \sigma_{\mathcal{H}_p}$$

for some $p \leqslant n$ and thus

$$\varphi = \sigma_{\mathcal{H}} \circ \sigma_{\mathcal{H}_1} \circ \cdots \circ \sigma_{\mathcal{H}_p}$$

is the composition of $p+1$ ($\leqslant n+1$) reflections. □

Remarks 2.3

- These theorems are interesting from the theoretical viewpoint (it never does harm to know a system of generators of a group); they are also very concrete as they allow us, for instance, to make a list of *all* the isometries of the Euclidean spaces of few dimensions (especially 2 and 3). We have seen this at the beginning of this section for dimension 2; we will come back to this (§§ III-2 and IV-1).
- Looking carefully at the proofs given above, the number of needed reflections can be estimated more precisely in terms of the dimension of the fixed subspace of the isometry under consideration, but I do not find this essential (see, *e.g.*, [**Ber94**]).

Rigid motions. The number of reflections showing up in the decomposition of a given isometry can be made arbitrarily large: $s_H \circ s_H \circ f = f$! But its parity is well defined, as we shall see.

Definition 2.4. An (affine) isometry is a *rigid motion* if its determinant (namely that of the associated linear mapping) is positive. An isometry that is not a rigid motion is a *negative isometry.*

Remarks 2.5

(1) Isometries are bijections and thus their determinant is nonzero.
(2) The set $\mathrm{Isom}^+(\mathcal{E})$ of rigid motions is a subgroup of $\mathrm{Isom}(\mathcal{E})$.
(3) The rigid motions are the isometries that preserve the orientations of the space (see Exercise II.11).

Proposition 2.6. *The number of reflections appearing in the decomposition of an isometry is even if and only if this isometry is a rigid motion.*

Proof. It is enough to prove that the determinant of a linear reflection is negative. And this is very easy. Let H be a hyperplane and $(e_1, \dots, e_{n-1})$ be a basis of H; let e_n denote a nonzero vector of the line orthogonal to H, so that $(e_1, \dots, e_n)$ is a basis of E in which the matrix of s_H is the diagonal matrix $(1, \dots, 1, -1)$ whose determinant is -1. □

Remark 2.7. This proof, added to the fact that the determinant is a group homomorphism and to the theorem of decomposition of isometries (Theorem 2.2) also shows that the determinant of an isometry is ± 1 ($+1$ for a rigid motion, -1 for a negative isometry).

It is deduced that $\mathrm{Isom}^+(\mathcal{E})$ is the kernel of the group homomorphism

$$\det : \mathrm{Isom}(\mathcal{E}) \longrightarrow \mathbf{R} - \{0\} \text{ or } \{1, -1\}$$

and thus that this is a normal subgroup of $\mathrm{Isom}(\mathcal{E})$.

Isometries and rigid motions in the plane. A plane linear isometry is either the identity or the composition of one or two reflections (about lines); an affine isometry of a Euclidean plane is either the identity or the composition of one, two or three reflections.

A rigid motion of the (linear or affine) plane is the composition of two reflections.

We will come back to a description of these rigid motions (they are rotations and translations) in the next chapter.

The reduced form of isometries. The next result is very useful, as it allows us to associate, with any affine isometry, another isometry that has the same associated linear mapping but has fixed points. This is simply Proposition I-2.22 in the special case of isometries.

Proposition 2.8. *Let φ be an affine isometry of the affine space $\mathcal{E}$. There exists an isometry ψ and a translation t_v of $\mathcal{E}$ such that*

– *the space $\mathcal{F}$ of fixed points of ψ is nonempty,*
– *the vector v of the translation belongs to the direction F of this subspace,*
– *we have $\varphi = t_v \circ \psi$.*

Moreover, the pair (v, ψ) is unique, t_v and ψ commute and $F = \operatorname{Ker}(\overrightarrow{\varphi} - \operatorname{Id}_{\mathcal{E}})$.

Proof. Let us prove first that, for any isometry φ, we have an orthogonal direct sum

$$E = \operatorname{Ker}(\overrightarrow{\varphi} - \operatorname{Id}_E) \oplus \operatorname{Im}(\overrightarrow{\varphi} - \operatorname{Id}_E).$$

The two subspaces have complementary dimensions, so that it suffices to prove that they are orthogonal. Let thus $x \in \operatorname{Ker}(\overrightarrow{\varphi} - \operatorname{Id}_E)$, that is, x is such that $\overrightarrow{\varphi}(x) = x$ and $y \in \operatorname{Im}(\overrightarrow{\varphi} - \operatorname{Id}_E)$, that is, such that $y = \overrightarrow{\varphi}(z) - z$ for some $z \in E$. We have indeed

$$x \cdot y = x \cdot (\overrightarrow{\varphi}(z) - z) = x \cdot \overrightarrow{\varphi}(z) - x \cdot z = \overrightarrow{\varphi}(x) \cdot \overrightarrow{\varphi}(z) - x \cdot z = 0.$$

Now we just have to apply Proposition I-2.22 or to copy its proof. Let O be any point of $\mathcal{E}$; we decompose

$$\overrightarrow{O\varphi(O)} = v + \overrightarrow{\varphi}(z) - z \text{ with } \overrightarrow{\varphi}(v) = v.$$

Let A be the point defined by $z = \overrightarrow{AO}$, so that

$$\overrightarrow{A\varphi(A)} = \overrightarrow{AO} + \overrightarrow{O\varphi(O)} + \overrightarrow{\varphi(O)\varphi(A)} = z + v + \overrightarrow{\varphi}(z) - z + \overrightarrow{\varphi}(\overrightarrow{OA}) = v.$$

Let us thus put $\psi = t_{-v} \circ \varphi$. We have

$$\psi(A) = t_{-v}(\varphi(A)) = A$$

so that ψ has fixed points. Moreover

$$\psi \circ t_v \circ \psi^{-1} = t_{\overrightarrow{\psi}(v)} = t_{\overrightarrow{\varphi}(v)} = t_v$$

so that ψ and t_v commute. The only thing left is the uniqueness of the pair (v, ψ). Let us assume that

$$\varphi = t_v \circ \psi = t_{v'} \circ \psi'$$

where ψ and ψ' have fixed points, say $\psi(A) = A$ and $\psi'(A') = A'$, and v, v' are fixed by $\overrightarrow{\varphi}$. Then $\overrightarrow{A\varphi(A)} = v$ and $\overrightarrow{A'\varphi(A')} = v'$, thus

$$\overrightarrow{AA'} = \overrightarrow{A\varphi(A)} + \overrightarrow{\varphi(A)\varphi(A')} + \overrightarrow{\varphi(A')A'} = v + \overrightarrow{\varphi}(\overrightarrow{AA'}) - v'$$

and

$$\overrightarrow{AA'} - \overrightarrow{\varphi}(\overrightarrow{AA'}) = v - v'.$$

This vector is thus in $\operatorname{Ker}(\overrightarrow{\varphi} - \operatorname{Id}_E) \cap \operatorname{Im}(\overrightarrow{\varphi} - \operatorname{Id}_E) = 0$, therefore $v = v'$ and $\psi = \psi'$. □

3. The group of linear isometries

The orthogonal group. The isometry group $O(E)$ of the Euclidean vector space E is called the *orthogonal group of* E. An orthonormal basis[(3)] of E being chosen, this group identifies with the orthogonal group of the Euclidean space $\mathbf{R}^n$; it is then denoted $O(n)$. It can be considered as a matrix group. Notice in particular that it is endowed with a topology, that induced by the usual topology of the real vector space $M_n(\mathbf{R})$ of square matrices.

Proposition 3.1. *A matrix A belongs to the group $O(n)$ if and only if it satisfies the relation ${}^tAA = \mathrm{Id}$.*

Here ${}^tA$ denotes the transpose of the matrix A.

Proof. Say that A is the matrix of an isometry, that is, the images of the vectors of the canonical basis, in other words the column vectors of A, form an orthonormal basis. But this is exactly what the relation ${}^tAA = \mathrm{Id}$ expresses: the fact that entries outside the diagonal in the matrix ${}^tAA$ are zero says that two distinct column vectors are orthogonal, while the diagonal entries are the squares of the norms of the column vectors. □

Remark 3.2. More intrinsically, we could have said that an endomorphism f of E is an isometry if and only if it satisfies the relation ${}^tff = \mathrm{Id}$ (I suggest that the reader looks for a "matrix-free", or "basis-free" proof).

Corollary 3.3. *The group $O(n)$ is compact.*

Proof. As this is a subset of a finite-dimensional real vector space, it is enough to prove that it is closed and bounded. But the mapping

$$\begin{array}{ccc} M_n(\mathbf{R}) & \longrightarrow & M_n(\mathbf{R}) \\ A & \longmapsto & {}^tAA \end{array}$$

is continuous (the entries of ${}^tAA$ are polynomials in the entries of A) and $O(n)$ is the inverse image of the closed subset $\{\mathrm{Id}\}$; it is thus closed.

Since all the column vectors of the elements of $O(n)$ are unit vectors, all the entries have an absolute value less than or equal to 1, in particular $O(n)$ is a bounded subset of $M_n(\mathbf{R})$. □

---

(3) It is probably not too late to recall that in a Euclidean space, there are orthonormal bases.

Positive isometries. The subgroup of vectorial rigid motions, which it is probably better to call *positive isometries*, is denoted by $O^+(E)$, $O^+(n)$ in the case of $\mathbf{R}^n$. This is a closed subgroup[(4)] of $O(n)$, being the inverse image of $\{1\}$ by the "determinant mapping", a mapping that is continuous as it is polynomial in the entries of the matrices.

The $n = 2$ case. The case of the group of plane isometries is particularly important: the identification of the group $O^+(2)$ with the group of complex numbers of modulus 1 and to $\mathbf{R}/2\pi\mathbf{Z}$ is what will be used in Chapter III to *measure* the angles.

Let us denote by $\mathbf{U}$ the multiplicative group of complex numbers of modulus 1.

Proposition 3.4. *The group $O^+(2)$ is isomorphic and homeomorphic with the multiplicative group $\mathbf{U}$ of complex numbers of modulus 1.*

Proof. A matrix

$$A = \begin{pmatrix} a & c \\ b & d \end{pmatrix}$$

belongs to $O(2)$ if and only if its two columns form an orthonormal basis, that is, if and only if $a^2 + b^2 = 1$, $c^2 + d^2 = 1$ and $ac + bd = 0$, and this gives

$$A = \begin{pmatrix} a & -\varepsilon b \\ b & \varepsilon a \end{pmatrix} \text{ with } \varepsilon = \pm 1 \text{ and } a^2 + b^2 = 1.$$

A direct computation gives $\varepsilon = \det A$ and thus $\varepsilon = 1$ if $A \in O^+(2)$. Now we only have to consider the mapping

$$\varphi : \; O^+(2) \longrightarrow \mathbf{U}$$

$$\begin{pmatrix} a & -b \\ b & a \end{pmatrix} \longmapsto a + ib.$$

We check that this is a group homomorphism by the computation of $\varphi(AA')$:

$$\varphi \begin{pmatrix} aa' - bb' & -(ba' + ab') \\ ba' + ab' & aa' - bb' \end{pmatrix} = aa' - bb' + i(ba' + ab') = (a + ib)(a' + ib').$$

This is obviously a bijective and continuous mapping, and so is its inverse mapping

$$\varphi^{-1} : a + ib \longmapsto \begin{pmatrix} a & -b \\ b & a \end{pmatrix}$$

thus ending the proof. □

---

(4) This is also an open subgroup of $O(n)$, being the inverse image of $]0, +\infty[$ by the determinant.

Corollary 3.5. *The group $O^+(2)$ is commutative.*

Proof. It is isomorphic with the commutative group $\mathbf{U}$ of complex numbers of modulus 1. □

Corollary 3.6. *The group $O^+(2)$ is path connected. The group $O(2)$ has two path connected components.*

Proof. The group of complex numbers of modulus 1 is path connected (this is a circle). The same is true of $O^+(2)$, which is homeomorphic to it.

On the other hand, the determinant mapping

$$\det : O(2) \longrightarrow \{\pm 1\}$$

is continuous as we have already said; it takes its values in a discrete space and it is surjective. Thus $O(2)$ is not connected; it can be written as the disjoint union

$$O(2) = O^+(2) \cup O^-(2)$$

denoting by $O^-(2)$ the set of isometries that have a negative determinant. We have already said that $O^+(2)$ is path connected. But $O^-(2)$ also is path connected, simply because it is homeomorphic to it: we choose once and for all an element $g_0 \in O^-(2)$ (we know that $O^-(2)$ is nonempty, it contains the reflections) and we consider the mapping

$$\begin{aligned} O^-(2) &\longrightarrow O^+(2) \\ f &\longmapsto f \circ g_0. \end{aligned}$$

It is continuous (why?), bijective (same question) and its inverse mapping is also continuous (is this clear?), hence this is a homeomorphism, and thus $O^-(2)$ is as path connected as $O^+(2)$. □

Remark 3.7. The end of the proof, namely the fact that $O^-(2)$ is homeomorphic with $O^+(2)$, works in any dimension n; we are going to use this in the same way... as soon as we have proved that $O^+(n)$ is path connected in general (Proposition 3.15 below).

Rotations, measures of the angles of rotations. The complex exponential defines a mapping

$$\begin{array}{ccccc} \mathbf{R} & \longrightarrow & \mathbf{U} & \longrightarrow & O^+(2) \\ \theta & \longmapsto & e^{i\theta} & \longmapsto & \begin{pmatrix} \cos\theta & -\sin\theta \\ \sin\theta & \cos\theta \end{pmatrix}, \end{array}$$

that is, of course, surjective and periodic of period 2π and that defines a group isomorphism

$$\mathbf{R}/2\pi\mathbf{Z} \longrightarrow O^+(2).$$

Definition 3.8. The image of the real number θ is called a *rotation of angle* θ. The positive isometries of a Euclidean plane are called *rotations*.

From what we have just done follows the fact that any element of $O(2)$ has the form

$$\begin{pmatrix} \cos\theta & -\varepsilon\sin\theta \\ \sin\theta & \varepsilon\cos\theta \end{pmatrix} \text{ with } \varepsilon = \pm 1.$$

Remark 3.9. The existence and periodicity of the functions cos and sin, even if familiar to the reader since she was very young, are not trivial facts. They come:

- either from the properties of the complex exponential, defined by its expansion as a power series (see [**Car95**] or the astounding prologue of [**Rud87**]),
- or from a precise definition of the measure of the length of an arc of circle (evoked here in § VII-4).

Change of basis, conjugacy. Let us consider now a Euclidean (vector) plane E. Choosing an orthonormal basis, we get an isomorphism (of Euclidean vector spaces) from E to $\mathbf{R}^2$ and group isomorphisms from $O(E)$ to $O(2)$ and from $O^+(E)$ to $O^+(2)$.

It is important to notice that the writing of plane rotations in an orthonormal basis does not depend much on this basis. More precisely:

Proposition 3.10. *Let E be a Euclidean plane and f be a positive isometry of E. We have*

$$g \circ f \circ g^{-1} = \begin{cases} f & \text{if } g \in O^+(E) \\ f^{-1} & \text{if } g \in O^-(E). \end{cases}$$

Proof. The first equality is an immediate consequence of the commutativity of $O^+(E)$. To prove the second, the simplest thing to do is to use matrices as above. □

Remark 3.11. This is another version of the conjugacy principle (here the statement I-2.19).

Corollary 3.12. *Let E be an oriented plane. The matrix of a rotation does not depend on the choice of the direct orthonormal basis used to write it.*

An orientation of the plane being fixed, let us choose a direct orthonormal basis. The changes of direct orthonormal bases belong to $O^+(E)$ (see if necessary Exercise II.11). It suffices now to apply the first equality of Proposition 3.10. □

Reduction, the general case. Let us now consider a Euclidean vector space of dimension n. The elements of the orthogonal group $O(n)$ have a little more intricate form than those of $O(2)$ but it is nevertheless possible to write them quite explicitly and deduce results that are analogous to the previous ones.

Proposition 3.13. *Let f be a linear isometry of E. The vector space E is an orthogonal direct sum*

$$E = V \oplus W \oplus P_1 \oplus \cdots \oplus P_r,$$

an equality in which the subspaces V, W and P_i are stable under f and where $f|_V = \mathrm{Id}_V$, $f|_W = -\mathrm{Id}_W$ and every P_i is a plane, the restriction of f to which is a rotation.

In terms of matrices, this says that there exists an orthonormal basis of E in which the matrix of f is written

$$\begin{pmatrix} 1 & & & & & & & \\ & \ddots & & & & & & \\ & & 1 & & & & & \\ & & & -1 & & & & \\ & & & & \ddots & & & \\ & & & & & -1 & & \\ & & & & & & \begin{pmatrix} \cos\theta_1 & -\sin\theta_1 \\ \sin\theta_1 & \cos\theta_1 \end{pmatrix} & & \\ & & & & & & & \ddots & \\ & & & & & & & & \begin{pmatrix} \cos\theta_r & -\sin\theta_r \\ \sin\theta_r & \cos\theta_r \end{pmatrix} \end{pmatrix}.$$

Notice also that $\det f = (-1)^{\dim W}$ and thus that W has even dimension when f is a positive isometry, and that W is not reduced to 0 when f is a negative isometry.

Proof. This is by induction on n. For $n = 1$, the positive isometries are the identity and the central symmetry $-\mathrm{Id}$ and thus the result is true; for $n = 2$, we have already said that an isometry is

- a reflection, which corresponds to $\dim V = 1$, $\dim W = 1$ and $P_i = 0$
- or a rotation, the case where $V = W = 0$ and $r = 1$

and the result also holds. We assume now (this is the induction hypothesis) that the result is true for all the isometries of the Euclidean vector spaces of dimension $\leqslant n - 1$.

Let us consider an isometry f of a space E of dimension $n \geqslant 3$. It suffices to find a nontrivial subspace F of E that is stable by f. Its orthogonal $F^\perp$ will also be stable under f and we will apply the induction hypothesis to F and $F^\perp$.

If f has a real eigenvalue, we simply take for F the line generated by a nonzero eigenvector.

Otherwise, let λ be a complex eigenvalue (assumed to be nonreal). Then $\overline{\lambda}$ is also an eigenvalue. Let x be a (complex) eigenvector[5] of f for the eigenvalue λ, so that $\overline{x}$ is an eigenvector for $\overline{\lambda}$. The complex plane spanned by x and $\overline{x}$ is stable by f. But the two vectors $(x+\overline{x})/2$ and $(x-\overline{x})/2i$ are real. They span a plane in E and this plane is stable by f and we can use it for F. □

Remark 3.14. The reader should now determine why we have initialized the induction by the consideration of both the cases $n = 1$ *and* $n = 2$.

Proposition 3.15. *The group $O^+(n)$ is path connected. The group $O(n)$ has two path connected components that are homeomorphic to $O^+(n)$.*

Proof. We proceed as in the $n = 2$ case (proof of Corollary 3.6): it suffices to prove that $O^+(n)$ is path connected, that is to say that any two elements can be connected by a path in $O^+(n)$. To do this, it suffices of course to prove that it is possible to connect any element of $O^+(n)$ to the identity.

Let thus A be an element of $O^+(n)$. Proposition 3.13 tells us that there exists an orthogonal matrix P such that ${}^tPAP$ has the form above. As $A \in O^+(n)$, the number of -1 is even and so we can group them by pairs and consider that

$$\begin{pmatrix} -1 & 0 \\ 0 & -1 \end{pmatrix} = \begin{pmatrix} \cos\theta & -\sin\theta \\ \sin\theta & \cos\theta \end{pmatrix} \text{ for } \theta = \pi,$$

in other words, the matrix $B = {}^tPAP$ contains only 1's and diagonal rotation blocks.

For all $t \in [0,1]$, let us form now a matrix $B(t) \in O^+(n)$ by replacing the θ's that appear in B by $t\theta$'s, so that $t \mapsto B(t)$ is a continuous path from Id to B in $O^+(n)$... and, more interesting, that $t \mapsto PB(t){}^tP$ is a continuous path from Id to A in $O^+(n)$, and this is what we were looking for. □

---

(5) It is possible to choose a basis of E, then we can assume that we are in $\mathbf{R}^n$ and consider the latter to be a subset of $\mathbf{C}^n$. It is also possible to construct a stable subspace F geometrically (without complexification), see, *e.g.*, [**Per96**, Exercise VI.5].

Why rigid motion? From the previous results, we deduce:

Proposition 3.16. *The group* $\operatorname{Isom}^+(\mathcal{E})$ *of rigid motions of a Euclidean affine space* $\mathcal{E}$ *is path connected.*

Remark 3.17. This proposition justifies the terminology "rigid motions" for these isometries: a path from the identity to the rigid motion f allows us to move (in the usual sense) an object.

Proof. The choice of an origin O in $\mathcal{E}$ allows us to construct a homeomorphism from $\operatorname{Isom}^+(\mathcal{E})$ to the product space $O^+(E) \times E$. This is simply the mapping defined by

$$\begin{array}{ccc} \operatorname{Isom}^+(\mathcal{E}) & \longrightarrow & O^+(E) \times E \\ \varphi & \longmapsto & (\overrightarrow{\varphi}, \overrightarrow{O\varphi(O)}). \end{array}$$

It is continuous and its inverse is also (think of their expressions in coordinates). Since $O^+(E)$ and E are path connected, their product is path connected as well. The space $\operatorname{Isom}^+(\mathcal{E})$, which is homeomorphic to it, is thus path connected too. □

Remark 3.18. Complementary information on the properties and the topology of $O^+(n)$ and other classical groups can be found in Exercises II.22, IV.35, IV.44, IV.45, V.48 and in the books [**Per96**], [**Ber94**], [**MT86**].

Exercises and problems

Direct applications and complements

Exercise II.1. State and prove the Cauchy–Schwarz inequality (taking care of the equality case). Prove that $\|\cdot\|$ is a norm and that d is a distance. When is there equality in the triangle inequality? Prove that the straight line is the shortest way (by broken lines) from a point to another.

Exercise II.2. Let a, b and c be three nonnegative numbers such that

$$|b - c| < a < b + c.$$

Prove that there exists a triangle of sides a, b, c.

Exercise II.3. Prove that G is the barycenter of $((A, \alpha), (B, 1 - \alpha))$ if and only if for every point O,

$$\alpha OA^2 + (1 - \alpha)OB^2 = OG^2 + \alpha GA^2 + (1 - \alpha)GB^2.$$

Exercise II.4. Let $\varphi : \mathcal{E} \to \mathcal{E}$ be a mapping defined on a Euclidean affine space. Assume that φ preserves the distances:

$$d(\varphi(A), \varphi(B)) = d(A, B).$$

Prove that φ is affine (and hence is an isometry)[(6)].

Exercise II.5. Let E be a Euclidean vector space and let $f : E \to E$ be a (set theoretical) mapping that preserves the scalar product. Prove that f is linear (and thus is an isometry).

Exercise II.6. Prove that a symmetry is an isometry only if this is an orthogonal symmetry.

Exercise II.7 (Reflections, perpendicular bisector). Let H be a hyperplane of a Euclidean vector space E and let x_0 be a nonzero vector of $H^\perp$. Prove that

$$s_H(x) = x - 2\frac{x \cdot x_0}{\|x_0\|^2} x_0.$$

Prove that, if x and y are two vectors of the same norm in the Euclidean vector space E, there exists a hyperplane H such that $s_H(x) = y$ (and that H is unique if $x \neq y$). Prove similarly that, if A and B are two points of an affine space $\mathcal{E}$, there exists an affine hyperplane $\mathcal{H}$ such that $\sigma_{\mathcal{H}}(A) = B$ (and that $\mathcal{H}$ is unique if $A \neq B$).

Exercise II.8 (Partition of the plane by the perpendicular bisector). Let A and B be two points of a Euclidean affine plane $\mathcal{P}$. Prove that the set

$$\{M \in \mathcal{P} \mid MA < MB\}$$

is the half-plane defined by the perpendicular bisector of the segment AB that contains A (see Exercise I.46).

Exercise II.9 (The "scalar function" of Leibniz). This is a name given to the function F, defined by the system of punctual masses $((A_1, \alpha_1), \ldots, (A_k, \alpha_k))$ and that, with a point M of the Euclidean affine space $\mathcal{E}$, associates the scalar

$$F(M) = \sum_{i=1}^{k} \alpha_i MA_i^2.$$

Assume the sum $\sum \alpha_i$ is nonzero. Prove that there exists a fixed vector v such that, for any point M' of $\mathcal{E}$,

$$F(M') = F(M) + 2\overrightarrow{MM'} \cdot v.$$

---

(6) This is not that easy...

If the sum $\sum \alpha_i$ is nonzero, call the barycenter of the system G. Check that

$$F(M) = F(G) + \left(\sum \alpha_i\right) MG^2.$$

Applications. A real number k is given. Determine, according to the values of k,

- the set of points M satisfying the equation $MA^2 + MB^2 = k$,
- the set of points M satisfying the equation $MA^2 - MB^2 = k$,
- the set of points M satisfying the equation $\dfrac{MA}{MB} = k$.

Exercise II.10 (A summary of orientations). Let E be a real vector space of finite dimension n. Let $\mathcal{B}$ be the set of all the bases of the vector space E. If b and b' are two bases (two elements of $\mathcal{B}$), it is said that they are equivalent if $\det_b b' > 0$. Prove that this is indeed an equivalence relation and that there are two equivalence classes. *To orient* the space E is to choose one of these two classes. The bases of the chosen class are said to be *direct*.

Exercise II.11. Prove that rigid motions are the isometries that preserve an orientation (all the orientations) of a space. Prove that a reflections reverses the orientation.

Exercise II.12. What are the positive isometries of the Euclidean plane that are involutional?

Exercises

For all the following exercises, we are in a Euclidean affine plane.

Exercise II.13. Let ABC be a triangle. Construct a square $MNPQ$ such that $M, N \in BC$, $Q \in AB$, $P \in AC$.

Exercise II.14. Let $\mathcal{D}$ and $\mathcal{D}'$ be two parallel lines and let A and B be two points on both sides of the strip determined by $\mathcal{D}$ and $\mathcal{D}'$. Construct M on $\mathcal{D}$ and M' on $\mathcal{D}'$ such that MM' has a given direction and:

- such that $AM = BM'$,
- then such that AM and BM' are perpendicular,
- finally such that $AM + MM' + M'B$ is minimal.

Exercise II.15. Given two circles $\mathcal{C}$ and $\mathcal{C}'$, find all the dilatations that map $\mathcal{C}$ to $\mathcal{C}'$.

Exercise II.16. Three circles $\mathcal{C}_1$, $\mathcal{C}_2$ and $\mathcal{C}_3$ whose centers are not collinear are given. Let I_1, I_2 and I_3 (*resp.* J_1, J_2 and J_3) be the centers of the dilatations of positive (*resp.* negative) ratios mapping $\mathcal{C}_2$ to $\mathcal{C}_3$, $\mathcal{C}_3$ to $\mathcal{C}_1$ and $\mathcal{C}_1$ to $\mathcal{C}_2$. Prove that I_1, J_2 and J_3 (and similarly I_2, J_3 and J_1, I_3, J_1 and J_2)

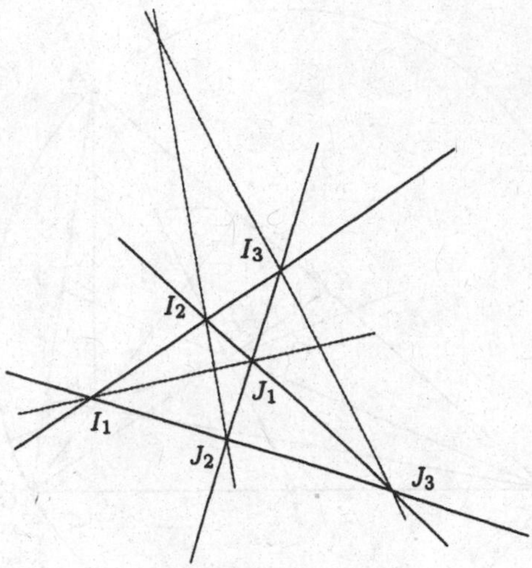

Fig. 6

and I_1, I_2 and I_3 are collinear (see Figure 6, on which one should mark the centers of the three circles).

Exercise II.17. Construct a circle tangent to two given lines and passing through a given point.

Exercise II.18 (Orthocenter). Why are the perpendicular bisectors of the sides of a triangle concurrent?[7] Look at Figure 19 of Chapter I with a Euclidean eye and prove that the three altitudes of a triangle are concurrent[8].

Exercise II.19. Let AB be a chord of a circle $\mathcal{C}$. Prove that the locus of the orthocenters H of the triangles AMB (when M describes $\mathcal{C}$) is the circle $\mathcal{C}'$, the image of $\mathcal{C}$ by reflection about AB.

Exercise II.20 (The Euler circle). Let ABC be a triangle, and let G, O and H be its centroid, circumcenter and orthocenter (respectively). Thinking again of Figure 19 of Chapter I, prove that O, G and H are collinear and more precisely that

$$\overrightarrow{GO} = -\frac{1}{2}\overrightarrow{GH}.$$

[7]The intersection point is the *circumcenter*, center of the *circumcircle*.
[8]The intersection point is the *orthocenter*.

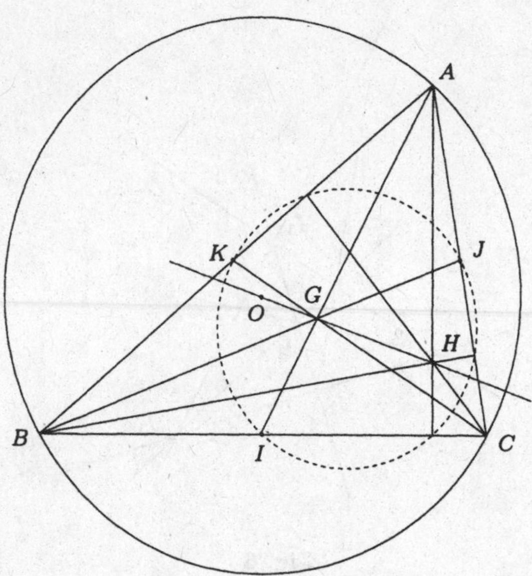

Fig. 7. The Euler circle

Let I, J and K be the midpoints of BC, CA and AB and $\mathcal{C}$ the circumcircle of the triangle IJK (Figure 7). Prove that it also passes through[9] the feet of the altitudes of ABC and the midpoints of AH, BH and CH.

More theoretical exercises

Exercise II.21 (Centers of $O(n)$, $O^+(n)$). Let g be an isometry of $\mathbf{R}^n$ that commutes with all the elements of $O(n)$. Prove that g preserves all the lines. Deduce that the center of $O(n)$ consists of Id and $-$ Id. What is the center of $O^+(n)$?

Exercise II.22 (The Gram–Schmidt orthonormalization process). Let $(x_1, \ldots, x_n)$ be a basis of a Euclidean vector space E. Prove that there exists a unique orthonormal basis of E, denoted by $(e_1, \ldots, e_n)$, such that

- the subspace spanned by $(e_1, \ldots, e_k)$ coincides with that spanned by $(x_1, \ldots, x_k)$ (for all k),
- the bases $(e_1, \ldots, e_k)$ and $(x_1, \ldots, x_k)$ define the same orientation of this subspace.

Let A be an $n \times n$ invertible real matrix. Prove that there exists an upper triangular matrix $P = P(A)$, depending continuously of A, the diagonal

(9) This is why this circle is also called the *nine point circle*, although it contains much more than nine points... and even many more than nine "remarkable" points.

entries of which are all nonnegative and such that AP belongs to $O(n)$. Prove that the linear group $GL(n;\mathbf{R})$ has two connected components.

Exercise II.23 (The Lorentz group). Consider the quadratic form[10]

$$q(x) = x_1^2 - x_2^2 - \cdots - x_n^2$$

on $\mathbf{R}^n$. Characterize the matrices of the "isometries" of the form q, namely of the linear isomorphisms f such that $q \circ f(x) = q(x)$ for all x in $\mathbf{R}^n$. Let O_q be the isometry group.

Determine the elements of O_q when $n = 2$. Do you think O_q is a compact group?

Assume that $n = 2$. Let O_q^+ be the group of isometries of q with nonnegative determinant. How many connected components does it have? Same question for O_q.

(10) See if necessary § VI-8 for a summary of quadratic forms.

Chapter III

Euclidean Geometry in the Plane

In this chapter, there are plane isometries, triangles and angles at the circumference, similarities, inversions and even pencils of circles. But there is also, and we are forced to begin with this, a discussion of what an angle is and how to measure it. The proofs are of course very simple but the statements and their precision are subtle and important.

1. Angles

The intuitive idea we have of an angle as being the subset of the plane limited by two half-lines of the same origin is hard to use rigorously. It seems that the best definition of this "gap" between the two half-lines is *via* the rigid motion that allows us to pass from one to the other.

It allows us in any case to endow the set of angles with the structure of a group and to *measure* the angles, namely to associate with any angle a real number (nonunique, alas) in an additive way.

Notation and reminder. We are in a Euclidean plane P. We use the same notation as in the previous chapter. The positive linear isometries of the plane are, as we have seen in § II-3, the rotations. We shall keep the word "rigid motion" to describe an *affine* isometry.

The basic remark is the next proposition.

Proposition 1.1. *Two unit vectors of a plane being given, there exists a unique rotation that maps one to the other.*

Proof. Let u and u' be two unit vectors of P. Let v be a unit vector such that (u, v) is an orthonormal basis. Then u' can be written in this basis:

$$u' = au + bv$$

with $a^2 + b^2 = 1$ since u' is a unit vector. The rotation f whose matrix in the basis (u, v) is $\begin{pmatrix} a & -b \\ b & a \end{pmatrix}$ satisfies $f(u) = u'$ and is the only one to do so (it is determined by a and b). □

The next question is of course: given *two* ordered pairs of unit vectors (u, v), (u', v'), does there exist a rotation that maps u to u' and v to v'? The answer is negative in general and this is where angles are hidden.

Oriented angles of vectors. An equivalence relation on the set of ordered pairs of unit vectors is defined by $(u, v)\mathcal{R}_1(u', v')$ if and only if there exists a rotation f of P such that $f(u) = u'$ and $f(v) = v'$.

The equivalence class of (u, v) is called the *oriented angle* of u and v, but also the oriented angle of u' and v' for all the vectors $u' = \lambda v$, $v' = \mu v$ (with λ, $\mu > 0$)[1]. Since we want to save on notation, we denote by (u, v) the class of (u, v), and no confusion should arise.

We call $\widehat{\mathcal{A}}$ the set of ordered pairs of unit vectors and $\mathcal{A}$ the set of oriented angles of vectors (the equivalence classes for the relation $\mathcal{R}_1$). Our next aim is to define a group structure on $\mathcal{A}$. In fact, this set looks very much like $O^+(P)$, which is a group. Proposition 1.1 allows us, indeed, to define a mapping

$$\widehat{\Phi} : \widehat{\mathcal{A}} \longrightarrow O^+(P)$$

that, with the ordered pair (u, u'), associates the unique rotation f that maps u to u'. We have:

Lemma 1.2. *The ordered pairs (u, u') and (v, v') have the same image by $\widehat{\Phi}$ if and only if they define the same oriented angle.*

Proof. Let f and g be the rotations defined (thanks to Proposition 1.1) by

$$f(u) = u', \quad g(v) = v'$$

and let r and r' those defined by

$$r(u) = v, \quad r'(u') = v'$$

(Figure 1). To say that $\widehat{\Phi}(u, u') = \widehat{\Phi}(v, v')$ is to say that the rotations r and r' are the same. To say that the oriented angles (u, u') and (v, v') are equal is to say that they satisfy $(u, u')\mathcal{R}_1(v, v')$, namely that f and g are equal.

The rotation $r' \circ f$ maps u to v', as does $g \circ r$. The uniqueness assertion in Proposition 1.1 gives $r' \circ f = r \circ g$ and, $O^+(P)$ being commutative (Proposition II-3.5), $r' \circ f = r \circ g$. This gives indeed that $r = r'$ if and only if $f = g$. □

(1) So that the dilatations preserve the oriented angles of vectors, by definition.

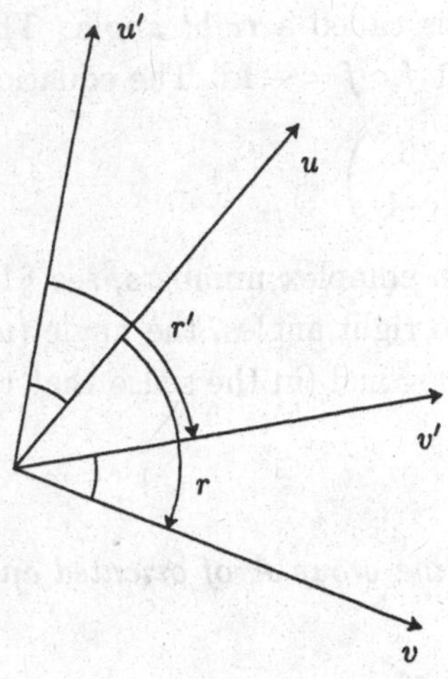

Fig. 1

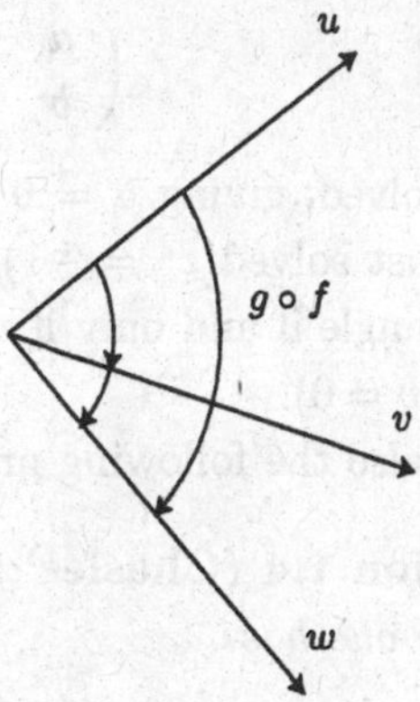

Fig. 2. Chasles' relation

The lemma asserts that the mapping $\widehat{\Phi}$ defines a *bijective* mapping

$$\Phi : \mathcal{A} \longrightarrow O^+(P),$$

the inverse mapping associating with the rotation f the angle $(u, f(u))$ (for any unit vector u). Because $O^+(P)$ is a (commutative) group, this gives $\mathcal{A}$ the structure of a (commutative) group, denoted additively:

$$(u,v) + (u',v') = (u'',v'')$$

where u'' is an arbitrary unit vector and $v'' = r' \circ r(u'')$, r and r' being defined by

$$v = r(u) \text{ and } v' = r'(u').$$

Remark 1.3. This operation is called a *transport of structure*: the mapping Φ is used to transport the group structure of $O^+(P)$. The given formula is equivalent to:

$$(u,v) + (u',v') = \Phi^{-1}\left(\Phi(u,v) \circ \Phi(u',v')\right).$$

It metamorphoses the bijection Φ into a group isomorphism.

The zero angle is the one corresponding to the rotation identity, namely (u,u). The angle $(u,-u)$, which corresponds to the central symmetry $-\mathrm{Id}$ is called the *flat angle*. This is an order-2 element in the group $\mathcal{A}$. We have indeed

$$2(u,-u) = 0 \text{ since } (-\mathrm{Id}) \circ (-\mathrm{Id}) = \mathrm{Id}.$$

An angle (u, v) such that $2(u, v) = (u, -u)$ is called a *right angle*. These angles correspond to the rotations f such that $f \circ f = -\operatorname{Id}$. The equation

$$\begin{pmatrix} a & -b \\ b & a \end{pmatrix}^2 = \begin{pmatrix} -1 & 0 \\ 0 & -1 \end{pmatrix}$$

is easily solved, giving $a = 0$ and $b = \pm 1$ (in complex numbers, see § II-3, we have just solved $z^2 = -1$). There are two right angles, the angle (u, v) is a right angle if and only if u and v are orthogonal (in the sense that they satisfy $u \cdot v = 0$).

Notice also the following property:

Proposition 1.4 (Chasles' relation). *In the group $\mathcal{A}$ of oriented angles of vectors, one has*

$$(u, v) + (v, w) = (u, w)$$

for all unit vectors u, v and w.

Proof. If the rotations f and g satisfy the relations $f(u) = v$ and $g(v) = w$, then the composition $g \circ f$ satisfies $g \circ f(u) = w$ (Figure 2). □

Remark 1.5. One could have *defined* the group structure by the Chasles relation.

Oriented angles of lines. We would like to call the oriented angle of the two affine lines (in an affine Euclidean plane $\mathcal{P}$) or of the two vectorial lines (in the Euclidean linear plane P) D and D' the oriented angle of two unit vectors. Unfortunately, there are two unit vectors on a line. Therefore, the oriented angle (D, D') is defined as the equivalence class of (u, u') (where u directs D and u' directs D') under the relation $\mathcal{R}_2$ defined by:

$$(u, u')\mathcal{R}_2(v, v') \iff \begin{cases} (u, u')\mathcal{R}_1(v, v') \\ \text{or} \\ (u, u')\mathcal{R}_1(-v, v'). \end{cases}$$

The set of oriented angles of lines is a group, simply because this is the quotient of the group $\mathcal{A}$ by the index-2 subgroup generated by the flat angle. We have, by definition:

Proposition 1.6. *Let D, D', Δ and Δ' be lines with spanning unit vectors u, u', v and v' respectively. We have the equality $2(u, u') = 2(v, v')$ in $\mathcal{A}$ if and only if the oriented angles of lines (D, D') and (Δ, Δ') are equal.* □

This property is sometimes expressed as "when you divide an equality of oriented angles of vectors by 2, what you get is an equality of oriented angles of lines".

Notice that, if (D, D') is an angle of lines, $2(D, D')$ is a well-defined element of the group $\mathcal{A}$ of oriented angles of vectors.

Orientations of the plane and measures of the oriented angles. Let us choose an orientation of the plane P. The matrix of a rotation f is the same in all the *direct* orthonormal bases (see Corollary II-3.12). With the angle (u, v) is associated, by Φ, a matrix of the form

$$\begin{pmatrix} \cos\theta & -\sin\theta \\ \sin\theta & \cos\theta \end{pmatrix}$$

where θ is a real number that is well-defined only modulo 2π, which we have called the angle of the rotation f and that we call here a *measure of the angle* (u, v).

Remarks 1.7

- Any oriented angle of vectors has measures: that is, the composed mapping
$$\mathbf{R} \to \mathbf{R}/2\pi\mathbf{Z} \to O^+(P) \to \mathcal{A}$$
is surjective.
- If θ is a measure of the angle (u, v), all the other measures of this angle are the $\theta + 2k\pi$, for $k \in \mathbf{Z}$: that is, the mapping
$$\mathbf{R}/2\pi\mathbf{Z} \to O^+(P) \to \mathcal{A}$$
is injective.
- The measures of an angle depend on the chosen orientation: if θ is a measure of the angle (u, v) and if we choose the other orientation, these are the $-\theta + 2k\pi$ that are now its measures.
- The number π is a measure of the flat angle.
- The right angles have measures $\dfrac{\pi}{2} + 2k\pi$ or $-\dfrac{\pi}{2} + 2k\pi$.
- The basis (u, u') is orthonormal and direct if and only if $\pi/2$ is a measure of the oriented angle of vectors (u, u').

One can also measure the angles of lines. As we have just said that π is a measure of the flat angle, the measures are, for the angles of lines, the elements of $\mathbf{R}/\pi\mathbf{Z}$.

Remark 1.8. Proposition 1.6 can be translated very simply in terms of measures of angles by the equivalence

$$2\theta \equiv 2\varphi \mod 2\pi \quad \Longleftrightarrow \quad \theta \equiv \varphi \mod \pi.$$

This means that there exists an integer k such that the equality $2\theta = 2\varphi + 2k\pi$ holds if and only if there exists an integer ℓ such that $\theta = \varphi + \ell\pi$. When you divide a congruence, you need to divide everything!

If we speak of *measuring* the angles, this is because there is an additivity property. From the previous definitions and constructions, we deduce:

Proposition 1.9. *The mapping that, with an oriented angle of vectors (resp. of lines), associates one of its measures, defines a morphism from the group of oriented angles to* $\mathbf{R}/2\pi\mathbf{Z}$ *(resp. in* $\mathbf{R}/\pi\mathbf{Z}$*). This morphism depends on the chosen orientation. This is an isomorphism.* □

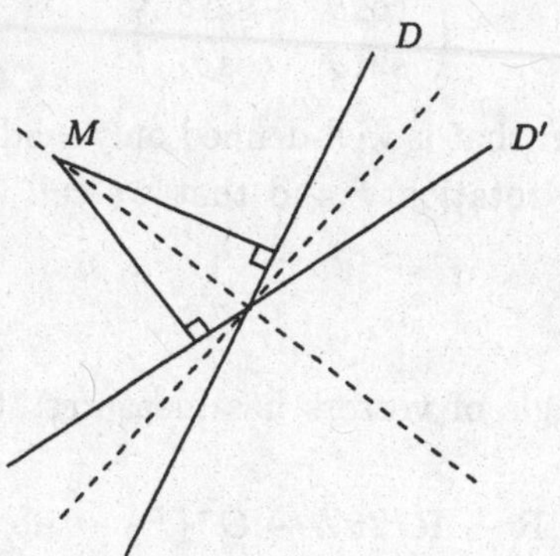

Fig. 3. Bisectors

Bisectors. The figure consisting of two distinct (either vectorial or affine and secant) lines D and D' has two axes of symmetry, the two lines spanned by the vectors $u+u'$ and $u-u'$ for any choice of unit vectors $u \in D$, $u' \in D'$. These two lines are orthogonal, and their union is the set of points M that are equidistant from D and D' (Figure 3).

Moreover, these are *bisectors* of the angle (D, D') in the sense that, if Δ is one of these two lines, $(D, \Delta) = (\Delta, D')$. This is a consequence of the following proposition.

Proposition 1.10. *Reflections reverse oriented angles (of vectors or of lines).*

Proof. Let us consider a line D and two unit vectors u, v. We want to prove that

$$(s_D(u), s_D(v)) = (v, u).$$

We consider the perpendicular bisector D' of u and v, namely the line $(u-v)^\perp$. The composition $s_D \circ s_{D'}$ is a positive isometry, hence this is a rotation, so that

$$(v, u) = (s_D \circ s_{D'}(v), s_D \circ s_{D'}(u)) = (s_D(u), s_D(v)).$$

□

The reader should remember that an angle of lines has two bisectors and there is no way to distinguish one from the other.

Geometric angles and internal bisector. To define a geometric angle, we simply identify the angles (u, v) and (v, u). Notice that we lose the group structure (but the notions of flat or right angle persist). As a consequence of the previous results, we have:

Proposition 1.11. *Isometries preserve geometric angles.* □

One could think of a geometric angle as being the convex subset of the plane contained between the two half-lines generated by the vectors u and u', more precisely as being the convex hull of the union of the two half-lines (see § I-5 and Figure 4). I will abuse the language and write "geometric angle" to mean this convex hull.

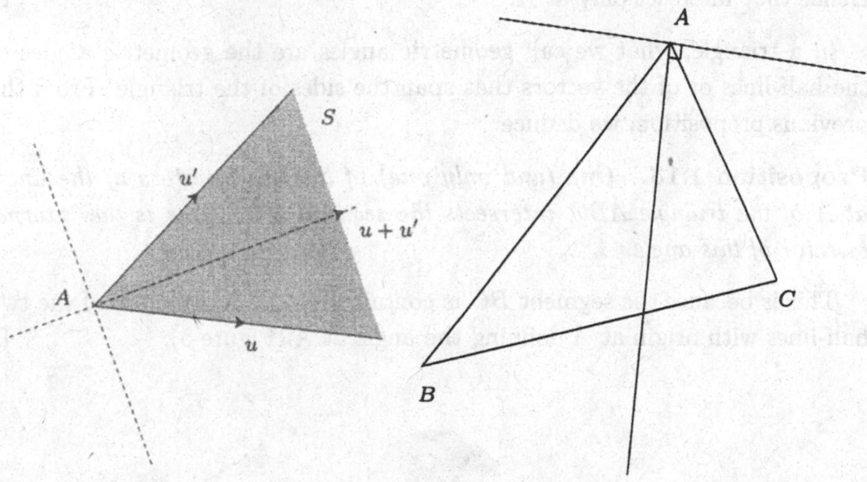

Fig. 4. The bisectors of a geometric angle

Fig. 5. The bisectors of an angle of a triangle

Notice finally that it is also possible to define an *internal* bisector for a nonflat geometric angle.

Proposition 1.12. *If d and d' are two half-lines of origin A that form a nonflat geometric angle, only one of the two bisectors of the angle of lines supporting d and d' intersects the geometric angle determined by d and d' at other points than A.*

Proof. Let S be the convex hull of the union of the two half-lines d and d'. Let u and u' be unit vectors directing d and d'. The bisectors of the two lines generated by u and u' are the lines through A that are spanned by $u + u'$

and $u - u'$. As S is the convex hull of $d \cup d'$, all the points M of the half-line of origin A spanned by $u + u'$, satisfying

$$\overrightarrow{AM} = \lambda(u + u') \text{ with } \lambda \geqslant 0,$$

belong to S (see Proposition I-5.6). Therefore, the bisector defined by $u + u'$ intersects S along a half-line (see Figure 4).

Let us prove that the other bisector (the *external* bisector) intersects S only at A. Recall that the convex hull S is the set of barycenters of the points of $d \cup d'$ endowed with nonnegative coefficients (still following Proposition I-5.6), here of the points M such that

$$\overrightarrow{AM} = \lambda u + \lambda' u' \text{ with } \lambda \geqslant 0 \text{ and } \lambda' \geqslant 0.$$

The line through A spanned by $u - u'$ consists of the points M satisfying

$$\overrightarrow{AM} = \mu(u - u') \text{ with } \mu \in \mathbf{R}.$$

Hence they intersect only at A. □

In a triangle, what we call geometric angles are the geometric angles of the half-lines or of the vectors that span the sides of the triangle. From the previous proposition, we deduce:

Proposition 1.13. *One (and only one) of the two bisectors of the angle at A of the triangle ABC intersects the segment BC. This is the internal bisector of this angle.*

This is because the segment BC is contained in the convex hull of the two half-lines with origin at A defining the angle at A (Figure 5). □

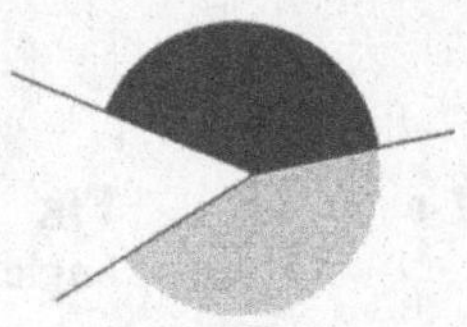

Fig. 6

"Measure of geometric angles". With a geometric angle defined by two vectors u and v, a number can be associated; this is the unique real number $\theta \in [0, \pi]$ where θ is a measure of one of the two oriented angles (u, v) or (v, u). This is what you read on a protractor[(2)]. One also speaks of *measure*

[(2)] On a different scale in which the value of π is 180.

in this case, although the qualification is somewhat improper: the measure thus defined is not additive, and moreover, it is not possible to add overlarge geometric angles. See Figure 6 in which two geometric angles of measure $\pi - \varepsilon$ add (?) to give a geometric angle of measure 2ε!

Until the end of this chapter, even if we do not state it explicitly, we will be in an affine Euclidean plane, which we will assume to be oriented when we will use measures of angles.

A few basic theorems on angles. After the dryness of the previous general considerations, let us refresh ourselves with a few geometry theorems.

Proposition 1.14. *Let A, B and C be three distinct points in an affine Euclidean plane. The sum of the oriented angles of vectors*

$$(\overrightarrow{AB}, \overrightarrow{AC}) + (\overrightarrow{BC}, \overrightarrow{BA}) + (\overrightarrow{CA}, \overrightarrow{CB})$$

is a flat angle.

Proof. We use a central symmetry to get:

$$\begin{aligned}(\overrightarrow{AB}, \overrightarrow{AC}) + (\overrightarrow{BC}, \overrightarrow{BA}) + (\overrightarrow{CA}, \overrightarrow{CB}) &= (\overrightarrow{AB}, \overrightarrow{AC}) + (\overrightarrow{BC}, \overrightarrow{BA}) + (\overrightarrow{AC}, \overrightarrow{BC})\\ &= (\overrightarrow{AB}, \overrightarrow{BA})\end{aligned}$$

thanks to Chasles' relation. □

There is a slightly more complicated proof, which has a historical interest (see Remark 1.16) and allows us to prove a more precise result.

We draw the parallel to BC through A (Figure 7), on which we define a point D by $\overrightarrow{AD} = \overrightarrow{BC}$. If E satisfies $\overrightarrow{CE} = \overrightarrow{BC}$, we have $(\overrightarrow{CE}, \overrightarrow{CD}) = (\overrightarrow{BC}, \overrightarrow{BA})$ (using the translation of vector $\overrightarrow{BC}$).

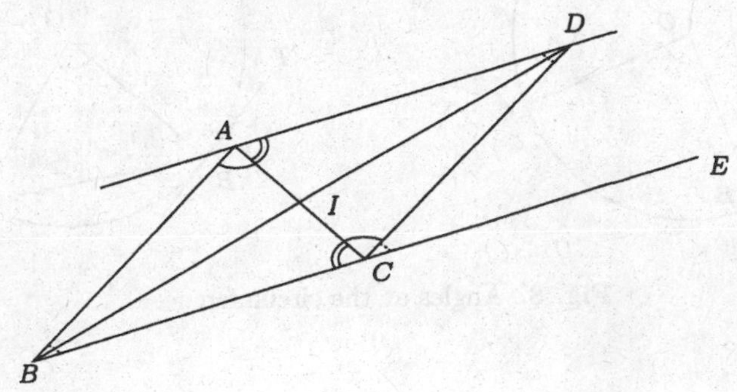

Fig. 7

Let I be the common midpoint of BD and AC. The symmetry of center I maps A to C and B to D, so that $(\overrightarrow{CD}, \overrightarrow{CA}) = (\overrightarrow{AB}, \overrightarrow{AC})$.

Hence the flat angle $(\overrightarrow{CE}, \overrightarrow{CB})$ is the sum $(\overrightarrow{AB}, \overrightarrow{AC}) + (\overrightarrow{BC}, \overrightarrow{BA}) + (\overrightarrow{CA}, \overrightarrow{CB})$. □

Notice that the translation and the central symmetry, as all isometries, preserve the geometric angles, so that the geometric flat angle $\widehat{BCE}$ is also the sum of the three geometric angles of the triangle. We have thus proven:

Corollary 1.15. *Let A, B and C be three distinct points of an oriented affine Euclidean plane. If α, β and γ are measures of the angles of vectors* $(\overrightarrow{AB}, \overrightarrow{AC})$, $(\overrightarrow{BC}, \overrightarrow{BA})$ *and* $(\overrightarrow{CA}, \overrightarrow{CB})$, *they satisfy the equality*

$$\alpha + \beta + \gamma \equiv \pi \quad (2\pi)$$

and the sum of the measures of the geometric angles is exactly π. □

Remark 1.16. This is a *Euclidean* result in all senses (including the most classical one). It is related to the parallel axiom, since *the unique* parallel to BC through A was used to prove it. We shall mention below (in Corollary IV-3.2 and Exercise V.50) other "geometries" in which this statement is not true.

Proposition 1.17 (Angles at the circumference). *If A, B and C are three distinct points on a circle of center O, the equality*

$$(\overrightarrow{OA}, \overrightarrow{OB}) = 2(\overrightarrow{CA}, \overrightarrow{CB})$$

holds.

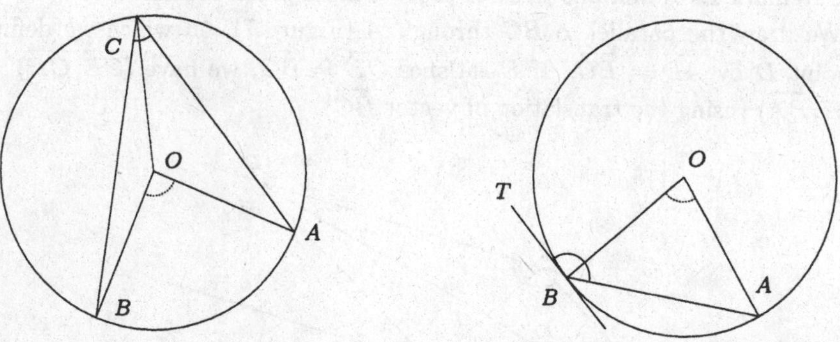

Fig. 8. Angles at the circumference

Proof. Since O is on the perpendicular bisector of AC, the reflection about this bisector gives $(\overrightarrow{CA}, \overrightarrow{CO}) = (\overrightarrow{AO}, \overrightarrow{AC})$ (this is Exercise III.5) and the proposition about the sum of the angles tells us that

$$(\overrightarrow{OA}, \overrightarrow{OC}) + 2(\overrightarrow{CO}, \overrightarrow{CA})$$

is a flat angle. But O belongs to the perpendicular bisector of BC, and this gives similarly that

$$(\overrightarrow{OC}, \overrightarrow{OB}) + 2(\overrightarrow{CB}, \overrightarrow{CO})$$

is also a flat angle. Adding these two relations, we get

$$(\overrightarrow{OA}, \overrightarrow{OC}) + (\overrightarrow{OC}, \overrightarrow{OB}) + 2\left((\overrightarrow{CO}, \overrightarrow{CA}) + (\overrightarrow{CB}, \overrightarrow{CO})\right) = 0,$$

that is,

$$(\overrightarrow{OA}, \overrightarrow{OB}) + 2(\overrightarrow{CB}, \overrightarrow{CA}) = 0,$$

and this is what we wanted to prove. □

In this proof, the point C may coincide with B; the line BC must then be replaced by the tangent to the $\mathcal{C}$ at B (see Figure 8). The vector $\overrightarrow{CB}$ can be replaced by any directing vector $\overrightarrow{TB}$ of this line: the angles of vectors in which $\overrightarrow{CB}$ appear in the proof are all endowed with a factor 2, so that the direction of the vector $\overrightarrow{TB}$ does not matter. We thus get:

Proposition 1.18. *If $\mathcal{D}$ is the tangent at B to the circle $\mathcal{C}$ of center O and if A is a point of $\mathcal{C}$ distinct of B, the equality of angles*

$$(\overrightarrow{OA}, \overrightarrow{OB}) = 2(AB, \mathcal{D})$$

holds. □

A criterion of cyclicity follows from these results:

Corollary 1.19 (Cyclicity). *Let A, B and C be three noncollinear points. A point D belongs to the circle they determine if and only if the angles of lines (CA, CB) and (DA, DB) are equal.*

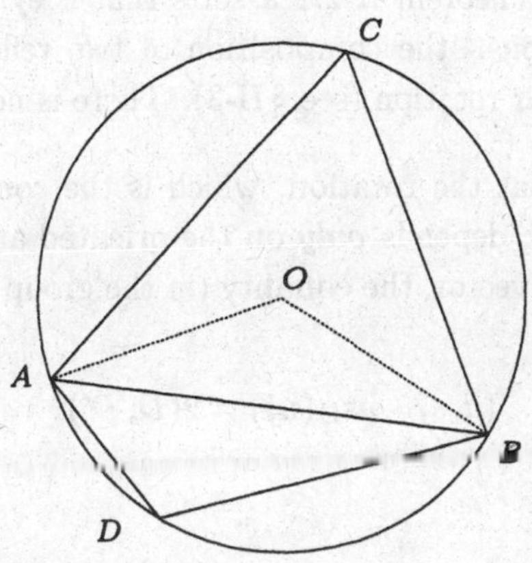

Fig. 9

Proof. Assume firstly that D belongs to the circumcircle to the triangle ABC. We apply Proposition 1.17 twice, in other words, we write the equalities

$$2(\overrightarrow{CA}, \overrightarrow{CB}) = (\overrightarrow{OA}, \overrightarrow{OB}) = 2(\overrightarrow{DA}, \overrightarrow{DB})$$

where O is the center of the circle. We then divide by 2 the equality of angles of vectors... and this transforms it into an equality of angles of lines (this is Proposition 1.6).

Conversely, we assume the equality of angles to hold. Let O be the center of the circumcircle to the triangle ABC and O' that of the circumcircle to ABD (because of the equality of angles, A, B and D are not collinear). Thanks to Proposition 1.18, the tangents at B to these two circles coincide. But then, the points O and O' also coincide: they are at the intersection of this tangent with the perpendicular bisector of the segment AB. Thus the two circles coincide and the four points are cyclic. □

This corollary is often used in the following form.

Corollary 1.20. *The angles of lines (CA, CB) and (DA, DB) are equal if and only if the four points A, B, C and D are cyclic or collinear.* □

2. Isometries and rigid motions in the plane

Equipped with these new tools, let us come back to the structure theorems on isometries of § II-2 in the case of the plane.

Linear isometries. Theorem II-2.1 asserts that they are the composition of one or two reflections. the composition of two reflections is a positive isometry, hence a linear rotation (see § II-3). There is not much to add!

Notice, however, that the rotation, which is the composition of the two reflections $s_{D'}$ and s_D, depends *only* on the oriented angle of lines (D, D'). Actually, if u is a unit vector, the equality (in the group of oriented angles of vectors)

$$(u, s_{D'} \circ s_D(u)) = 2(D, D')$$

holds.

Proof. By definition of the angles, that formed by u and its image does not depend on the choice of u. Thus it can be chosen on D. One chooses also a unit vector u' on D'. Thus

$$\begin{aligned}(u, s_{D'} \circ s_D(u)) &= (u, s_{D'}(u)) \\ &= (u, u') + (u', s_{D'}(u)) \\ &= (u, u') + (u, u') \\ &\quad \text{car } (u', s_{D'}(u)) = (u, s_{D'}(u')) \\ &= 2(u, u')\end{aligned}$$

using Proposition 1.10. □

Affine isometries. Thanks to Theorem II-2.2, they are the compositions of one, two or three reflections. Let φ be an affine isometry of the plane and let $\overrightarrow{\varphi}$ be the associated linear isometry. Then, according to the linear study above, $\overrightarrow{\varphi}$ is:

- the identity
- or a reflection
- or a linear rotation.

Let us investigate these three cases.

- If $\overrightarrow{\varphi}$ is the identity, φ is a translation.
- If $\overrightarrow{\varphi}$ is a reflection about a line D, then:
  - either φ has a fixed point A; we vectorialize the affine plane at A and we apply the corresponding linear result to get that φ is the reflection about the line through A and directed by D;
  - or φ has no fixed point; we can then look at the general results on affine isometries. Here, $\overrightarrow{\varphi}$ being a reflection, it has nonzero fixed vectors (the vectors of D) and we can apply Proposition II-2.8. There exists thus a unique vector v in D and a unique reflection ψ about an affine line directed by D such that
  $$\varphi = t_v \circ \psi = \psi \circ t_v.$$
  The isometry φ is a *glide reflection* (depicted in Figure 10; see

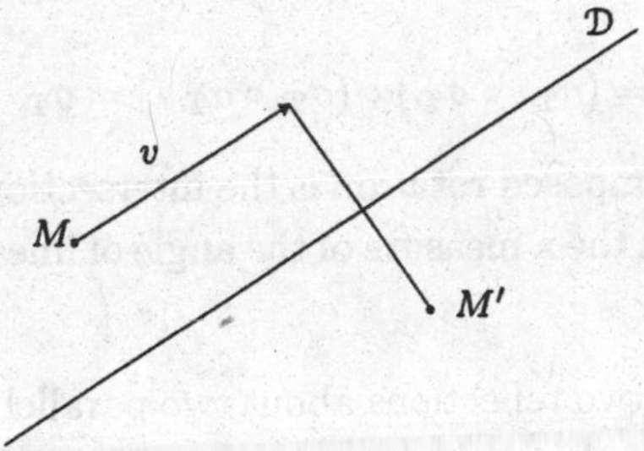

Fig. 10. A glide reflection

also Exercise I.15).

– If $\overrightarrow{\varphi}$ is a linear rotation (different from the identity), it has no nonzero fixed vector. According to Proposition I-2.20, φ has a unique fixed point A. We just have to vectorialize the affine plane at A for φ to act like $\overrightarrow{\varphi}$. It is said that this is an affine *rotation* of center A. If the plane is oriented and if $\overrightarrow{\varphi}$ is a rotation of angle θ, it is said that φ is a rotation of center A and angle θ (notation $\rho_{A,\theta}$).

To summarize. The affine isometries of the affine Euclidean plane are the translations, the rotations, the reflections and the glide reflections. One has the following table:

	translations	rotations	reflections	glide reflections
fixed points	no fixed point	a unique fixed point	a line of fixed points	no fixed point
invariant lines	a direction of lines	no invariant line	and a direction of lines	a unique invariant line
decomposition in reflections	2 parallel lines	2 secant lines	1 line	3 lines

Complements. The linear isometry associated with the composition of two affine rotations is a linear rotation (they form a group!). The composition is thus a translation (when the sum of the angles of the two rotations is zero modulo 2π) or a rotation (otherwise). Here is a geometric construction of its center.

We write the two rotations $\rho_{A,\alpha}$ and $\rho_{B,\beta}$ as the composition of two reflections. For $\rho_{A,\alpha}$, the two lines pass through A and one can be chosen arbitrarily. Similarly for $\rho_{B,\beta}$, with two lines through B. The solution is obvious: let $\mathcal{D}$ be the line AB, $\mathcal{D}'$ the line through A such that $(\mathcal{D}, \mathcal{D}') \equiv \alpha/2$ modulo π, and $\mathcal{D}''$ the line through B such that $(\mathcal{D}'', \mathcal{D}) \equiv \beta/2$ modulo π, so that

$$\rho_{A,\alpha} \circ \rho_{B,\beta} = (\sigma_{\mathcal{D}'} \circ \sigma_{\mathcal{D}}) \circ (\sigma_{\mathcal{D}} \circ \sigma_{\mathcal{D}''}) = \sigma_{\mathcal{D}'} \circ \sigma_{\mathcal{D}''}.$$

The center of the composed rotation is the intersection point of $\mathcal{D}'$ and $\mathcal{D}''$. It is easily checked that the a measure of the angle of lines $(\mathcal{D}'', \mathcal{D}')$ is $(\alpha+\beta)/2$ (Figure 11).

The composition of two reflections about two parallel lines $\mathcal{D}$ and $\mathcal{D}'$ is, we have said, a translation. Let us determine the vector of this translation. We choose a point A of $\mathcal{D}$ and a point A' of $\mathcal{D}'$ so that AA' is perpendicular to $\mathcal{D}$. The composition of the two reflections $\sigma_{\mathcal{D}'} \circ \sigma_{\mathcal{D}}$ is the translation of vector $2\overrightarrow{AA'}$ (Figure 12). Notice, by the way, that this translation depends only

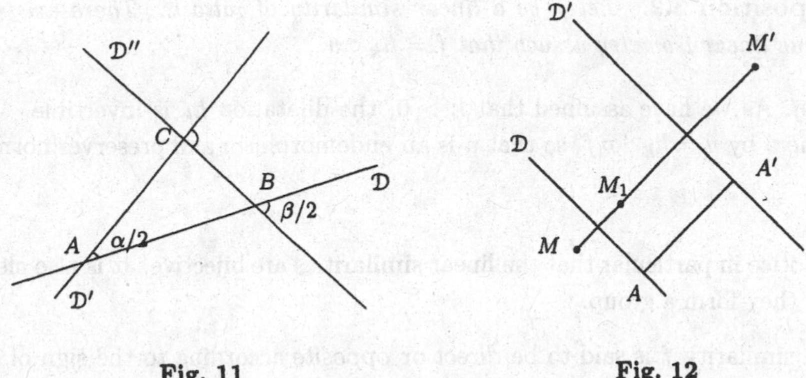

Fig. 11 Fig. 12

on the vector $\overrightarrow{AA'}$: one of the two lines $\mathcal{D}$ and $\mathcal{D}'$ can be chosen arbitrarily (except that it must be perpendicular to AA').

The last complement is the way to write the rotations using complex numbers. An orthonormal frame and an origin being chosen, the affine Euclidean plane $\mathcal{P}$ can be identified with $\mathbf{C}$. With the point of coordinates (x, y) is associated the complex number $x + iy$, its *affix*.

To save on notation, the simplest thing to do is to use small letters for the affixes of the points denoted by the corresponding capital letters, in other words, as usual, a is the affix of A, b that of B... and z that of M (!).

This way, the formula

$$z' = a + e^{i\theta}(z - a)$$

gives the affix of the image M' of M under the rotation of center A and angle θ.

3. Plane similarities

We introduce now transformations that are slightly more general than the isometries, as they contain the dilatations as well; these are the *similarities*.

Definition 3.1. An endomorphism f of a Euclidean vector space is a linear *similarity* if there exists a positive real number k, the *ratio* of the similarity, such that, for any vector x of E,

$$\|f(x)\| = k\,\|x\|.$$

We have said that the isometries and the dilatations are similarities (be careful to note that the ratio of the dilatation of ratio λ as a similarity is $|\lambda|$). And, in some sense, this is all:

Proposition 3.2. *Let f be a linear similarity of ratio k. There exists a unique linear isometry u such that $f = h_k \circ u$.*

Proof. As we have assumed that $k > 0$, the dilatation h_k is invertible. We define u by $u = h_{k^{-1}} \circ f$, so that u is an endomorphism and preserves norms. □

Notice in particular that the linear similarities are bijective. It is also clear that they form a group.

A similarity f is said to be *direct* or *opposite* according to the sign of its determinant.

Remark 3.3. The definition 3.1 of similarities given here is meaningful in any dimension and Proposition 3.2 is true in full generality. However, starting from here, we must restrict ourselves to the case of the plane.

Proposition 3.4. *Any linear direct similarity is the composition of a dilatation of positive ratio and a linear rotation. Any linear opposite similarity is the composition of a dilatation of positive ratio and a reflection.*

Proof. This is a direct consequence of the list of linear plane isometries, once we have noticed that

$$\det(h_k \circ u) = k^2 \det(u)$$

and thus that u is a positive isometry if and only if the similarity f is direct. □

Let us come now to affine similarities. They are defined as the affine mappings(3) $\varphi : \mathcal{E} \to \mathcal{E}$ for which there exists a nonnegative real number k such that, for all points M and N of images M' and N', we have

$$M'N' = kMN,$$

or, equivalently, they are the affine mappings that have a linear similarity as their associated linear mapping. They are of course bijective and they form a group. In the same way, there are direct and opposite affine similarities.

Proposition 3.5. *A similarity of $\mathcal{E}$ that is not an isometry has a unique fixed point, called its center.*

(3) Using Exercise II.4, it is easily seen that it is not necessary to require that the mappings under consideration are affine.

Proof. Let φ be a similarity (that is not an isometry) and $\overrightarrow{\varphi}$ the associated endomorphism. As the linear similarity $\overrightarrow{\varphi}$ is not an isometry, its ratio k is different from 1 so that 1 is not one of its eigenvalues. We apply Proposition I-2.20 to conclude that φ has, indeed, a unique fixed point. □

Vectorializing the affine plane at the center O of the similarity, it is seen that a direct similarity that is not an isometry is the composition of a dilatation of center O and ratio k (the ratio of the similarity) and a rotation of center O.

If the angle of the rotation is θ, the similarity is denoted $\sigma_{O,k,\theta}$. We have of course

$$\sigma_{O,k,\theta} \circ \sigma_{O',k',\theta'} = \sigma_{O'',kk',\theta+\theta'} \text{ if } kk' \neq 1.$$

The reciprocal similarity of a similarity is expressed by the relation

$$(\sigma_{O,k,\theta})^{-1} = \sigma_{O,k^{-1},-\theta}.$$

As for the rotations, the number θ is called the *angle* of the similarity.

The group of direct similarities consists thus of the similarities $\sigma_{O,k,\theta}$ (including the rotations $\rho_{O,\theta} = \sigma_{O,1,\theta}$ and the dilatations $h(O,k) = \sigma_{O,k,0}$) and the translations.

Properties of direct similarities

(1) Direct similarities preserve oriented angles (as this is the case for dilatations and rotations).
(2) They map a line $\mathcal{D}$ to a line $\mathcal{D}'$ such that the angle $(\mathcal{D}, \mathcal{D}')$ is the angle of the similarity (modulo π).
(3) A direct similarity of ratio k maps a circle of radius R to a circle of radius kR whose center is the image of the center.
(4) Given two ordered pairs of points (A, B) and (A', B') (with $A \neq B$ and $A' \neq B'$), there exists a unique direct similarity mapping A to A' and B to B'. The linear similarity is well determined, its ratio is $A'B'/AB$ and its angle is $(\overrightarrow{AB}, \overrightarrow{A'B'})$. Adding the fact that A' is the image of A, a unique direct (affine) similarity is determined.

Construction of the center. Let us come back to more geometrical constructions and show a way to determine the center of the direct similarity defined by the two distinct points A and B together with their images A' and B'.

Notice first that, if $\overrightarrow{AB} = \overrightarrow{A'B'}$, the similarity is a translation. Assume thus that $\overrightarrow{AB} \neq \overrightarrow{A'B'}$. Notice also that, if the lines AB and $A'B'$ are parallel,

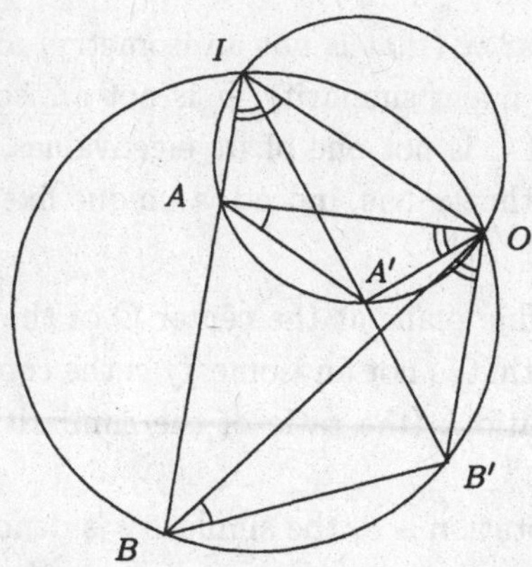

Fig. 13

this is a dilatation of center $AA' \cap BB'$. We thus assume now that the lines AB and $A'B'$ intersect at a point I. The center O must satisfy

$(OA, OA') \equiv \alpha \mod \pi$ (α is a measure of the angle of the similarity).

But the point I satisfies

$$(IA, IA') = (AB, A'B') \equiv \alpha \mod \pi.$$

Therefore O is cyclic with I, A and A' (on a circle $\mathcal{C}$) and similarly with I, B and B' (on a circle $\mathcal{C}'$). The two circles intersect at I and at another point O (Figure 13) which may coincide with I. The triangles OAA' and OBB' are similar (the equalities of angles shown on Figure 13 hold by cyclicity). Hence the triangles OAB and $OA'B'$ are similar too (see, *e.g.*, Proposition 3.7 below). □

Remarks 3.6

- The careful reader will have noticed that I have assumed that $I \neq A$, A', B, B'. The remaining cases are (of course) left as an exercise.
- It can be shown that the circles $\mathcal{C}$ and $\mathcal{C}'$ are tangent if and only if the lines AA' and BB' are parallel (exercise).

Using complex numbers. As above, we identify the Euclidean affine plane with **C**, using an orthonormal frame. The direct similarities are the mappings of the form

$$z \longmapsto az + b.$$

The ratio of this similarity is $|a|$, and its angle is an argument of a.

This writing may be extremely convenient. Let us give for instance a very short proof of the next useful property:

Proposition 3.7. *There exists a direct similarity s of center A mapping B to B' and C to C' if and only if there exists a direct similarity s' of center A mapping B to C and B' to C'.*

Proof. We fix the origin at the point A. With obvious notation for the affixes of the points, the statement of the proposition is simply the equivalence:

$$\frac{b'}{b} = \frac{c'}{c} \iff \frac{c'}{b'} = \frac{c}{b}.$$

□

Remark 3.8. This is of course easy to prove even without complex numbers. For instance, we notice that, if τ is the similarity of center A that maps B' to C, we have

$$s' = \tau \circ s = s \circ \tau$$

(similarities with the same center commute).

Characterization of the similarities. They are the only affine transformations that preserve the angles, as the next proposition shows.

Proposition 3.9. *Let P be a Euclidean (vectorial) plane and let $f : P \to P$ be a linear mapping that preserves the angles (resp. that reverses the angles). Then f is a direct (resp. opposite) similarity.*

Proof. Let us fix a vector $u \neq 0$. Composing f with a direct similarity, we can assume that $f(u) = u$. Since f preserves the angles, we have $(u, v) = (u, f(v))$ for all v and thus $f(v) = \lambda v$ (for some $\lambda > 0$, which, *a priori* depends on v). As usual (see the "trick" of Exercise I.23), we conclude that f is a dilatation.

If f reverses the angles, we compose it with a reflection and we apply what we have just done. □

Remark 3.10. There is a much more general result: any bijection of class $\mathcal{C}^1$ of P to itself that preserves the angles is a similarity (this is a theorem of Liouville, see Exercise III.66).

4. Inversions and pencils of circles

We are going to study now transformations that are *not* isometries (and not even affine transformations), the inversions. They have the property that they transform certain circles in lines and that they preserve the angles, which can be very useful to transform a complicated figure in a simpler one[4].

This will take place in an affine Euclidean plane $\mathcal{E}$ but the definitions keep their meaning in any affine Euclidean space; the properties are proved as well,

(4) This is a good opportunity to recommend reading [**CG67**, Chapter 5].

with the same proofs, replacing everywhere "circle" by "sphere" and "line" by "hyperplane".

Definition 4.1. Let O be a point of the plane and k be a nonzero real number. The *inversion* of *pole* O and *power* k is the transformation

$$I(O,k) : \mathcal{E} - \{O\} \longrightarrow \mathcal{E} - \{O\}$$

that associates, with any point M, the point M' of the line OM satisfying the equality $\overrightarrow{OM} \cdot \overrightarrow{OM'} = k$.

Remark 4.2. As O, M and M' are collinear, the same thing can be expressed by:

$$\overrightarrow{OM'} = \frac{k}{OM^2}\overrightarrow{OM}.$$

It is clear that the inversion $I(O,k)$ is an involution, that it has fixed points only if $k > 0$, in which case its fixed points are the points of the circle of center O and radius $\sqrt{k}$. This circle is called the *inversion circle*[(5)]. It should be noticed, and this is important, that the inversion exchanges the interior and the exterior of the inversion circle.

Effect on lengths and angles. We are going to see that the inversions are not isometries. Here is the way they transform distances and angles.

Proposition 4.3 (Distance of the images). *If A' and B' are the images of A and B by an inversion of pole O and power k, their distance is expressed by the formula*

$$A'B' = \frac{|k|\,AB}{OA \cdot OB}.$$

Proof. We have, by definition of A' and B'

$$\overrightarrow{A'B'} = k\left[\frac{\overrightarrow{OB}}{OB^2} - \frac{\overrightarrow{OA}}{OA^2}\right],$$

this allowing to compute

$$\left\|\overrightarrow{A'B'}\right\|^2 = k^2\left[\frac{OB^2}{OB^4} + \frac{OA^2}{OA^4} - 2\frac{\overrightarrow{OA}\cdot\overrightarrow{OB}}{OA^2OB^2}\right] = \frac{k^2}{OA^2OB^2}\left\|\overrightarrow{AB}\right\|^2.$$

□

(5) It can be considered that the inversions of negative power have an inversion circle of imaginary radius. This point of view will be used in § VI-7.

Proposition 4.4. *An inversion of pole O is a diffeomorphism*[(6)] *from the open subset $\mathcal{E}-\{O\}$ to itself. The differential of $I=I(O,k)$ at the point M is the linear mapping $dI_M : E \to E$ defined by*

$$dI_M(u)=\frac{k}{OM^2}\left[u-2\frac{\overrightarrow{OM}\cdot u}{OM^2}\overrightarrow{OM}\right].$$

Proof. This is a direct computation. To save on writing, let us use the notation "$M+u$" as defined in Remark I-1.3.

$$\begin{aligned}
\overrightarrow{OI(M+u)} &= \frac{k}{\left\|\overrightarrow{OM}+u\right\|^2}\left(\overrightarrow{OM}+u\right)\\
&= \frac{k}{OM^2}\left(1+2\frac{\overrightarrow{OM}\cdot u}{OM^2}+\frac{\|u\|^2}{OM^2}\right)^{-1}\left(\overrightarrow{OM}+u\right)\\
&= \frac{k}{OM^2}\left(1-2\frac{\overrightarrow{OM}\cdot u}{OM^2}+\mathrm{o}(\|u\|)\right)\left(\overrightarrow{OM}+u\right)\\
&= \frac{k}{OM^2}\overrightarrow{OM}+\frac{k}{OM^2}\left[u-2\frac{\overrightarrow{OM}\cdot u}{OM^2}\overrightarrow{OM}\right]+\mathrm{o}(\|u\|)\\
&= \overrightarrow{OI(M)}+dI_M(u)+\mathrm{o}(\|u\|).
\end{aligned}$$

The differential dI_M is thus the composition of the reflection about the line orthogonal to OM (see if necessary Exercise II.7) and the dilatation of ratio k/OM^2. Since I is an involution, this is enough. □

In particular, the linear mapping dI_M transforms an oriented angle into the opposite angle.

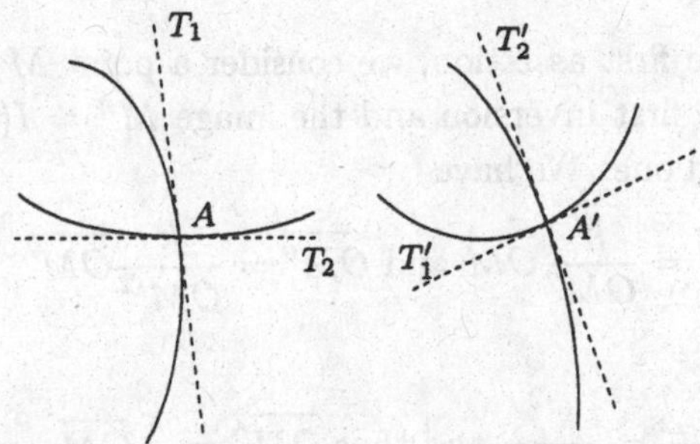

Fig. 14

Let us imagine now two plane curves intersecting at a point A. The *angle* of these two curves is the angle of their tangents at A (Figure 14). Consider the images of these curves under an inversion whose pole is not at A. These

(6) Recall that a *diffeomorphism* is a bijective and differentiable mapping whose inverse mapping is differentiable.

are two curves intersecting at the image A' of A. The tangents at A' to these two curves are the images of the tangents at A to the original curves by the differential of the inversion (see if necessary Chapter VII for the notions used here).

In particular, the angle of the image curves is the opposite of the angle of the original curves, a property that can be expressed in the following way:

Corollary 4.5 ("Inversions preserve the angles"). *An inversion transforms an oriented angle into its opposite.* □

Remark 4.6. I insist that angles are preserved but not straight lines: we are going to see that the lines are in general transformed into circles.

The special case of right angles is used quite often: inversions preserve the orthogonality. The special case of zero or flat angles also: inversions transform two tangent curves into two tangent curves.

Inversions and dilatations. It is often very convenient to change the power of an inversion in order to replace it by another inversion that is better adapted to the problem under consideration. This is made possible by the following result.

Proposition 4.7. *The composition $I(O,k)\circ I(O,k')$ of two inversions of the same pole is the restriction to $\mathcal{E}-\{O\}$ of the dilatation of center O and ratio k/k'. The composition $h(O,\lambda)\circ I(O,k)$ is the inversion of pole O and power λk.*

Proof. To prove the first assertion, we consider a point M, its image $M' = I(O,k')(M)$ by the first inversion and the image $M'' = I(O,k)(M')$ of the latter by the second one. We have

$$\overrightarrow{OM'} = \frac{k'}{OM^2}\overrightarrow{OM} \text{ and } \overrightarrow{OM''} = \frac{k}{OM'^2}\overrightarrow{OM'}$$

so that

$$OM'^2 = \frac{k'^2}{OM^2} \text{ and thus } \overrightarrow{OM''} = \frac{k}{k'}\overrightarrow{OM}.$$

Hence $h(O,\lambda) = I(O,\lambda k)\circ I(O,k)$. We just need to compose (on the right) the two sides with $I(O,k)$ to get the second assertion. □

This proposition can be used to replace an inversion of negative power by the composition of an inversion of positive power and a symmetry about the pole:

$$I(O,k) = h(O,-1)\circ I(O,-k).$$

Images of lines and circles, first part. We are going to use the previous remark to study the image of a line or of a circle by an inversion.

It is clear that lines passing through the pole are globally invariant. Let us thus consider lines that do not pass through the pole.

Proposition 4.8. *The image of a line not through the pole is a circle through the pole.*

Remark 4.9. This is an abuse of language: the reader should read this statement and its proof with a critical eye to understand that, of course, the pole is not in the image. The image of the line is, in fact, the circle punctured at the pole.

Proof of Proposition 4.8. Let thus $I(O,k)$ be the inversion of pole O and power k and let $\mathcal{D}$ be a line not through O.

To begin with, we reduce to the case of an inversion of pole O having a fixed point on $\mathcal{D}$. Let H be the orthogonal projection of O on $\mathcal{D}$, so that H is fixed by $I(O,OH^2)$ (see Figure 15). If $h(O,\lambda)$ is a dilatation of center O, we know (Proposition 4.7) that

$$h(O,\lambda)\circ I(O,k)=I(O,k\lambda).$$

We choose $\lambda=k/OH^2$, so that

$$I(O,k)=h(O,\lambda)\circ I(O,OH^2)$$

and we consider the image $I(O,OH^2)(\mathcal{D})$ of $\mathcal{D}$ by this inversion.

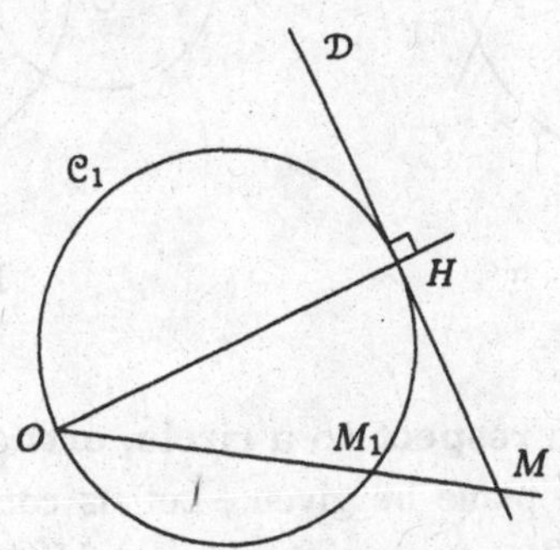

Fig. 15

A point M of the plane belongs to $\mathcal{D}$ if and only if it satisfies the equation $\overrightarrow{OM}\cdot\overrightarrow{OH}=OH^2$. If M_1 is the image of M, we have the equality

$$\overrightarrow{OM}=\frac{OH^2}{OM_1^2}\overrightarrow{OM_1}.$$

Therefore M belongs to $\mathcal{D}$ if and only if

$$\frac{OH^2}{OM_1^2}\overrightarrow{OM_1}\cdot\overrightarrow{OH} = OH^2,$$

namely if and only if

$$\overrightarrow{OM_1}\cdot\left(\overrightarrow{OH}-\overrightarrow{OM_1}\right) = 0,$$

in other words if and only if $\overrightarrow{OM_1}\cdot\overrightarrow{M_1H} = 0$. But this last relation expresses the fact that M_1 belongs to the circle $\mathcal{C}_1$ of diameter OH. The latter is thus the image of the line $\mathcal{D}$ by the inversion $I(O, OH^2)$. See Figure 15.

We have eventually

$$I(O,k)(\mathcal{D}) = h(O,\lambda)\circ I(O,OH^2)(\mathcal{D}) = h(O,\lambda)(\mathcal{C}_1).$$

Now the image under a dilatation of center O of a circle through O is a circle through O, so that we have proved, indeed, that the image of $\mathcal{D}$ is a circle through O. □

Before we go further, it could be useful to recall a few facts about circles.

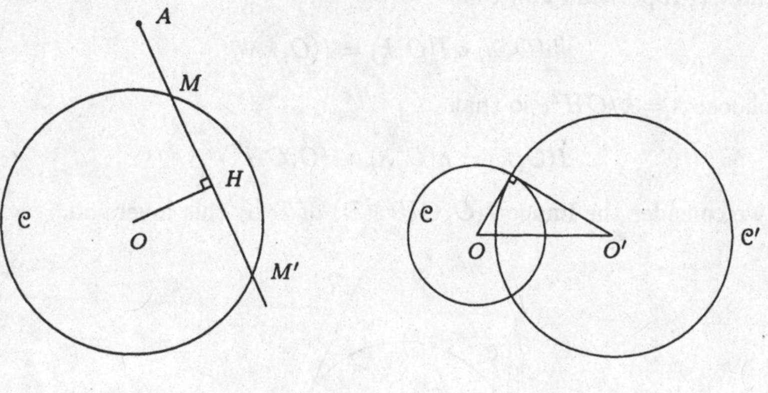

Fig. 16 Fig. 17

Power of a point with respect to a circle, orthogonality. Let a circle $\mathcal{C}$ and a point A of the plane be given. Let us consider an arbitrary line through A intersecting $\mathcal{C}$ at two points M and M' (Figure 16). If H is the orthogonal projection of the center O of $\mathcal{C}$ on $\mathcal{D}$, we compute

$$\begin{aligned}\overrightarrow{AM}\cdot\overrightarrow{AM'} &= (\overrightarrow{AH}-\overrightarrow{MH})\cdot(\overrightarrow{AH}+\overrightarrow{MH})\\ &= AH^2 - MH^2\\ &= AO^2 - OH^2 - MH^2\\ &= AO^2 - R^2\end{aligned}$$

where R is the radius of the circle $\mathcal{C}$.

The quantity $\overrightarrow{AM} \cdot \overrightarrow{AM'}$ does not depend on the secant line $\mathcal{D}$ we have chosen; it is called the *power of A with respect to $\mathcal{C}$* and it is denoted by $P_{\mathcal{C}}(A)$.

Remark 4.10. The point A is on $\mathcal{C}$ if and only if its power with respect to $\mathcal{C}$ is zero ($P_{\mathcal{C}}(A) = 0$); it is outside $\mathcal{C}$ if and only if the power is nonnegative ($P_{\mathcal{C}}(A) > 0$).

If A is outside $\mathcal{C}$, the secant line may as well be a tangent line and the power of A with respect to $\mathcal{C}$ is the square of the distance from A to the contact point of the tangent.

Definition 4.11. Two circles $\mathcal{C}$ and $\mathcal{C}'$ are said to be *orthogonal* when their angle is a right angle, that is, when they are secant and the tangents at the intersection points are orthogonal (Figure 17). This relation is denoted by $\mathcal{C} \perp \mathcal{C}'$.

Letting O, O' and R, R' be the centers and radii of $\mathcal{C}$ and $\mathcal{C}'$, a simple application of Pythagoras' theorem proves that the circles $\mathcal{C}$ and $\mathcal{C}'$ are orthogonal if and only if we have

$$OO'^2 = R^2 + R'^2,$$

which is equivalent both to $P_{\mathcal{C}'}(O) = R^2$ and $P_{\mathcal{C}}(O') = R'^2$.

Let us come back to inversions.

Proposition 4.12. *Let I be an inversion of circle $\mathcal{C}$. Let M and M' be a point of the plane and its image. Then all the circles through M and M' are orthogonal to $\mathcal{C}$.*

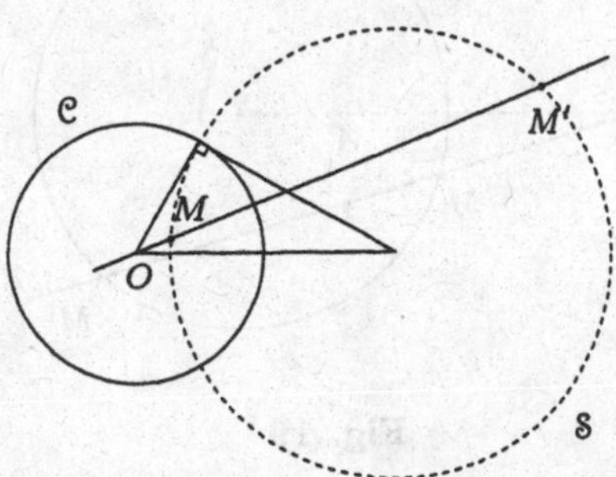

Fig. 18

Proof. Let $\mathcal{S}$ be a circle through M and M' (see Figure 18). Then the power of the pole O of the inversion with respect to $\mathcal{S}$ is

$$P_{\mathcal{S}}(O) = \overrightarrow{OM} \cdot \overrightarrow{OM'} = k,$$

namely the square of the radius of $\mathcal{C}$ (whose center is O). Therefore the circles $\mathcal{S}$ and $\mathcal{C}$ are orthogonal. □

Remark 4.13. It can be said that a line is *orthogonal to a circle* when it is a diameter of the circle (it is then orthogonal to the tangents to the circle at the intersection points). The statement of the proposition still holds when "inversion" is replace by "reflection" and (simultaneously) "circle" $\mathcal{C}$ by "line" $\mathcal{C}$.

Images of lines and circles, continuation. We now have all the necessary tools to proceed in our study of the image of the circles and lines by the inversions.

Proposition 4.14. *The image of a circle not through the pole by an inversion is a circle not through the pole.*

Proof. Let $I(O,k)$ be an inversion of pole O and of power k and let $\mathcal{C}$ be a circle not through O. Let $p = P_{\mathcal{C}}(O)$ be the power of the pole with respect to the circle under consideration. We choose λ so that

$$I(O,k) = h(O,\lambda) \circ I(O,p)$$

(namely $\lambda = k/p$). But the power p has been chosen in such a way that the inversion of pole O and of power p keeps the circle $\mathcal{C}$ globally invariant, exchanging the two intersection points of the lines through O with $\mathcal{C}$.

Finally, $I(O,k)(\mathcal{C}) = h(O,\lambda)(\mathcal{C})$ is indeed a circle not through O. □

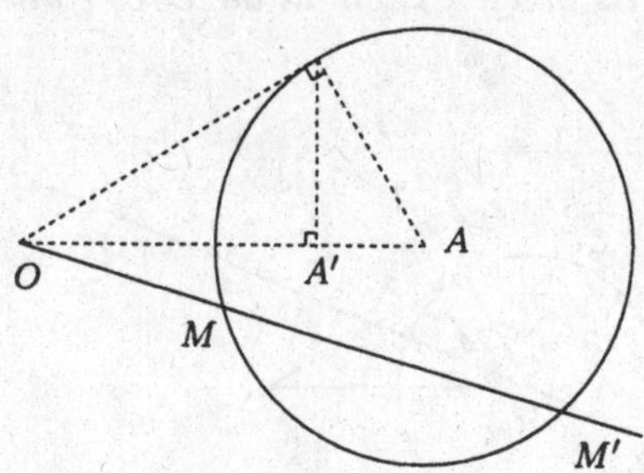

Fig. 19

Remark 4.15. Notice that the image of the center of the circle is *never* the center of the image circle. Let us consider for instance Figure 19, which shows a circle $\mathcal{C}$ globally invariant by an inversion of pole O. The image of its center A is the point A' (see Exercise III.20).

Let us summarize the information contained in the previous propositions:

Theorem 4.16. *Let I be an inversion of pole O. The image by I:*

- *of a line through O is this line itself,*
- *of a line not through O is a circle through O,*
- *of a circle through O is a line not through O,*
- *of a circle not through O is a circle not through O.*

We have already proved everything (the third assertion is a consequence of the second and of the involutivity of I). □

Remark 4.17. It would be simpler to remember that "the image of a circle is a circle", which is true... with a small, easy to justify, convention. We decide that there is, somewhere, very far, a[7] *point at infinity*, denoted ∞, and that the straight lines are the circles through this point.

The relation $\overrightarrow{OM} \cdot \overrightarrow{OM'} = k$ sends M' very far when M is very close to O; it is thus natural to decide that ∞ is the image of the pole of any inversion, and, of course, that the pole is the image of ∞. We have thus extended $I(O, k)$ as a mapping

$$\mathcal{E} \cup \{\infty\} \longrightarrow \mathcal{E} \cup \{\infty\}.$$

The assertions of the previous theorem can then be read as follows: let I be an inversion of pole O; the image by I:

- of a circle through ∞ and through O is a circle through O (image of ∞) and ∞ (image of O),
- of a circle through ∞ but not through O is a circle through O but not through ∞,
- of a circle through O but not through ∞ is a circle through ∞ but not through O,
- of a circle neither through ∞ nor through O is a circle neither through O nor through ∞.

In other words, as we know that O and ∞ are images of each other: the image of a circle is a circle.

Remark 4.18. It is of course possible to formalize the existence of the point ∞ and to justify its name (see Exercises IV.24 and IV.43 and §§ V-7 and VI-7).

---

(7) A unique point at infinity; we will come back to this in §§ V-7 and VI-7.

Radical axis of two circles. Let two nonconcentric circles $\mathcal{C}$ and $\mathcal{C}'$ be given. Then the set of points of the plane that have the same power with respect to the two circles $\mathcal{C}$ and $\mathcal{C}'$ is a line, the *radical axis* of $\mathcal{C}$ and $\mathcal{C}'$, perpendicular to the line of centers of $\mathcal{C}$ and $\mathcal{C}'$.

Indeed, as $P_{\mathcal{C}}(M) = MO^2 - R^2$ (the reader will have already guessed who O and R are and who O' and R' will be), M has the same power with respect to $\mathcal{C}$ and $\mathcal{C}'$ if and only if it satisfies

$$MO^2 - MO'^2 = R^2 - R'^2$$

this is to say, if and only if we have

$$\overrightarrow{O'O} \cdot \left(\overrightarrow{MO} + \overrightarrow{MO'}\right) = R^2 - R'^2,$$

and this, the vector $\overrightarrow{O'O}$ being nonzero, is the equation of a line perpendicular to the line of centers[(8)]. □

Remark 4.19. If $\mathcal{C}$ and $\mathcal{C}'$ are secant, the two intersection points obviously have the same power with respect to the two circles. The radical axis is then the line through these two points (Figure 20). A more general construction is sketched in Exercise III.49.

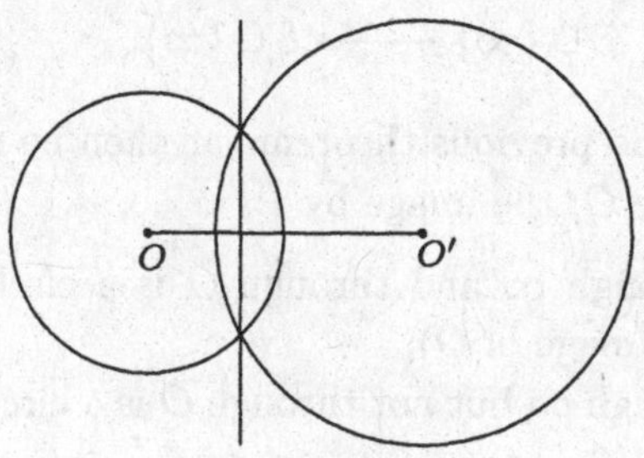

Fig. 20

Pencils of circles. Two nonconcentric circles $\mathcal{C}$ and $\mathcal{C}'$ and their radical axis Δ being given, one can consider the set $\mathcal{F}$ of all the circles C such that Δ is the radical axis of C and $\mathcal{C}$... or, but this is equivalent, of C and $\mathcal{C}'$.

Such a family $\mathcal{F}$ of circles is a *pencil of coaxal circles*, more precisely the pencil of circles generated by $\mathcal{C}$ and $\mathcal{C}'$. The pencil $\mathcal{F}$ contains the two circles $\mathcal{C}$ and $\mathcal{C}'$ that span it, but it is also spanned by any pair of circles it contains (in the same way as a line is spanned by any pair of points).

---

(8) See, *e.g.*, Exercise II.9. Another proof, using a computation in coordinates, is given below.

The tradition in mathematics is to use the word "pencil" for a *linear* family. Here, this is the "line generated by $\mathcal{C}$ and $\mathcal{C}'$ in the space of circles", something I explain now[9].

Let us choose an orthonormal affine frame in such a way that the line of centers of $\mathcal{C}$ and $\mathcal{C}'$ is the x-axis. The coordinates of O and O' are denoted by $(a,0)$ and $(a',0)$. The equations of $\mathcal{C}$ and $\mathcal{C}'$ are respectively:

$$(x-a)^2+y^2-R^2=0 \quad \text{and} \quad (x-a')^2+y^2-R'^2=0.$$

If λ and λ' are two real numbers (not both zero), we consider the curve $\mathcal{C}_{\lambda,\lambda'}$ of equation

$$\lambda\left((x-a)^2+y^2-R^2\right)+\lambda'\left((x-a')^2+y^2-R'^2\right)=0.$$

If $\lambda+\lambda'$ is nonzero, $\mathcal{C}_{\lambda,\lambda'}$ has an equation of the form

$$x^2+y^2-2bx+c=0,$$

and hence is a circle (which may be empty or consist of a single point).

If $\lambda+\lambda'=0$, $\mathcal{C}_{\lambda,\lambda'}$ has an equation of the form $(a-a')x=b$, and hence is a line ($a-a'\neq 0$ since the two circles are not concentric) orthogonal to the line of centers of $\mathcal{C}$ and $\mathcal{C}'$. This equation also has the form

$$P_{\mathcal{C}}(M)-P_{\mathcal{C}'}(M)=0$$

since we have noticed above that OM^2-R^2 (here $(x-a)^2+y^2-R^2$) is the power of M with respect to $\mathcal{C}$. In other words, the line $\mathcal{C}_{\lambda,-\lambda}$ is the radical axis of $\mathcal{C}$ and $\mathcal{C}'$.

But this is also the radical axis of $\mathcal{C}_{\lambda,\lambda'}$ and $\mathcal{C}$ for all λ and λ'. Indeed, writing again $\mathcal{C}$ and $\mathcal{C}'$ for the equations of $\mathcal{C}$ and $\mathcal{C}'$, an equation of the radical axis of $\mathcal{C}_{\lambda,\lambda'}$ and $\mathcal{C}$ is, according to what we have just seen,

$$\frac{\lambda\mathcal{C}+\lambda'\mathcal{C}'}{\lambda+\lambda'}-\mathcal{C}=0,$$

that is, $\lambda'\mathcal{C}'-\lambda'\mathcal{C}=0$, which is indeed the equation of the radical axis Δ of $\mathcal{C}$ and $\mathcal{C}'$.

The equations $\mathcal{C}_{\lambda,\lambda'}$ thus describe circles of the pencil spanned by $\mathcal{C}$ and $\mathcal{C}'$.

Conversely, to say that Γ is a circle of the pencil $\mathcal{F}$ is to say that the radical axis of Γ and $\mathcal{C}$ is the line Δ, in other words that we have the equivalence

$$P_{\Gamma}(M)=P_{\mathcal{C}}(M) \iff P_{\mathcal{C}}(M)-P_{\mathcal{C}'}(M)=0$$

since the last equation is that of Δ. All these equations having degree 1 or 2, this is also equivalent to the fact that the two equations are proportional and

[9] *Hic et nunc*, but also in § VI-7, using the equations of the circles in the plane.

eventually to the fact that there exists a scalar λ such that, for all point M, we have:

$$P_\Gamma(M) = P_{\mathcal{C}}(M) + \lambda(P_{\mathcal{C}}(M) - P_{\mathcal{C}'}(M)).$$

In short, Γ has an equation of the required form and is a circle of the pencil.

Types of pencils. There are different types of pencils, according to the relative positions of the circles $\mathcal{C}$ and $\mathcal{C}'$:

Fig. 21. An intersecting pencil

Fig. 22. A nonintersecting pencil

– If $\mathcal{C}$ and $\mathcal{C}'$ intersect at two (distinct) points A and B, the radical axis is the line AB, it contains the two points A and B whose power with respect to $\mathcal{C}$ and $\mathcal{C}'$ is zero. Their power with respect to all the circles of the pencil is thus zero, in other words, all the circles of the pencil pass through A and B, and all the circles through A and B belong to the pencil: the pencil is the *intersecting pencil of circles* consisting of all the circles through A and B, and the points A and B are its *base points* (Figure 21).

  From the point of view of the equations above, in this case, all the $\mathcal{C}_{\lambda,\lambda'}$ with $\lambda + \lambda' \neq 0$ are circles (of positive radius).

– If $\mathcal{C}$ and $\mathcal{C}'$ do not intersect, their radical axis cannot intersect them (there is no point with a zero power with respect to $\mathcal{C}$ and $\mathcal{C}'$); it consists of points that all have a positive power with respect to $\mathcal{C}$, $\mathcal{C}'$, and all the circles of the pencil (all the points of the radical axis are outside the circles $\mathcal{C}$ and $\mathcal{C}'$). It intersects no circle of the pencil.

  In this case, there are two circles of zero radius (point-circles) and a lot of circles of imaginary radius in the pencil. The radical axis is the perpendicular bisector of the segment defined by the two point-circles. The pencil is said to be a *nonintersecting pencil of circles*. The two point-circles it contains are its *limiting points* (Figure 22).

  These two point-circles are indeed circles of the pencil and in particular, they can be used to span it.

– The case where $\mathcal{C}$ and $\mathcal{C}'$ are tangent is a limiting case of the two previous ones. It is easily checked that all the circles of the pencil are tangent to each other, with a common tangent that is the radical axis.

With a pencil $\mathcal{F}$ defined by two circles $\mathcal{C}$ and $\mathcal{C}'$, one associates the family $\mathcal{F}^\perp$ of the circles orthogonal to $\mathcal{C}$ and $\mathcal{C}'$ (see Figure 23).

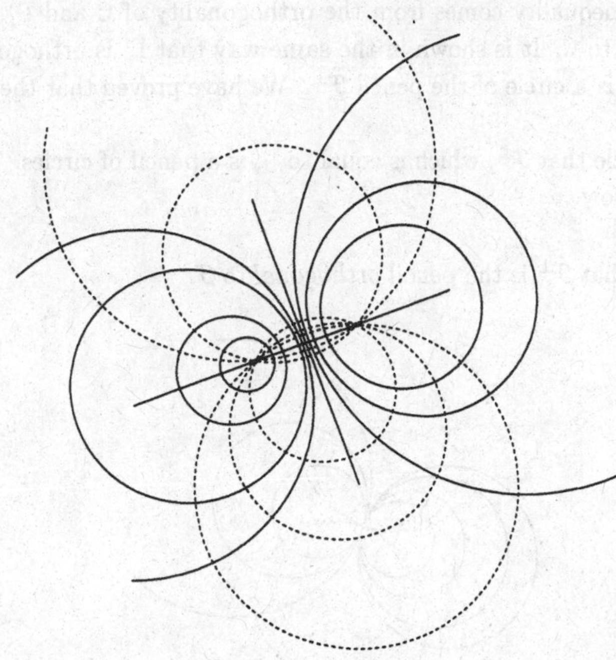

Fig. 23. Two orthogonal pencils

Proposition 4.20. *The family $\mathcal{F}^\perp$ is a pencil of circles, the radical axis of which is the line of centers of $\mathcal{F}$, and that depends only of $\mathcal{F}$. All the circles of $\mathcal{F}^\perp$ are orthogonal to all the circles of $\mathcal{F}$. Moreover, we have $(\mathcal{F}^\perp)^\perp = \mathcal{F}$.*

Proof. Let us call O and O' respectively the centers of the circles $\mathcal{C}$ and $\mathcal{C}'$, R and R' their radii. The line OO' is a diameter of $\mathcal{C}$ and $\mathcal{C}'$; it is thus orthogonal to these two circles. Let us fix a circle Γ orthogonal to $\mathcal{C}$ and $\mathcal{C}'$ and let us consider the pencil $\mathcal{G}$ generated by Γ and OO'. Let us prove that $\mathcal{F}^\perp = \mathcal{G}$.

Let Γ' be a circle orthogonal to $\mathcal{C}$ and $\mathcal{C}'$. We have

$$\mathcal{P}_\Gamma(O) = R^2 = \mathcal{P}_{\Gamma'}(O)$$

since Γ and Γ' are orthogonal to $\mathcal{C}$ and similarly, using $\mathcal{C}'$,

$$\mathcal{P}_\Gamma(O') = R'^2 = \mathcal{P}_{\Gamma'}(O')$$

hence the line OO' is the radical axis of Γ and Γ'. Therefore Γ' belongs to $\mathcal{G}$: we have shown the inclusion $\mathcal{F}^{\perp} \subset \mathcal{G}$.

Conversely, if Γ' is a circle of the pencil $\mathcal{G}$, O belongs to the radical axis of Γ and Γ', we thus have

$$\mathcal{P}_{\Gamma'}(O) = \mathcal{P}_{\Gamma}(O) = R^2$$

(the second inequality comes from the orthogonality of $\mathcal{C}$ and Γ), hence Γ' is orthogonal to $\mathcal{C}$. It is shown in the same way that Γ' is orthogonal to $\mathcal{C}'$, hence that Γ is a circle of the pencil $\mathcal{F}^{\perp}$. We have proved that the inclusion $\mathcal{G} \subset \mathcal{F}^{\perp}$ holds.

We conclude that $\mathcal{F}^{\perp}$, which is equal to $\mathcal{G}$, is a pencil of circles. The other assertions follow. □

It is said that $\mathcal{F}^{\perp}$ is the *pencil orthogonal* to $\mathcal{F}$.

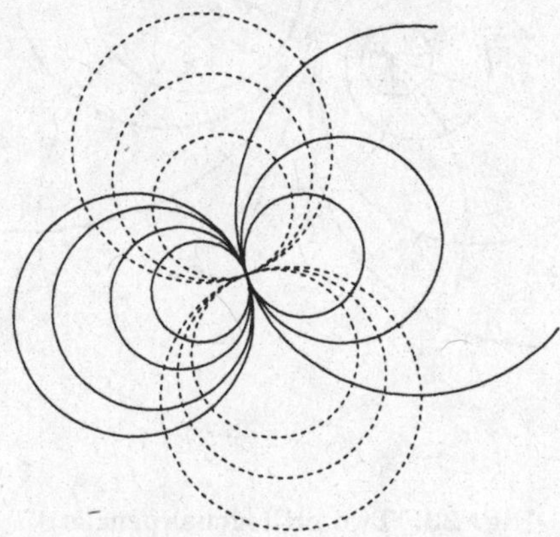

Fig. 24. Two orthogonal tangent pencils

Noticing that, for any point A of $\mathcal{C}$, the point-circle A is orthogonal to $\mathcal{C}$ (the power of its center with respect to $\mathcal{C}$ is the square of its radius), it is deduced that:

- the orthogonal of the intersecting pencil with base points A and B is the nonintersecting pencil with limiting points A and B (Figure 23)
- and conversely, the orthogonal of the pencil with limiting points I and J is the pencil with base points I and J.
- Similarly, the orthogonal of a pencil of tangent circles is the pencil of circles tangent to the line of centers of the center at the same point (Figure 24).

Another description of this orthogonality relation will be given in § VI-7.

Inverse of a pencil. In general, according to Theorem 4.16, the inverse of a pencil must be a pencil of the same nature. Here are a few comments and special cases (using Theorem 4.16 and the fact that inversions preserve the orthogonality, namely Corollary 4.5).

- If the pole is general enough, one of the circles of the pencil passes through this point[10]. It is transformed into a line, the radical axis of the new pencil. The radical axis is transformed into a circle of the new pencil.
- If the pole belongs to the radical axis and if the power of the inversion is the power of the pole with respect to the circles of the pencil, then the pencil is invariant (in the sense that any of the circles is globally invariant).

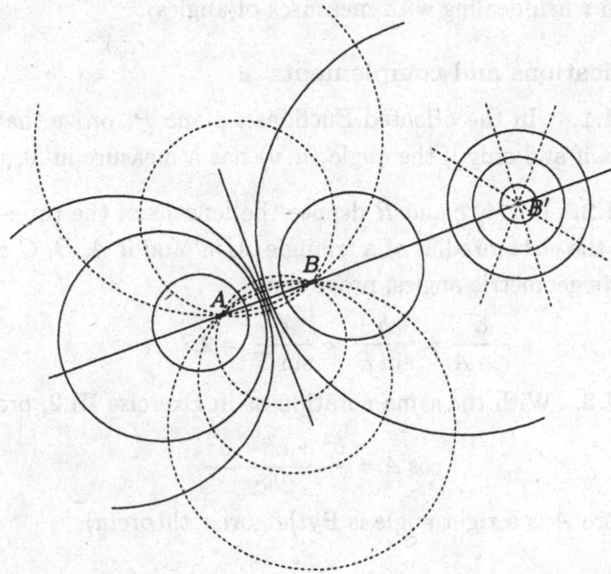

Fig. 25. Two orthogonal pencils and their inverses

- If $\mathcal{F}$ is a pencil with base points A and B and if the pole of the inversion is at A, the pencil $\mathcal{F}$ is transformed into the family of all lines through the inverse B' of B, and the orthogonal pencil $\mathcal{F}^{\perp}$ into the family of circles orthogonal to these lines, that is, into the family of all circles of center B' (Figure 25).
- In other words, if $\mathcal{G}$ is a nonintersecting pencil with limiting points I and J, and if the pole of the inversion is at I, $\mathcal{G}$ is transformed into the

(10) See if necessary Exercise III.60.

family of circles centered at the image J' of J and $\mathcal{G}^{\perp}$ into the family of lines through J'. Remember from this that it is always possible to transform by an inversion a nonintersecting pencil into a pencil of concentric circles, allowing us to simplify notably some problems (see, *e.g.*, Exercises III.61 and III.62).

– If $\mathcal{F}$ is a pencil of tangent circles at A and if the pole of the inversion is at A, $\mathcal{F}$ is transformed into the family of lines parallel to its radical axis and $\mathcal{F}^{\perp}$ into the family of orthogonal lines.

Exercises and problems

For all these exercises, we are in a Euclidean affine plane (which we assume to be oriented when dealing with measures of angles).

Direct applications and complements

Exercise III.1. In the oriented Euclidean plane P, prove that the basis (u, v) is direct if and only if the angle (u, v) has a measure in $[0, \pi]$.

Exercise III.2. If a, b, c and R denote the lengths of the three sides BC, CA, AB and the circumradius of a triangle ABC and if A, B, C also denote measures of its geometric angles, prove that

$$\frac{a}{\sin A} = \frac{b}{\sin B} = \frac{c}{\sin C} = 2R.$$

Exercise III.3. With the same notation as in Exercise III.2, prove that

$$\cos A = \frac{b^2 + c^2 - a^2}{2bc}$$

(the case where A is a right angle is Pythagoras' theorem).

Exercise III.4 (Incircle, excircles). Prove that the internal bisectors of a triangle ABC are concurrent at a point I that is equidistant from the three sides of the triangle and (thus) is the center of a circle tangent to the three sides of the triangle, the incircle.

Prove that the internal bisector of the angle at A and the two external bisectors of the angles at B and C are concurrent at a point J that is equidistant form the three sides of the triangle and (thus) that is the center of a circle tangent to the three sides of the triangle, the excircle of the angle A (Figure 26)[11].

---

[11] The reader is requested to check that his proof does *not* allow to prove that the three external bisectors are concurrent (this is wrong and thus the proof would be wrong too).

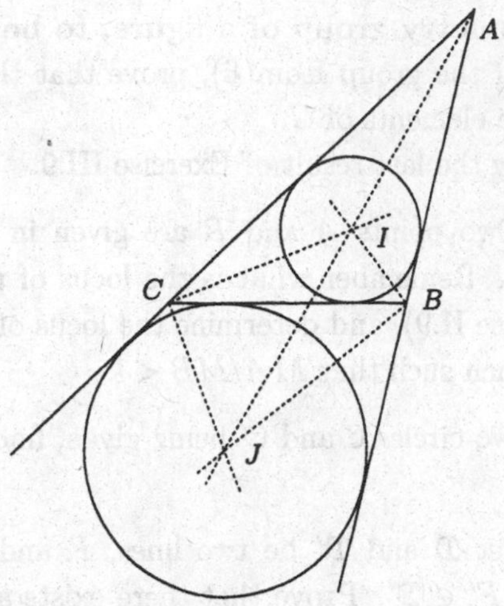

Fig. 26. The incircle and an excircle

Exercise III.5 (Isosceles triangles). If C belongs to the perpendicular bisector of the segment AB, we have the equality of angles $(\overrightarrow{AB}, \overrightarrow{AC}) = (\overrightarrow{BC}, \overrightarrow{BA})$.

Exercise III.6. What is the composition of the two rotations $\rho_{B,-\theta} \circ \rho_{A,\theta}$?

Exercise III.7. Plane isometries can be written, in complex numbers, as shown in the following table.

	translations	rotations	reflections	glide reflections
writing in complex numbers	$z \mapsto z + b$	$z \mapsto az + b$ $\lvert a\rvert = 1$ $a \neq 1$	$z \mapsto a\bar{z} + b$ $\lvert a\rvert = 1$ $a\bar{b} + b = 0$	$z \mapsto a\bar{z} + b$ $\lvert a\rvert = 1$ $a\bar{b} + b \neq 0$

Exercise III.8. By the way, how is an affine transformation written in complex numbers?

Exercise III.9 (Isometry group of a figure, to be continued). Consider two rigid motions φ and ψ of the plane. What can be said of the commutator $\varphi \circ \psi \circ \varphi^{-1} \circ \psi^{-1}$?

Prove that the group of rigid motions that preserve a bounded subset of the plane is commutative.

Prove that any finite subgroup of the group of affine rigid motions of the plane is commutative.

Exercise III.10 (Isometry group of a figure, to be continued). If G is a finite subgroup of the group $\mathrm{Isom}(\mathcal{E})$, prove that there is a point of $\mathcal{E}$ that is fixed by all the elements of G.

Find again this way the last result of Exercise III.9.

Exercise III.11. Two points A and B are given in the plane together with a real number k. Remember what is the locus of points M such that $MA/MB = k$ (Exercise II.9), and determine the locus of the points M such that $MA/MB > k$, then such that $MA/MB < k$.

Exercise III.12. Two circles $\mathcal{C}$ and $\mathcal{C}'$ being given, find all the similarities that map $\mathcal{C}$ to $\mathcal{C}'$.

Exercise III.13. Let $\mathcal{D}$ and $\mathcal{D}'$ be two lines, F and F' be two points. Assume that $F \notin \mathcal{D}$, $F' \notin \mathcal{D}'$. Prove that there exists a direct similarity σ such that $\sigma(\mathcal{D}) = \mathcal{D}'$ and $\sigma(F) = F'$.

Exercise III.14. The plane is identified with $\mathbf{C}$. If A is a point of affix a, prove that the inversion $I(A, k)$ maps the point of affix z to the point of affix

$$z' = a + \frac{k}{\overline{z} - \overline{a}}.$$

Exercise III.15. Do the inversions constitute a group? Prove that the conjugate of an inversion by another inversion is again (the restriction of) an inversion (determine its pole and power)[12].

Exercise III.16. Let A and B be two points in the plane, and A' and B' their images by an inversion. Prove that A, B, A' and B' are cyclic or collinear.

Exercise III.17. What are the circles preserved by an inversion?

Exercise III.18. What is the image of a circle of center O by an inversion of pole O?

Exercise III.19. Transform by the inversion of pole A and of power AC^2 the figure consisting of the square $ABCD$ and its circumcircle.

Exercise III.20. What is the image of the center of a circle by an inversion whose pole is not on the circle (see Figure 19)?

Exercise III.21. Is the mapping defined by the formula $z \mapsto \dfrac{z-i}{z+i}$ affine? Is this an inversion?

Exercises

Exercise III.22. Let ABC be a triangle. The perpendicular bisector of

[12] Another avatar of the conjugacy principle I-2.19.

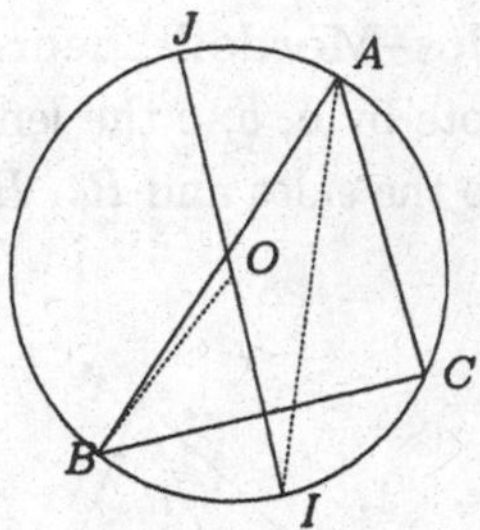

Fig. 27

BC intersects the circumcircle at two points I and J. Let J be the one that is on the same side of BC as A (Figure 27). Prove that AI and AJ are the (*resp.* internal and external) bisectors of the angle at A of the triangle.

Exercise III.23. Let D, E and F be the symmetric points of the orthocenter of a triangle ABC with respect to its three sides. Prove that these points belong to the circumcircle of ABC. Assume that the angles of the triangle ABC are acute. Prove that the altitudes of ABC are the internal bisectors of the triangle DEF.

A triangle DEF is given. Construct a triangle ABC whose altitudes are the internal bisectors of DEF.

Exercise III.24 (Fagnano's problem). Let ABC be an acute-angled triangle. We look for three points P, Q and R on its three sides such that the perimeter of PQR is minimal. Prove that there exists a solution, then construct three points that give a solution. One can firstly consider an arbitrary

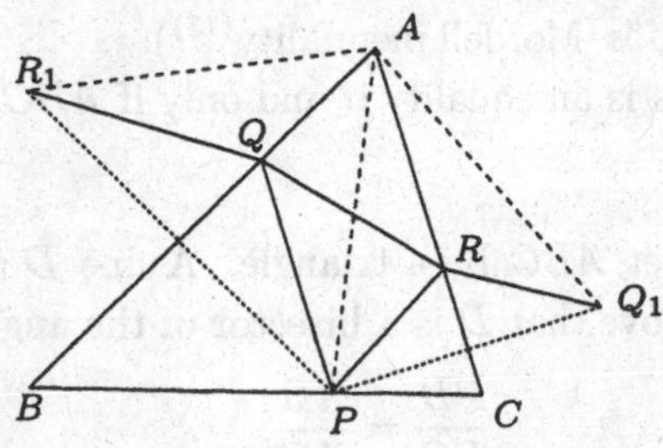

Fig. 28. Fagnano's problem

point P of the segment BC and its symmetric points Q_1 and R_1 with respect to the two other sides (Figure 28) to minimize the perimeter of PQR, with P fixed, then let P vary.

Prove that the altitudes of ABC are the internal bisectors of PQR.

Exercise III.25 (The Erdős–Mordell theorem). Let P be a point inside a triangle ABC. Denote by a, b, c the lengths of the three sides, r_a, r_b, r_c the distances from P to the sides and R_a, R_b, R_c the lengths PA, PB, PC (Figure 29).

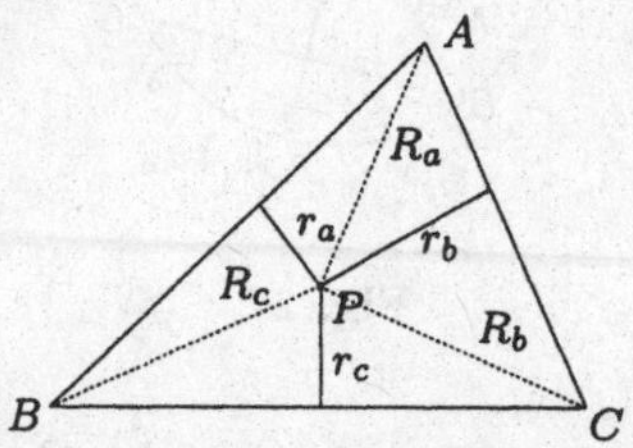

Fig. 29. The Erdős–Mordell theorem

(1) Assume that $P \in BC$. Prove that the area of ABC is equal to $\frac{1}{2}(br_b + cr_c)$. Deduce that $aR_a \geqslant br_b + cr_c$.
(2) Using a dilatation of center A, prove that this inequality holds for any P.
(3) Using the image of P by the reflection about the bisector of the angle at A, prove that $aR_a \geqslant br_c + cr_b$.
(4) Deduce that

$$R_a + R_b + R_c \geqslant \frac{b^2 + c^2}{bc} r_a + \frac{c^2 + a^2}{ac} r_b + \frac{a^2 + b^2}{ac} r_c$$

and eventually that

$$R_a + R_b + R_c \geqslant 2(r_a + r_b + r_c)$$

(this is the Erdős–Mordell inequality[13]).
(5) Prove that this is an equality if and only if ABC is equilateral and P is its center.

Exercise III.26. Let ABC be a triangle. A line D through A intersects the line BC at P. Prove that D is a bisector of the angle at A if and only if

$$\frac{PB}{PC} = \frac{AB}{AC}.$$

Exercise III.27. Let A and B be two distinct points of the oriented Euclidean affine plane $\mathcal{P}$ and let α be a real number. Find the locus of points M such that $(\overrightarrow{MA}, \overrightarrow{MB})$ (*resp.* (MA, MB), *resp.* the geometric angle $\widehat{AMB}$) has measure α.

---

[13] This short proof is due to Vilmos Komornik **[Kom97]**.

Exercise III.28 (Criteria for congruence of triangles). Let ABC and $A'B'C'$ be two triangles. Under each of the following hypotheses, prove that ABC and $A'B'C'$ are isometric (namely, that there exists an isometry φ such that $A' = \varphi(A)$, $B' = \varphi(B)$ and $C' = \varphi(C)$).

(1) First criterion, "an equal angle between two respectively equal sides", in clear:
$$AB = A'B', \quad AC = A'C' \quad \text{and} \quad A = A'$$
(A denotes the measure of the angle at A).

(2) Second criterion, "an equal side between two respectively equal angles", in clear:
$$AB = A'B', \quad A = A' \quad \text{and} \quad B = B'.$$

(3) Third criterion, "three respectively equal sides", in clear:
$$AB = A'B', \quad BC = B'C', \quad \text{and} \quad CA = C'A'.$$

Exercise III.29. Let ABC be a triangle and let I be its incenter, K, K' and K'' the orthogonal projections of I on the sides BC, CA, AB. Let a, b and c be as usual the lengths of the three sides BC, CA, AB and p the half-perimeter. Prove[14] that $CK = p - c$, $AK' = p - a$, $BK'' = p - b$.

Exercise III.30. Let ABC be a triangle. Let α, β and γ be some measures of the angles $(\overrightarrow{AB}, \overrightarrow{AC})$, $(\overrightarrow{BC}, \overrightarrow{BA})$ and $(\overrightarrow{CA}, \overrightarrow{CB})$. Identify the composition $\rho_{A,\alpha} \circ \rho_{B,\beta} \circ \rho_{C,\gamma}$.

Exercise III.31. Let $\mathcal{C}$ be a circle[15] and D_1, D_2, D_3 be three directions of lines. Let M_0 be a point of $\mathcal{C}$, M_1 be the other intersection point of the

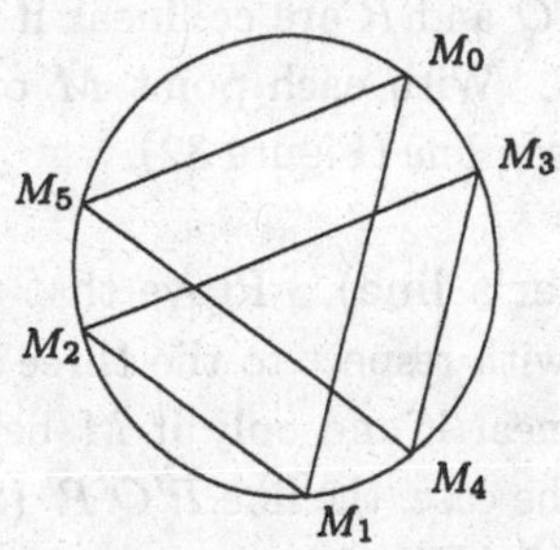

Fig. 30

(14) This exercise is not terribly exciting, but its result is useful; it will be used in Exercise III.64.

(15) The circle can be replaced by any conic in this exercise; this becomes a variant of Pascal's theorem (here Theorem VI-4.4; see also Exercise VI.49 and another variant of Pascal's theorem on a circle in Exercise III.51).

parallel to D_1 through M_0 with $\mathcal{C}$, M_2 the other intersection point of the parallel to D_2 through M_1 and $\mathcal{C}$, *etc.* We define this way points M_i for $i \geqslant 0$. Prove that $M_6 = M_0$ (Figure 30).

Exercise III.32 (Miquel's theorem). Four circles $\mathcal{C}_1$, $\mathcal{C}_2$, $\mathcal{C}_3$ and $\mathcal{C}_4$ are given, such that $\mathcal{C}_4$ and $\mathcal{C}_1$ intersect at A and A', $\mathcal{C}_1$ and $\mathcal{C}_2$ intersect at B and B', $\mathcal{C}_2$ and $\mathcal{C}_3$ intersect at C and C' and eventually $\mathcal{C}_3$ and $\mathcal{C}_4$ intersect at D and D'. Prove that A, B, C and D are cyclic if and only if A', B', C'

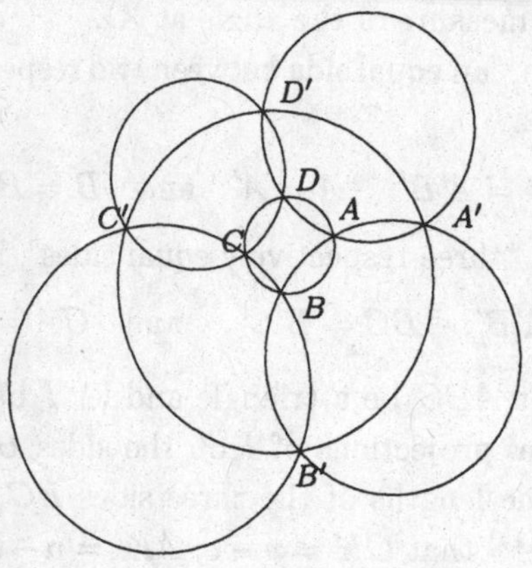

Fig. 31. Miquel's theorem

and D' are (Figure 31[16]).

Exercise III.33 (Simson's line). Let ABC be a triangle. With any point M of the plane, associate its orthogonal projections P, Q and R on BC, CA, AB. Prove that P, Q and R are collinear if and only if M belongs to the circumcircle to ABC. With each point M of the circumcircle is thus associated a line, its *Simson line* (Figure 32).

Exercise III.34 (Steiner's line). Prove that the symmetric points P', Q' and R' of a point M with respect to the three sides BC, CA and AB of a triangle ABC are collinear if and only if M belongs to the circumcircle. Prove that, when this is the case, the line $P'Q'R'$ (*Steiner's line* of M) passes through the orthocenter of ABC.

Exercise III.35 (The pivot). Let ABC be a triangle and let A', B' and C' be three points (distinct from A, B and C) on its sides BC, CA and AB. Prove that the circumcircles of $AB'C'$, $BC'A'$ and $CA'B'$ have a common point, the *pivot* (see Figure 33).

(16) If this figure reminds you of a cube, do not hesitate to find out why in Exercise V.38.

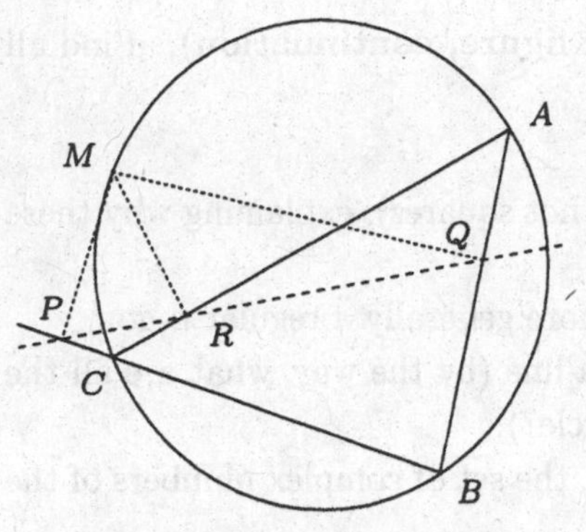

Fig. 32. Simson's line

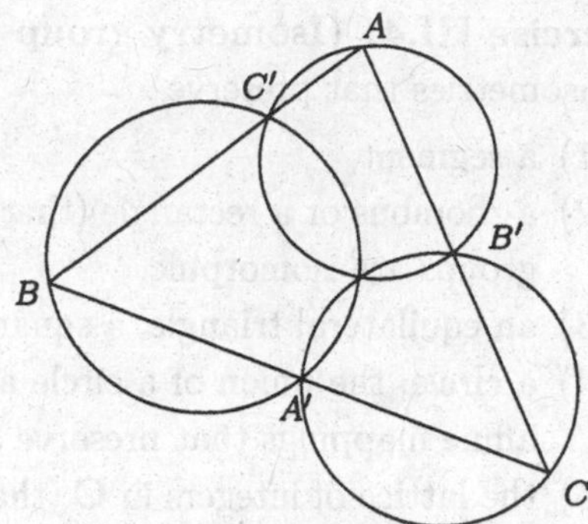

Fig. 33. The pivot

Exercise III.36. Three parallel lines D_1, D_2 and D_3 are given. Construct an equilateral triangle ABC with A on D_1, B on D_2 and C on D_3.

Exercise III.37. Let ABC be a triangle and let β and γ be the two points such that βAB and γAC are right-angled isosceles triangles exterior to ABC of hypotenuses AB, AC respectively. Let I be the midpoint of BC. Prove that $\beta I \gamma$ is right-angled isosceles at I.

Exercise III.38 (A theorem attributed to Napoléon). Let $ABCD$ be a convex quadrilateral, and P, Q, R and S the four points such that APB, BQC, CRD and DSA are four right-angled isosceles triangles (at P *etc.*) exterior to the quadrilateral. Prove that $PR = QS$ and $PR \perp QS$. Prove that $PQRS$ is a square if and only if $ABCD$ is a parallelogram.

Exercise III.39. On a circle of center O, three arcs AB, CD, EF intercepted by angles at the center of measure $\pi/3$ are given. Let M, N and P be the midpoints of the chords BC, DE and FA, and B', E' the midpoints of OB and OE. Prove that $PB'E'$ is equilateral and MNP too.

Exercise III.40 (Fermat's point). On the three sides AB, BC and CA of a triangle (and outside the triangle), three equilateral triangles ABC', BCA' and CAB' are constructed.

Prove that AA', BB' and CC' are concurrent at a point F, that they make angles of $2\pi/3$ and that the segments AA', BB' and CC' have the same length.

Assume now that the angles of the triangle have measures less than $2\pi/3$. Prove that the function

$$M \longmapsto MA + MB + MC$$

has a minimum, reached at F.

Exercise III.41 (Isometry group of a figure, continuation). Find all the isometries that preserve

(1) a segment,
(2) a rhombus or a rectangle (that are not squares), explaining why these groups are isomorphic,
(3) an equilateral triangle, a square, more generally a regular n-gon,
(4) a circle, the union of a circle and a line (by the way, what are all the affine mappings that preserve a circle?),
(5) the lattice of integers in $\mathbf{C}$ (that is, the set of complex numbers of the form $m + in$ with m and n in $\mathbf{Z}$).

Exercise III.42 (Isometry group of a figure, continuation). Does there exist a plane figure whose isometry (*resp.* rigid motion) group is isomorphic with the alternating group $\mathfrak{A}_4$?

Exercise III.43. Two (distinct) points A and A' and two real numbers θ and k (with $k > 0$) are given. Where is the center of the direct similarity of angle θ and ratio k that maps A to A'?

Exercise III.44 (Criteria for similarity of triangles). Let ABC and $A'B'C'$ be two triangles. Under each of the following hypotheses, prove that ABC and $A'B'C'$ are directly similar (that is, that there exists a direct similarity φ such that $A' = \varphi(A)$, $B' = \varphi(B)$ and $C' = \varphi(C)$).

(1) First criterion, "an equal angle between two respectively proportional sides", in clear:

$$\frac{A'B'}{AB} = \frac{A'C'}{AC} \text{ and } (\overrightarrow{A'B'}, \overrightarrow{A'C'}) = (\overrightarrow{AB}, \overrightarrow{AC}).$$

(2) Second criterion, "two respectively equal angles", in clear:

$$(B'C', B'A') = (BC, BA) \text{ and } (C'A', C'B') = (CA, CB)$$

(equality of angles of lines).

Exercise III.45. A line $\mathcal{D}$ and a point A not on $\mathcal{D}$ are fixed. With a point B that varies on $\mathcal{D}$, associate the unique point C such that ABC remains directly similar to a fixed triangle. Find the locus of C and that of the orthocenter of ABC when B describes $\mathcal{D}$.

Exercise III.46. Let $\mathcal{D}$ and $\mathcal{D}'$ be two parallel lines and S be a point outside the strip they determine. A variable line through S intersects $\mathcal{D}$ at M and $\mathcal{D}'$ at M'. Determine the locus of the contact points T and T' of the tangents through S to the circle of diameter MM'.

Exercise III.47. An arc of circle Γ of extremities A and B is given.

- With each point M of Γ distinct from A and B is associated the point M' of the half-line BM of origin B such that $BM' = AM$. What is the locus of points M' when M describes Γ?
- With each point M of Γ distinct from A and B is associated the point M'' of the half-line opposite to the half-line MA of origin M such that $MM'' = BM$. What is the locus of points M'' when M describes Γ?

Exercise III.48. Let $\mathcal{C}$ and $\mathcal{C}'$ be two circles of centers O and O' intersecting at two points I and J and let σ be the direct similarity of center I such that $\sigma(O) = O'$. Prove that $\sigma(\mathcal{C}) = \mathcal{C}'$ and that, for any point M of $\mathcal{C}$, the points M, $\sigma(M)$ and J are collinear.

Determine the locus of the orthogonal projection P of I on $M\sigma(M)$ when M describes $\mathcal{C}$.

Determine the locus of the centroid (and of the circumcenter) of the triangle $IM\sigma(M)$ when M describes $\mathcal{C}$.

Exercise III.49 (Construction of the radical axis). Let $\mathcal{C}$ and $\mathcal{C}'$ be two circles (assumed to be nonconcentric). Let Γ be a circle that intersects $\mathcal{C}$ at A and B and $\mathcal{C}'$ at A' and B'. Prove that the radical axis of $\mathcal{C}$ and $\mathcal{C}'$ is the perpendicular to the line of centers through the intersection point of AB and $A'B'$.

Exercise III.50. Let A', B' and C' be the feet of the altitudes through A, B and C in the triangle ABC. The lines BC and $B'C'$ intersect at α, CA and $C'A'$ at β, AB and $A'B'$ at γ. Prove that

$$\overrightarrow{\alpha B} \cdot \overrightarrow{\alpha C} = \overrightarrow{\alpha B'} \cdot \overrightarrow{\alpha C'}.$$

What can be said of α, β and γ? What is the radical axis of the circumcircle to ABC and its Euler circle (recall this is the circumcircle to $A'B'C'$; see Exercise II.20)?

Exercise III.51 (Pascal's theorem for circles). Consider six points A, B, C, D, E and F on a circle $\mathcal{C}$. Assume that the hexagon $ABCDEF$ has no pair of parallel sides. Consider the intersection points

$$S = AB \cap DE, \quad T = CD \cap AF \quad \text{and} \quad U = BC \cap EF.$$

We want to prove that S, T and U are collinear (Figure 34). Let P, Q and R be the intersection points

$$CD \cap FE, \quad FE \cap AB \quad \text{and} \quad AB \cap CD.$$

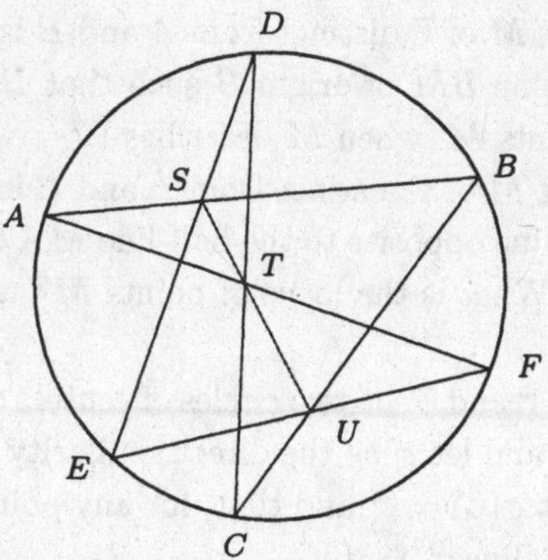

Fig. 34. Pascal's theorem

Considering the lines SDE, ATF and BCU as transversals to the sides of PQR, prove that the equality

$$\frac{\overrightarrow{SQ}}{\overrightarrow{SR}} \cdot \frac{\overrightarrow{TR}}{\overrightarrow{TP}} \cdot \frac{\overrightarrow{UP}}{\overrightarrow{UQ}} = 1$$

holds and conclude (the most general version of Pascal's theorem is Theorem VI-4.4).

Exercise III.52. Given two circles $\mathcal{C}$ and $\mathcal{C}'$, find all the inversions that map $\mathcal{C}$ to $\mathcal{C}'$.

Exercise III.53. Let A, B, C and D be four collinear points (in this order). Find an inversion that maps them to the four consecutive vertices of a rectangle.

Exercise III.54 ("Anallagmatic invariant" of two circles). Consider two intersecting circles[17] of radii R and R'. Let d be the distance between their centers. Prove that the ratio

$$\frac{R^2 + R'^2 - d^2}{2RR'}$$

is preserved by any inversion.

Exercise III.55 (Ptolemy's inequality). Let A, B, C and D be four points of a plane. Prove that

$$AB \cdot CD \leqslant AC \cdot BD + AD \cdot BC$$

(one can use a well-chosen inversion). In which cases does the equality hold?

---

[17] The result holds true for any two circles, but it is a little harder to prove directly; see [DC51] and Exercise VI.56.

Exercise III.56. Two circles $\mathcal{C}$ and $\mathcal{C}'$, a point A on $\mathcal{C}$ and a point A' on $\mathcal{C}'$ are given. Find a point P on the radical axis Δ of $\mathcal{C}$ and $\mathcal{C}'$ such that, if PA (*resp.* PA') intersects $\mathcal{C}$ at M (*resp.* M'), then $MM' \perp \Delta$.

Exercise III.57. What is the locus of centers of the circles of an intersecting pencil? of a nonintersecting pencil?

Exercise III.58. Find the locus of points such that the ratio of their powers with respect to two given nonconcentric circles has a given value k.

Exercise III.59. Prove that the three circles $\mathcal{C}$, $\mathcal{C}'$ and $\mathcal{C}''$ of centers and respective radii (O, R), (O', R') and (O'', R'') belong to the same pencil if and only if the three points O, O' and O'' are collinear and satisfy

$$R^2\overrightarrow{O'O''} + R'^2\overrightarrow{O''O} + R''^2\overrightarrow{OO'} + (\overrightarrow{O'O''} \cdot \overrightarrow{O''O}) \cdot \overrightarrow{OO'} = 0.$$

Exercise III.60. Let $\mathcal{F}$ be a pencil of circles. Prove that any point of the plane lies, in general, on a (unique) circle of $\mathcal{F}$. What are the exceptions?

Given a point in the plane, construct the circle of the pencil $\mathcal{F}$ that passes through this point.

Exercise III.61. Let $\mathcal{C}$ and $\mathcal{C}'$ be two circles. Prove that there exists an inversion that maps them to two lines or two concentric circles.

Exercise III.62 (Steiner's porism). Let $\mathcal{C}$ and $\mathcal{C}'$ be two circles. Assume that $\mathcal{C}$ is inside $\mathcal{C}'$. Let Γ_1 be a circle tangent (interiorly) to $\mathcal{C}'$ and (exteriorly) to $\mathcal{C}$. A sequence of circles Γ_i is constructed by induction so that Γ_{i+1} is

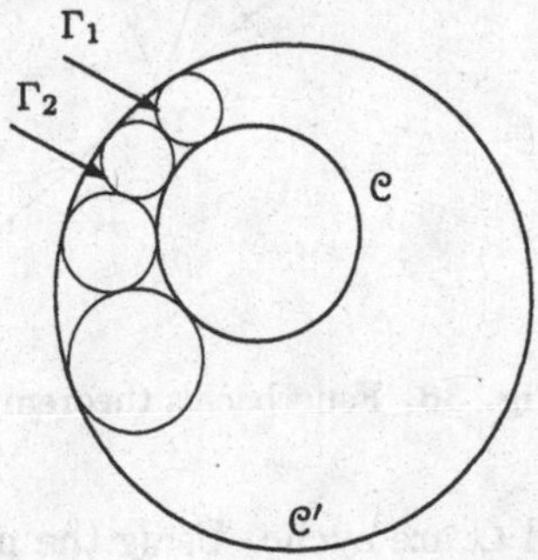

Fig. 35. Steiner's porism

tangent (interiorly) to $\mathcal{C}'$, and tangent (exteriorly) to $\mathcal{C}$ and Γ_i and different from Γ_{i-1} (Figure 35). Prove that, if for some n, $\Gamma_n = \Gamma_1$, then the same is true, for the same n, but starting from any choice of Γ_1.

Exercise III.63. Determine the locus of the poles of all the inversions that transform two secant lines into two circles of the same radius. Deduce the poles of the inversions that transform the three sides of a triangle into three circles of the same radius. Prove that these inversions transform the circumcircle of the triangle into a circle that has the same radius as the three others.

Let ABC be a triangle, O the center and R the radius of its circumcircle, I the center and r the radius of its incircle. Using an inversion of pole I that preserves the circumcircle, prove the "Euler relation":

$$R^2 - OI^2 = 2rR.$$

Exercise III.64 (Feuerbach's theorem). Let ABC be a triangle and let I be the center of its incircle $\mathcal{C}$, J the center of the excircle $\mathcal{C}'$ of the angle at A, and K and L the orthogonal projections of I and J on the side BC (Figure 36).

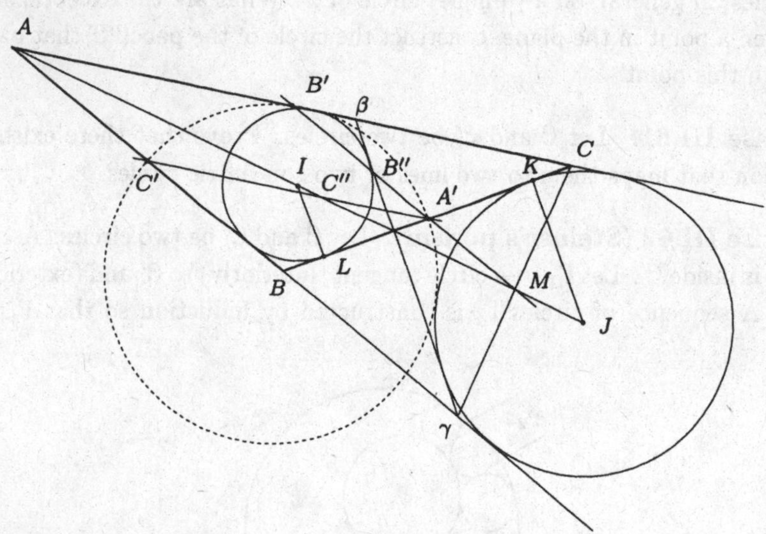

Fig. 36. Feuerbach's theorem

Prove that I, J, B and C are cyclic. Using the midpoint N of IJ, prove that the midpoint A' of BC is the midpoint of KL. Compute $A'K$ and $A'L$ (one can use Exercise III.29).

Let B' and C' be the respective midpoints of AC and AB. Let β and γ be the symmetric points of B and C with respect to AI, so that BC and $\beta\gamma$ are the two interior common tangents to $\mathcal{C}$ and $\mathcal{C}'$. Let M be the midpoint of γC. Prove that $M \in AI \cap A'B'$. Compute $A'M$.

Let $B'' = A'B' \cap \beta\gamma$, $C'' = A'C' \cap \beta\gamma$. Prove that the inversion of pole A' and of power $A'M^2$ exchanges B' and B'', C' and C''. Deduce that the Euler circle (circumcircle of $A'B'C'$, see Exercise II.20) is tangent to $\mathcal{C}$ and $\mathcal{C}'$.

More theoretical exercises

Exercise III.65. The differential at a point z_0 of a holomorphic mapping defined on an open subset of $\mathbf{C}$ is a linear mapping $\mathbf{C} \to \mathbf{C}$. What mapping is this? Prove that a nonconstant holomorphic mapping preserves the angles.

Exercise III.66 (Liouville's theorem). Let f be a bijection of class $\mathcal{C}^1$ from the Euclidean affine plane $\mathcal{P}$ to itself. Assuming that f preserves the angles (in the sense explained by Figure 14 in § 4), we want to prove that f is a similarity.

(1) For any point M of $\mathcal{P}$, the differential df_M is a linear mapping $P \to P$. Prove that df_M is, either a direct similarity for all M, or an opposite similarity for all M. Composing if necessary with a reflection, prove that it can be assumed that df_M is a direct similarity for all M.

(2) Choose an orthonormal frame and identify $\mathcal{P}$ with $\mathbf{C}$. Denote by df_z the differential of f at the point M of affix z. Thus df_z is a linear mapping, over $\mathbf{R}$, *a priori*, from $\mathbf{C}$ to $\mathbf{C}$, and we are assuming this is a direct similarity. Prove that f is a holomorphic function on $\mathbf{C}$ (what is called an *entire* function).

(3) Use the classical results on entire functions[18] and the assumption that f is a bijection to conclude that f is a degree-1 polynomial, namely a direct similarity.

(4) Why is this not contradictory to Corollary 4.5?

---

[18] See, *e.g.*, **[Car95]** or **[Sil72]**.

Chapter IV

Euclidean Geometry in Space

In this chapter, everything will take place in a Euclidean (affine or vector) space of dimension 3.

After having investigated the group of isometries and recalled a few facts about the vector product, we attack spheres and especially spherical triangles, the triangles drawn by great circles on the sphere. Having computed the area of these triangles, we deduce the Euler formula for convex polyhedra and the list of regular polyhedra.

1. Isometries and rigid motions in space

Vector isometries. Theorem II-2.1 asserts that they are the composition of one, two or three reflections and Proposition II-3.13 gives us a matrix description: there exists an orthonormal basis of E in which the matrix A of the isometry f has one of the following forms[(1)]:

– $A = \begin{pmatrix} 1 & & \\ & 1 & \\ & & -1 \end{pmatrix}$, where f is a reflection about the plane spanned by the two first basis vectors,

– $A = \begin{pmatrix} 1 & & \\ & \cos\theta & -\sin\theta \\ & \sin\theta & \cos\theta \end{pmatrix}$; it is said that the isometry is a *rotation* of axis the line spanned by the first basis vector; this is the form of all the positive isometries (the elements of $O^+(E)$),

(1) The entries that do not appear explicitly are zero.

- $A = \begin{pmatrix} -1 & & \\ & \cos\theta & -\sin\theta \\ & \sin\theta & \cos\theta \end{pmatrix}$; it is then said that the isometry is an *anti-rotation* or a *rotatory reflection*.

Rotations. Let us spend some time on rotations. They are the vector isometries composed of two reflections. If P_1 and P_2 are two planes intersecting along a line D, the isometry $s_{P_2} \circ s_{P_1}$ fixes D (pointwise) and the plane $Q = D^{\perp}$ (globally). In this plane, it operates as a plane rotation; this is the composed mapping $s_{D_2} \circ s_{D_1}$, where D_i denotes the line $P_i \cap Q$ (see Figure 1). Notice that, as in the case of plane rotations, one of the two planes used in the decomposition can be chosen arbitrarily (it must, of course, contain the fixed line D).

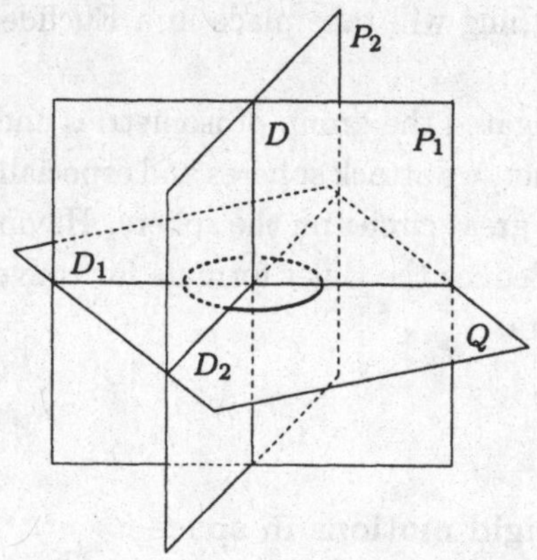

Fig. 1

In an orthonormal basis whose first vector, u, is a vector of D, the matrix of the rotation has the form above:

$$\begin{pmatrix} 1 & & \\ & \cos\theta & -\sin\theta \\ & \sin\theta & \cos\theta \end{pmatrix}.$$

We would like to say that the (or, better to say, a) number θ as written in this matrix is the *angle of the rotation*. A little care is needed: this number depends on the basis.

In the plane, we have seen (in § II-3) that we had to choose an orientation to define θ (modulo 2π). Here it is a little bit more subtle: we would need to have chosen orientations for all the planes in space, this being impossible[(2)].

[(2)] The meaning of the word "impossible" will be made precise in a statement given in Exercise V.46.

The solution is the following. We fix an orientation of the space. A rotation being given, we choose an orientation (or a unit vector u) of its fixed line D. The line thus becomes an *axis*. We complete u in an orthonormal basis of the space and write the matrix of the rotation in this basis. We thus get a number θ (at least modulo 2π). The rotation can then be denoted by $\rho_{u,\theta}$ and the number θ can be called its *angle*... but of course we have

$$\rho_{-u,\theta} = \rho_{u,-\theta}.$$

In brief, to measure the angle of a rotation, it is necessary to orient the (space *and*) the line of fixed points of this rotation. This is probably why this line is called the *axis* of the rotation. To change the orientation of the line changes the measure of the angle to its opposite, to change the orientation of the space too.

A confirmation of this difficulty can be found in the fact that the angle appears intrinsically (that is, without needing the choice of a basis or an orientation) only through its cosine, in the trace of the rotation f, *via* the formula

$$\operatorname{tr}(f) = 1 + 2\cos\theta.$$

An important special case is that of the *half-turns*, which are both the orthogonal symmetries about lines and the rotations of angle π, obtained as compositions of two reflections about orthogonal planes.

Notice also, as in the case of the plane, that the translations are the compositions of two reflections of parallel planes. The vector of the translation is obtained exactly as in the case of the plane (see § III-2).

Affine isometries. As in the case of the plane (§ III-2), the form of all the affine isometries of the Euclidean affine space is deduced from the results on linear isometries (the matrix form given by Proposition II-3.13) and the general results on the fixed points of affine isometries (as given in Proposition II-2.8).

Notice firstly, looking at the matrices above, that all the vectorial isometries except the anti-rotations have the eigenvalue 1. Let thus φ be an affine isometry of the space and let $\overrightarrow{\varphi}$ the associated linear isometry.

- If $\overrightarrow{\varphi}$ is the identity, φ is a translation.
- If $\overrightarrow{\varphi}$ is a reflection of plane P, then:
  - either φ has a fixed point A: we vectorialize the affine space at A and apply the corresponding linear result to obtain that φ is the reflection about the plane through A directed by P;
  - or φ has no fixed point: then, according to Proposition II-2.8, there exists a unique vector v of P and a unique reflection ψ of

plane directed by P such that

$$\varphi = t_v \circ \psi = \psi \circ t_v.$$

The isometry φ is said to be an orthogonal *glide reflection*.

- If $\overrightarrow{\varphi}$ is a rotation of axis D, then:
  - either φ has a fixed point A: we vectorialize the space at A and φ acts as $\overrightarrow{\varphi}$; it is said that φ is a *rotation* of axis the line through A directed by D;
  - or φ has no fixed point: then, still using Proposition II-2.8, we find a unique vector v of D and a unique rotation ψ of axis $\mathcal{D}$ directed by D such that

$$\varphi = t_v \circ \psi = \psi \circ t_v.$$

The isometry φ is a *screw displacement* of axis $\mathcal{D}$: we make the space turn about $\mathcal{D}$ and we push in the direction of $\mathcal{D}$ (this is indeed what you do when driving in a screw, Figure 2).

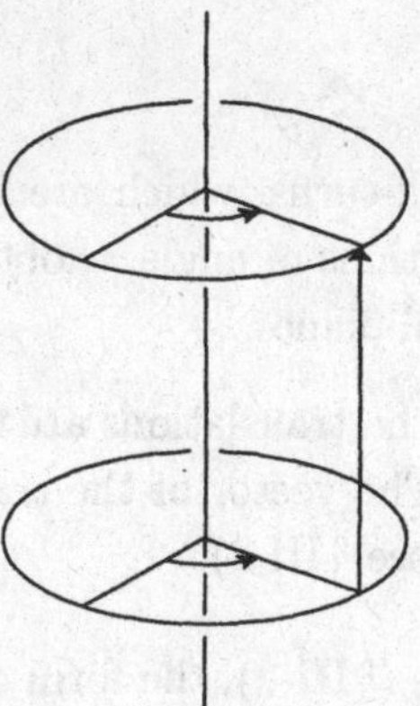

Fig. 2. A screw displacement

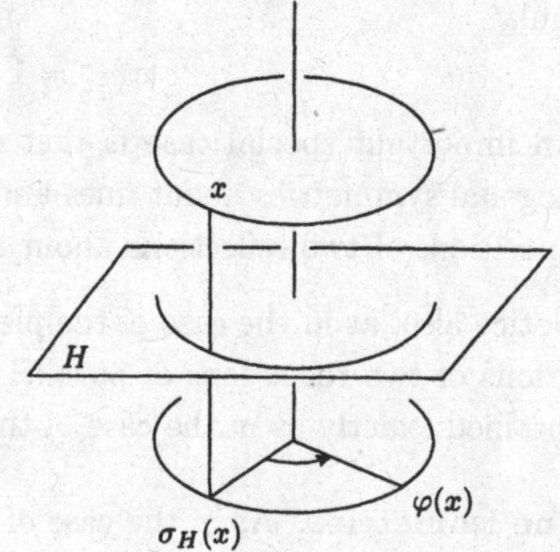

Fig. 3. An anti-rotation

- If $\overrightarrow{\varphi}$ is an anti-rotation, then it has no nonzero fixed vector; then (according to Proposition I-2.20), φ has a unique fixed point A. It thus works in the affine space as $\overrightarrow{\varphi}$ in the vector space. It is said that this is an *anti-rotation*. It is, for example, the composition of a rotation and a central symmetry, or of a rotation and a reflection (Figure 3); this is why it is also called a *rotatory reflection*.

2. The vector product, with area computations

In this section, the space E is an *oriented* Euclidean vector space of dimension 3.

Definition 2.1. The *vector product*, or *cross product*, a composition law on vectors, is defined by:

- $u \times v = 0$ if u and v are collinear,
- $u \times v = w$, the unique vector:
  - orthogonal to u and v,
  - of length $\|u\| \|v\| |\sin \alpha|$ where α is a measure of the angle (u, v) ($|\sin \alpha|$ does not depend on the choices made to define this measure),
  - and such that the basis (u, v, w) is a direct basis,

  otherwise.

Remarks 2.2

- For instance, if u and v are two orthogonal vectors of length 1, the basis $(u, v, u \times v)$ is a direct orthonormal basis of E.
- The vector product depends on the orientation of E. To choose the other orientation changes the vector product in its opposite.

The vector product is endowed with sympathetic properties, which are summarized in the following statement.

Proposition 2.3. *The vector product is an alternated (skew-symmetric) mapping*

$$E \times E \longrightarrow E.$$

Moreover, the product $u \times v$ is zero if and only if the vectors u and v are collinear.

Proof. Except for bilinearity, all the properties are obvious consequences of the definition. The word *alternated* means that $u \times u = 0$ for any vector u, or, equivalently, that all the vectors u and v satisfy $u \times v = -v \times u$, namely the vector product is skew-symmetric.

We still have to prove that this operation is bilinear. Fix a nonzero vector u and consider the mapping f_u defined by

$$f_u : v \longmapsto u \times v.$$

We want to prove that this is a linear mapping. To this aim, we prove that this is the composition

$$f_u = h_{\|u\|} \circ r_{(u,\pi/2)} \circ p_u$$

of three linear mappings, respectively a dilatation, a rotation and the projection onto the plane $u^\perp$.

To obtain the vector product $u \times v$, we begin by constructing a vector v_2 of the line $\langle u, v\rangle^\perp$ such that the basis (u, v, v_2) is direct. To do this, it suffices

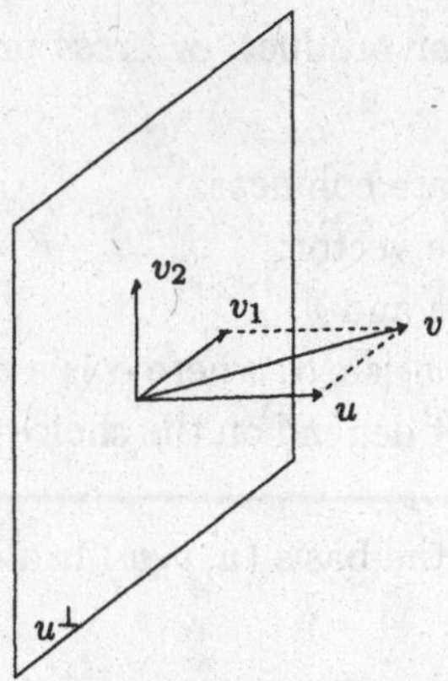

Fig. 4

to project v orthogonally on $u^\perp$, getting a vector v_1 orthogonal to u, then to perform a rotation of angle $\pi/2$ about u, getting v_2 with the expected properties (Figure 4).

We still have to dilate v_2 to transform it into a vector w with the suitable length. We have

$$\|v_2\| = \|v_1\| = \|v\| \, |\sin\alpha| \,,$$

so that it suffices to multiply v_2 by $\|u\|$. The mapping f_u is the composition of the three linear mappings, as announced. □

Calculations in coordinates. The vector product is very simply expressed in the *direct orthonormal* bases.

Proposition 2.4. *In a direct orthonormal basis, the coordinates of the vector product are given by the formulas*

$$\begin{pmatrix} a \\ b \\ c \end{pmatrix} \times \begin{pmatrix} x \\ y \\ z \end{pmatrix} = \begin{pmatrix} bz - cy \\ cx - az \\ ay - bx \end{pmatrix}.$$

Proof. Let (e_1, e_2, e_3) be a basis of the vector space E. We have just shown that the vector product is bilinear. We thus have

$$\begin{aligned}(ae_1 + be_2 + ce_3) &\times (xe_1 + ye_2 + ze_3) \\ &= (bz - cy)e_2 \times e_3 + (cx - az)e_3 \times e_1 + (ay - bx)e_1 \times e_2.\end{aligned}$$

If the basis (e_1, e_2, e_3) is orthonormal and direct, we also have $e_2 \times e_3 = e_1$, $e_3 \times e_1 = e_2$ and $e_1 \times e_2 = e_3$. The result follows. □

Remark 2.5. One could feel more comfortable, having these formulas. Notice, however, that they are complicated and can lead to rather tedious computations. To use them well means to *choose*, for each problem, the most suitable direct orthonormal basis, the one that gives the simplest formulas (one could for example solve Exercise IV.12 using this method).

Calculations of plane areas. Assume that we are now in an *oriented* Euclidean affine space $\mathcal{P}$. Let us consider it as a subspace of a 3-dimensional space. There exists a unique unit vector u, orthogonal to $\mathcal{P}$ and such that a direct basis of P completed by u is a direct basis of E.

Three points A, B and C being given in $\mathcal{P}$, the vector product $\overrightarrow{AB} \times \overrightarrow{AC}$ is orthogonal to $\mathcal{P}$, hence collinear with u:

$$\overrightarrow{AB} \times \overrightarrow{AC} = \lambda u.$$

If H is the orthogonal projection of C on AB, one has

$$\overrightarrow{AB} \times \overrightarrow{AC} = \overrightarrow{AB} \times \overrightarrow{HC},$$

thus allowing us to compute $|\lambda|$:

$$|\lambda| = AB \cdot HC.$$

This is double the area of the triangle ABC.

The unique real number $\mathcal{A}(ABC)$ satisfying the equality

$$\frac{1}{2}\overrightarrow{AB} \times \overrightarrow{AC} = \mathcal{A}(ABC)u$$

is called the *signed area* of the triangle ABC.

One can see that the signed areas of the triangles ABC and ACB do not coincide, that the signed area of the triangle ABC is positive if and only if the basis $(\overrightarrow{AB}, \overrightarrow{AC})$ is direct... and the signed area of the parallelogram constructed on $\overrightarrow{AB}$ and $\overrightarrow{AC}$ in this order is the number λ considered above.

Notice eventually that the signed area of the triangle ABC is half the determinant of $(\overrightarrow{AB}, \overrightarrow{AC})$ in a direct orthonormal basis.

This interpretation of the area by vector products allows us to prove very simply:

Proposition 2.6. *Let ABC be a triangle. The barycentric coordinates of a point M in the affine frame (A, B, C) are proportional to the signed areas of the triangles MBC, MCA, MAB.*

In other words (see Exercise I.28 for the definition of the barycentric coordinates),

$$\mathcal{A}(MBC)\overrightarrow{MA} + \mathcal{A}(MCA)\overrightarrow{MB} + \mathcal{A}(MAB)\overrightarrow{MC} = 0.$$

Proof. We write the equality

$$\alpha\overrightarrow{MA}+\beta\overrightarrow{MB}+\gamma\overrightarrow{MC}=0$$

in which α, β and γ are the quantities we want to compute. We then take the vector product of this relation with each of the vectors $\overrightarrow{MA}$, $\overrightarrow{MB}$ and $\overrightarrow{MC}$ and we obtain the linear system

$$\begin{cases} \qquad\qquad \beta\mathcal{A}(MAB) \; -\gamma\mathcal{A}(MCA) = 0 \\ \alpha\mathcal{A}(MAB) \qquad\qquad\qquad -\gamma\mathcal{A}(MBC) = 0 \\ \alpha\mathcal{A}(MCA) \; -\beta\mathcal{A}(MBC) \qquad\qquad\quad = 0. \end{cases}$$

Using Proposition 2.4, this system can be rewritten in the form:

$$\begin{pmatrix} \mathcal{A}(MBC) \\ \mathcal{A}(MCA) \\ \mathcal{A}(MAB) \end{pmatrix} \times \begin{pmatrix} \alpha \\ \beta \\ \gamma \end{pmatrix} = 0 \in \mathbf{R}^3.$$

According to Proposition 2.3, α, β and γ are solutions of the system if and only if the vectors

$$\begin{pmatrix} \mathcal{A}(MBC) \\ \mathcal{A}(MCA) \\ \mathcal{A}(MAB) \end{pmatrix} \text{ and } \begin{pmatrix} \alpha \\ \beta \\ \gamma \end{pmatrix}$$

are collinear, and this is what we wanted to prove. □

3. Spheres, spherical triangles

In this section, our space will be a sphere in the Euclidean space. Recall that the planes through the center of a sphere intersect it along *great* circles, circles whose radius is that of the sphere. The other planes also intersect the sphere along circles (that might be empty or point-circles) of smaller radius, the *small* circles[3].

The part of the sphere limited by two great circles (Figure 5) is called a time zone.. The figure limited by three arcs of great circles is called a *spherical triangle* (Figure 6).

The first result in this section is about the area of the spherical triangles. This is not the place to present the theory of what the area should be. I will only use the following properties. There exists an *area* function, defined on a set of subsets of the sphere that contains the time zones and the triangles, taking nonnegative values, and such that:

- the area is invariant by isometries,
- it is additive and

(3) The *meridians* in geography are great circles, while the *parallels* are in general small circles. See Exercises IV.25 and IV.26.

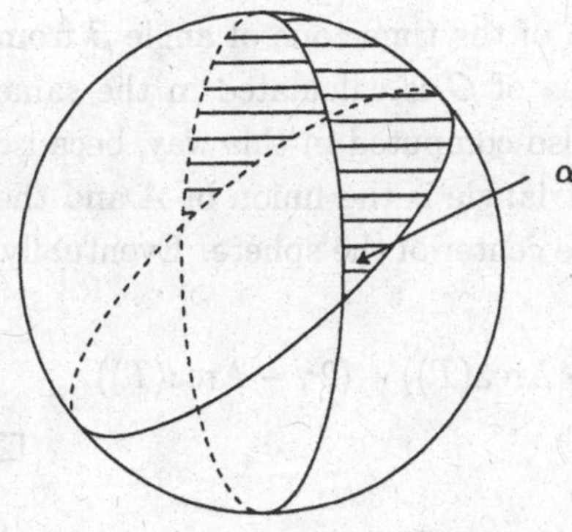

Fig. 5. A time zone

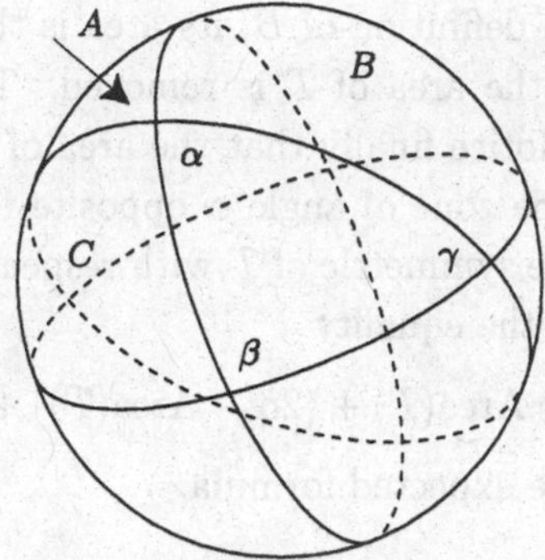

Fig. 6. Area of a triangle

– the total area of a sphere of radius R is $4\pi R^2$.

As a consequence of additivity, the area of a time zone of angle α (Figure 5), is $\alpha/2\pi$ times the area of the sphere, namely $2\alpha R^2$.

Area of the spherical triangle. Until the end of the chapter, we shall stay on a sphere of radius 1.

We shall express the area of a spherical triangle in terms of measures of its angles. Let us start with a precision on these "angles". The angle of two great circles is the angle of their planes. We should thus be able to measure the angle of two planes. We have seen (in § 1) that this is not completely obvious: if D is the line along which the two planes intersect, it is possible to measure the angle in the plane $D^\perp$... once D has been oriented. What stays is a geometric angle in this plane, with which a number $\alpha \in [0, \pi]$ can be associated, which will often be called the *angle of the two planes*.

Proposition 3.1 (Girard's Formula). *Let T be a spherical triangle of angles α, β and γ. Then*

$$\mathrm{Area}(T) = \alpha + \beta + \gamma - \pi.$$

Proof. Let D be the half-sphere containing T, determined by the great circle opposite to the angle α. Let us write D as the union of T and three other subsets:

– the subset A of D opposite to the angle α,
– the complement B of the triangle T in the time zone of angle β,
– and the complement C of this triangle in the time zone of angle γ

(see Figure 6). This way, the area of the half-sphere D, namely 2π, is also equal to

$$\mathrm{Area}(D) = \mathrm{Area}(T) + \mathrm{Area}(A) + \mathrm{Area}(B) + \mathrm{Area}(C).$$

By the definition of B, its area is the area of the time zone of angle β from which the area of T is removed. The area of C is calculated in the same way. Notice finally that the area of A is also computed in this way, because the time zone of angle α opposite to the triangle is the union of A and the triangle symmetric of T with respect to the center of the sphere. Eventually, we get the equality

$$2\pi = \text{Area}(T) + (2\alpha - \text{Area}(T)) + (2\beta - \text{Area}(T)) + (2\gamma - \text{Area}(T))$$

and the expected formula. □

There is no doubt that the area of the triangle T is a positive number, so that we see that a spherical triangle is something very different from a plane Euclidean triangle.

Corollary 3.2. *The sum of the angles of a spherical triangle is a number greater than π.* □

Remark 3.3. Imagine a "geometry" whose points are the points of the sphere and whose lines are the great circles: the plane is replaced by the sphere, the lines by the great circles. In such a geometry, there are no parallel "lines" since two great circles always intersect. In particular, through a point not on a "line", there is *no* parallel to this line. Recall that in Euclidean geometry, the sum of the angles of a triangle is a flat angle (Corollary 1.15) and that this property is related to the parallel axiom (Proposition I-1.17; see also Figure 7 of Chapter III).

The fact that, here, the sum of the angles of a triangle is always *greater* than π is related to the nonexistence of parallels. There are also "geometries" for which, through a point not on a line, there are infinitely many parallels to this line... and in which the sum of the angles of the "triangles" is *smaller* than π (Exercise V.50)[(4)].

Corollary 3.2 explains why it is impossible to draw a map of the Earth on which both the distances and angles are correct. See Exercises IV.36 and IV.39.

4. Polyhedra, Euler formula

Let us now consider convex polyhedra in 3-dimensional affine space.

Definition 4.1. The convex hull of finitely many noncoplanar points (see Figure 7) is called a *convex polyhedron*.

(4) These properties are a way the *curvature* of the spaces under consideration, here the sphere, shows up. See Chapter VIII.

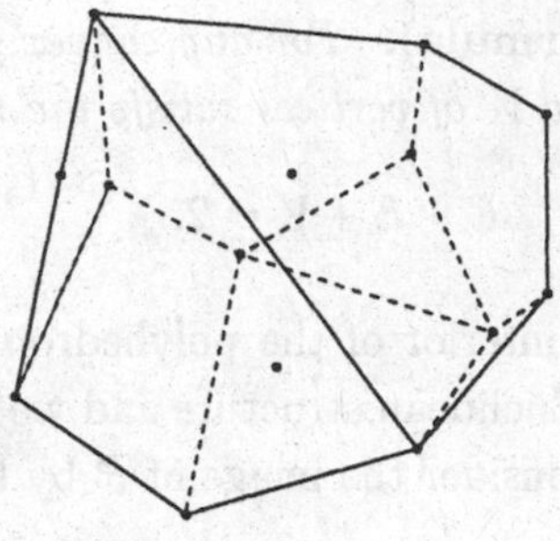

Fig. 7

Remarks 4.2

- The assumption that there are finitely many points ensures that the polyhedron is *compact* (see Exercise I.44).
- The assumption that the points are not coplanar ensures that the polyhedron is indeed of "dimension 3", more precisely that its interior is not empty (see Exercise I.45).

A polyhedron is limited by *faces*, plane polygons, themselves limited by *edges*, the sides of these polygons, the edges in their turn being limited by *vertices*, their ends. If this property seems to be clear, it deserves a proof that is not completely obvious[5]. It is based on the characterization of convex polyhedra expressed in the next proposition.

Proposition 4.3. *Any convex polyhedron is the intersection of a finite number of closed half-spaces. Conversely, any compact intersection of a finite number of half-spaces is a convex polyhedron.*

The first assertion is related to the existence of "supporting hyperplanes" (these are the faces), the second to the existence of extremal points (these are the vertices). I refer the reader to [**Ber77**] for a proof. We shall also accept (and this not either completely obvious) that a convex polyhedron is the convex hull of its vertices.

Euler's formula. Having accepted all this, let F be the number of faces, E the number of edges and V the number of vertices of a polyhedron P. The Euler formula is a statement both simple and remarkable:

[5]The reader that is not convinced that it is not obvious how to define a polyhedron, its faces, its vertices *etc.*, is requested to read [**Lak76**].

Theorem 4.4 (Euler formula). *For any convex polyhedron, the numbers F of faces, E of edges and V of vertices satisfy the relation*

$$F - E + V = 2.$$

Proof. A point O in the interior of the polyhedron P is chosen, the affine space is endowed with a Euclidean structure and we consider the sphere $\mathcal{S}$ of center O and radius 1. Consider the image of P by the radial projection

$$x \longmapsto \frac{x}{\|x\|}.$$

It maps:

- the vertices of P on points of $\mathcal{S}$,
- its edges on arcs of great circles: a segment MN is mapped on the intersection of $\mathcal{S}$ with the plane OMN,
- and the faces on spherical polygons.

In this way, we have constructed a curved version of our polyhedron, drawn on the sphere $\mathcal{S}$.

The proof consists in computing in two different ways the sum of all the angles of all the spherical polygons obtained.

- On the one hand, at each vertex, the sum is 2π, hence the total sum is $2\pi V$.
- On the other hand, this is also the sum of all the angles of all the faces. Now, triangulating a k-vertex polygon and applying Girard's formula (Proposition 3.1), it is noticed that the sum of angles of this polygon is its area to which $(k-2)\pi$ must be added. In this way, we have computed

$$2\pi V = \sum_{f \text{ face}} \left(\text{Area}(f) + (\text{number of vertices of } f - 2)\pi\right).$$

The sum of all the areas of all the faces is the area of the unit sphere, namely 4π. Each face having as many edges as it has vertices and each edge belonging to two faces, the sum of the number of vertices of all the faces is $2E$ and we get the equality

$$2\pi V = 4\pi + 2E\pi - 2F\pi,$$

which is the Euler formula when divided by 2π. □

Remark 4.5. As is seen in this proof, the Euler formula is related to Corollary 3.2, that is, eventually, to the curvature of the sphere. We shall come back to this in Chapter VIII.

Corollary 4.6. *Let P be a convex polyhedron, all the faces of which have the same number s of vertices and such that the same number r of edges end at each vertex. Then*

$$\{r, s\} \in \{\{3,3\}, \{3,4\}, \{3,5\}\}.$$

Proof. In this situation, we have:

- on the one hand, $sF = 2E$ (any edge belongs to two faces),
- on the other hand, $rV = 2E$ (any edge has two ends).

Therefore, from the Euler formula is deduced the relation

$$\frac{1}{s} + \frac{1}{r} = \frac{1}{2} + \frac{1}{E}.$$

As the number E of edges is nonzero, we get the inequality

$$\frac{1}{s} + \frac{1}{r} > \frac{1}{2}.$$

The convex polygons constituting the faces of P have at least three edges, in formulas $s \geqslant 3$. Similarly, at each vertex, at least three faces meet, thus we have also $r \geqslant 3$. The inequalities

$$\frac{1}{r} > \frac{1}{2} - \frac{1}{s} \geqslant \frac{1}{2} - \frac{1}{3} = \frac{1}{6}$$

are deduced, Hence, $r \leqslant 5$ and similarly $s \leqslant 5$. □

Remark 4.7. The numbers r and s determine E by the relation

$$\frac{1}{E} = \frac{1}{s} + \frac{1}{r} - \frac{1}{2},$$

therefore they also determine V and F as we have said that

$$V = \frac{2E}{r}, \quad F = \frac{2E}{s}.$$

Remark 4.8. The numbers r and s play exactly the same role in the statement 4.6. There is a geometric reason for that: given a polyhedron P corresponding to the numbers r (number of edges at each vertex) and s (number of edges of each face), a polyhedron corresponding to the numbers s and r, the *dual* of P can be constructed (see Exercises IV.28, IV.29, IV.30 and examples in Figures 10 and 11).

5. Regular polyhedra

In this section, I want to prove that there are five types of regular polyhedra[6] in 3-dimensional space. Let me just mention two problems:

- It is not at all obvious how to define what a *regular* polyhedron is.
- The word *type* also deserves a definition.

To have a more precise idea why these are problems, one could begin by considering their analogues in the case of the plane.

Regular polygons. A convex plane polygon is said to be *regular* if all its sides have the same length and all its angles have the same measure. Although this is completely obvious, notice that:

- neither is the equality of the sides enough (there are nonsquare rhombi),
- nor the equality of the angles (there are nonsquare rectangles)[7].

Triangulating a regular polygon with n sides (also called an n-gon), it is seen that the sum of its geometric angles is $(n-2)\pi$, and thus that its angles have measure $(n-2)\pi/n$.

Let us come now to the question of the *type*.

Proposition 5.1. *For any integer $n \geqslant 3$, there exists a regular polygon with n sides. Any two regular n-gons are similar.*

Proof. The existence is obvious: consider the convex hull of the n points $e^{2ik\pi/n}$ in $\mathbf{C}$.

If P and P' are two regular n-gons, using a similarity, you take an edge of P' to an edge of P. Up to the composition with a reflection (also a similarity), it can be assumed that P and the image of P' are on the same side of this edge. But then, by equality of angles and length, all the vertices coincide. □

The regular n-gon is thus unique *up to similarity*.

A definition of 3-dimensional regular polyhedra. I am not looking for the most perfect definition. As the definition given above for regular polygons, the one given here is an *ad hoc* definition. See, *e.g.*, [**Ber77**] for a more general definition and additional information.

---

(6) In my opinion, this statement is part of the cultural heritage of mankind, as are the *Odyssey*, the sonatas of Beethoven or the statues of Easter Island (without mentioning the Pyramids) and I cannot imagine how a citizen, *a fortiori* a maths teacher, could not know it.

(7) See, however, Exercise IV.14.

Definition 5.2. A convex polyhedron in the 3-dimensional Euclidean space is *regular* if all its faces are isometric regular polygons and if, at each vertex, the edges form regular isometric figures.

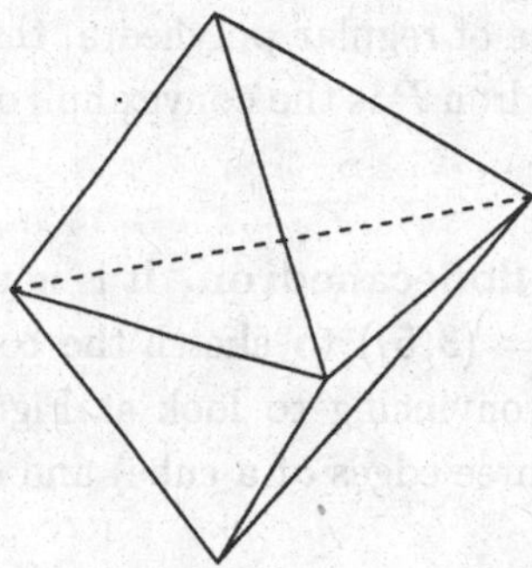

Fig. 8. This is not a regular polyhedron

Remarks 5.3

(1) In analogy with the case of the rhombi and rectangles in plane geometry, notice that there exist convex polyhedra whose faces are regular isometric polygons, but that are not regular polyhedra. This is the case, *e.g.*, for the polyhedron obtained by gluing two isometric regular tetrahedra along a face (Figure 8): all the faces are equilateral triangles, but at some of the vertices, three edges meet while at some other ones, four edges meet.

(2) In the case of a regular polyhedron, all the faces have the same number of vertices and, at any vertex, as many edges meet. We are thus limited, from the combinatorial viewpoint, by Corollary 4.6: the faces of a regular polyhedron are regular s-gons and r of them meet at each vertex.

It happens that any value of the ordered pair (r, s) that is not forbidden by Corollary 4.6 indeed corresponds to a regular polyhedron and that this one is unique up to similarity. For any of these pairs (r, s), it is enough to prove:

- That it is possible to glue r isometric regular s-gons along the edges. I will sketch a construction in each case.
- That the figure obtained is unique up to similarity. I will trust the readers' intuition, a quite complete proof being rather tedious.

All this seems to be very banal when one tries to glue three equilateral triangles ($(r,s) = (3,3)$) or three squares ($(r,s) = (3,4)$) at a vertex, simply because the obtained objects (regular tetrahedron, cube) are very familiar.

One goes from a regular polyhedron of type (r,s) to a regular polyhedron of type (s,r) by a duality operation, a description of which will be found in Exercise IV.28. In the case of regular polyhedra, this can be described very simply[8]: the dual polyhedron $\check{P}$ is the convex hull of the centers of the faces of P.

Gluing pentagons, the dodecahedron. It is not very hard to glue three regular pentagons ($(r,s) = (3,5)$) to sketch the construction of a dodecahedron. It will be very convincing to look at Figure 9 (that shows three pentagons nicely put on three edges of a cube) and to solve Exercise IV.40.

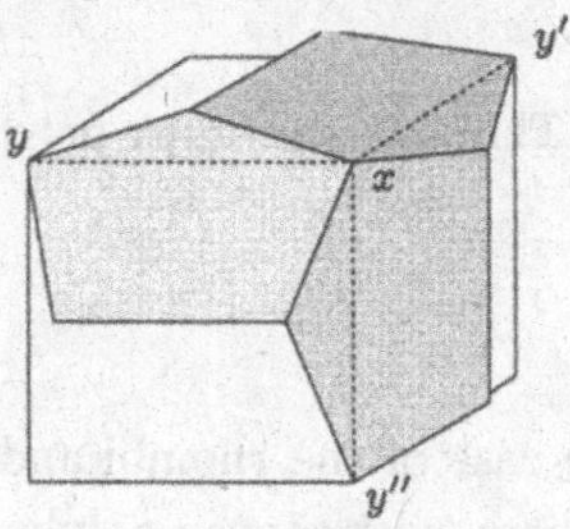

Fig. 9

Remark 5.4. Let us explain how to finish the construction of a regular dodecahedron: start from a cube and put three pentagons at the vertex x as shown on Figure 9. Use then the reflections about the perpendicular bisector planes of the faces to finish the construction: it is proved that the figure formed by these three pentagons is unique up to isometry (Exercise IV.40), from which it is deduced that the polyhedron is regular... and that implies also the uniqueness up to similarity.

The five regular polyhedra. There are thus five types of regular polyhedra[9], the *Platonic solids*, in the space of dimension 3. They are:

- the regular tetrahedron (Figure 10), consisting of four equilateral triangles, for which $(r,s) = (3,3)$,
- the cube, consisting of six squares, for which $(r,s) = (3,4)$ (Figure 11),

(8) For a first reading, this property can be used as a definition of the dual polyhedron. It is not very satisfactory in general, see Exercise IV.31.

(9) See the figures of Dürer in [**Dür95**] or those of Leonardo da Vinci as copied in [**Ber77**].

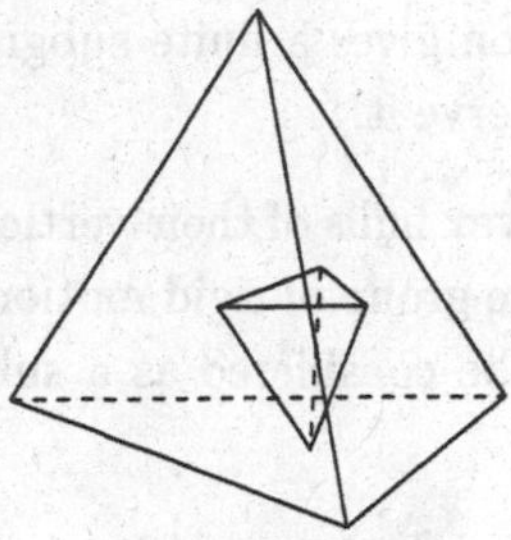

Fig. 10. Tetrahedra

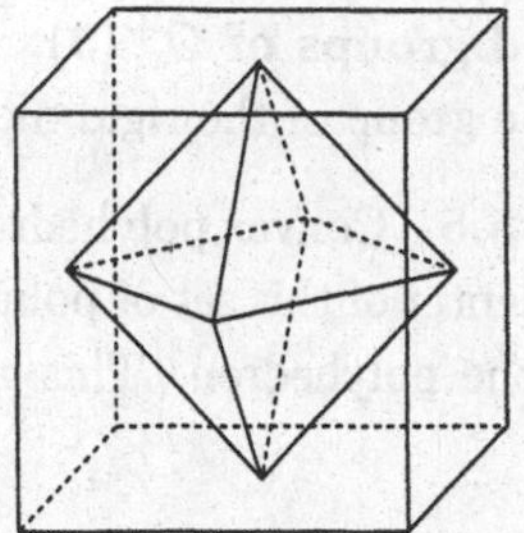

Fig. 11. A cube and an octahedron

– the octahedron, consisting of eight equilateral triangles, for which $(r, s) = (4, 3)$, of which Dürer writes that it is "like a diamond" and can be constructed as the convex hull of the centers of the faces of a cube (see Figure 11) or of the midpoints of the edges of a regular tetrahedron (see Figure 14 and Exercise IV.16),

– the dodecahedron, consisting of twelve regular pentagons, for which $(r, s) = (3, 5)$ (Figure 12),

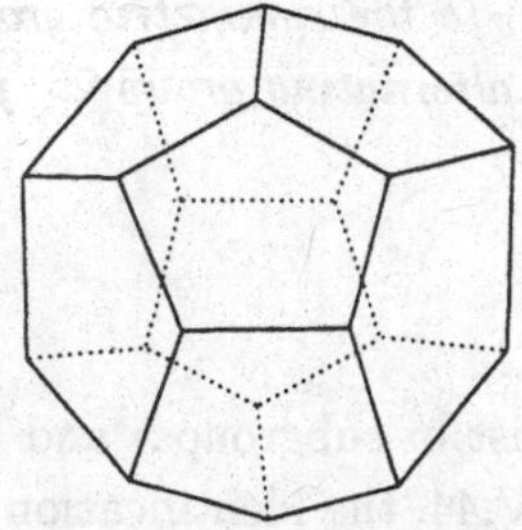

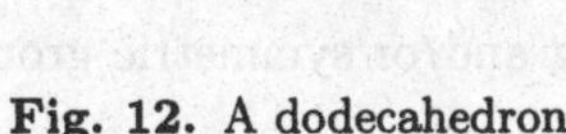

Fig. 12. A dodecahedron

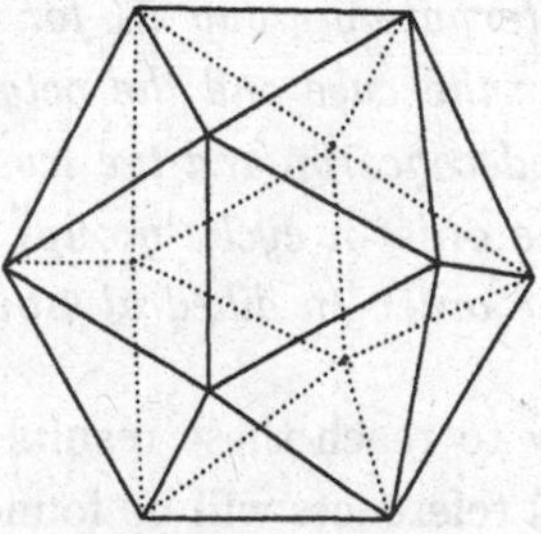

Fig. 13. An icosahedron

– the icosahedron, consisting of twenty equilateral triangles, for which $(r, s) = (5, 3)$ (Figure 13) and that can be constructed as the convex hull of the centers of the faces of the dodecahedron.

The combinatorial properties of the regular polyhedra can be summarized in a table. The order n of the group of rigid motions preserving the polyhedron has been added (see Exercises IV.18, IV.20, IV.42 and IV.44).

	r	s	S	A	F	n
tetrahedron	3	3	4	6	4	12
cube	3	4	8	12	6	24
octahedron	4	3	6	12	8	24
dodecahedron	3	5	20	30	12	60
icosahedron	5	3	12	30	20	60

Finite subgroups of $O^+(3)$. Each polyhedron gives a finite subgroup of $O^+(3)$, the group of the rigid motions that preserve it.

Remark 5.5. Convex polyhedra being the convex hulls of their vertices, the equibarycenter of this set of points is fixed by the group of rigid motions that preserve the polyhedron. This group can thus be considered as a subgroup of $O^+(3)$.

It is easy to believe (Exercise IV.29) that a polyhedron and its dual have the same group of rigid motions. We thus get three *types* of finite subgroups.

There are other finite subgroups: if P is a regular (plane) polygon with n sides, the group of rigid motions that preserve it is isomorphic with the dihedral group with $2n$ elements, denoted by D_{2n} (see, if necessary, Exercise IV.15), this group containing a cyclic group of order n.

It is indeed possible to prove:

Theorem 5.6. *Up to conjugacy in $O^+(3)$, there are five types of finite subgroups in $O^+(3)$:*

- *the groups of rigid motions of the regular polyhedra (isomorphic to the alternating group $\mathfrak{A}_4$ for the tetrahedron, to the symmetric group $\mathfrak{S}_4$ for the cube and the octahedron, to the alternating group $\mathfrak{A}_5$ for the dodecahedron and the icosahedron),*
- *the order-n cyclic groups,*
- *the order-$2n$ dihedral groups.*

A way to reach these results (namely, the list of subgroups) and bibliographical references will be found in Exercise IV.44, the identification of the groups of regular polyhedra to the alternating and/or symmetric groups in Exercises IV.18, IV.20 and IV.42.

Exercises and problems

Direct applications and complements

All these exercises take place in an affine Euclidean space of dimension 3.

Exercise IV.1. What can be said of the angle of the rotation

$$\begin{pmatrix} 0 & 0 & 1 \\ 1 & 0 & 0 \\ 0 & 1 & 0 \end{pmatrix}?$$

Exercise IV.2. Let f and g be two rotations. When is the angle of $g \circ f$ equal to the sum of the angles of f and g?

Exercise IV.3. What is the composition of three reflections about parallel planes?

Exercise IV.4. A translation and a rotation commute if and only if the vector of the translation is a director vector for the axis of the rotation.

Exercise IV.5. The composition of a rotation and a translation is, in general, a screw displacement.

Exercise IV.6. Investigate the composition of two affine rotations.

Exercise IV.7 (Half-turns). Let s_P be the (vectorial) reflection of plane P. Prove that the linear mapping $-s_P$ is a half-turn.

What can be said of the composition of two half-turns? When is the composition of two half-turns a half-turn?

Prove that $O^+(3)$ is generated by the half-turns and that any two half-turns are conjugated in $O^+(3)$.

Exercise IV.8. Describe the composition of three reflections about perpendicular planes.

Exercise IV.9. Prove that to give a screw displacement allows us (in general) to define an orientation of the space[10] (what are the exceptions?).

Exercise IV.10. Let u_1, u_2, v_1 and v_2 be unit vectors such that $\|u_1 - u_2\| = \|v_1 - v_2\|$. Prove that there exists a rotation f such that $f(u_1) = v_1$ and $f(u_2) = v_2$.

Exercise IV.11 (Common perpendicular). If the two lines $\mathcal{D}_1$ and $\mathcal{D}_2$ are not coplanar, prove that there is a unique line Δ which is perpendicular to $\mathcal{D}_1$ and $\mathcal{D}_2$. If M_1 and M_2 are the two intersection points of Δ with $\mathcal{D}_1$ and $\mathcal{D}_2$, prove that M_1M_2 is the distance[11] between $\mathcal{D}_1$ and $\mathcal{D}_2$.

Exercise IV.12 (Double vector product). In an oriented 3-dimensional Euclidean vector space, prove that, for all vectors u, v and w, the following equality holds:

$$u \times (v \times w) = (u \cdot w)v - (u \cdot v)w.$$

Is the vector product associative?

---

[10] If mankind consisted only of right-handed (or left-handed) people, the corkscrew could be considered as a natural (for human species) orientation of the space.

[11] Namely, the minimum of the distances $d(A_1, A_2)$ for points A_1 and A_2 of D_1 and D_2.

Exercise IV.13. In a 3-dimensional oriented Euclidean vector space E, endowed with a direct orthonormal basis (e_1, e_2, e_3), prove that

$$(u \times v) \cdot w = \det_{(e_1,e_2,e_3)}(u, v, w)$$

and that this number is the (signed) volume of the parallelepiped constructed on the three vectors u, v and w.

Exercise IV.14. Prove that, if a polygon inscribed in a circle has all its angles equal and an odd number of sides, then all its sides have the same length.

Exercise IV.15. Let P be a regular n-gon in a Euclidean affine plane. Prove that the group of rigid motions that preserve P is isomorphic with the cyclic group of order n, and that the group of isometries of the plane that preserve P is isomorphic with the dihedral group[12] D_{2n} of order $2n$.

Consider now that $\mathcal{P}$ is a plane in a Euclidean affine space of dimension 3. Prove that the group of rigid motions that preserve the polygon P is isomorphic with the dihedral group D_{2n}.

Let M be a point of $\mathcal{E}$ out of $\mathcal{P}$ that is projected on the center of P in $\mathcal{P}$. Consider the pyramid with vertex M built on the polygon P (if $n = 3$, choose M so that the pyramid is not a regular tetrahedron). Prove that the group of rigid motions of $\mathcal{E}$ that preserve the pyramid is isomorphic with the cyclic group of order n.

Exercise IV.16 (Regular tetrahedron). Prove that, in a regular tetrahedron, two opposite edges are orthogonal and that their common perpendicular passes through their midpoints.

Prove that the figure formed by the midpoints of the edges of a regular tetrahedron is a regular octahedron (Figure 14).

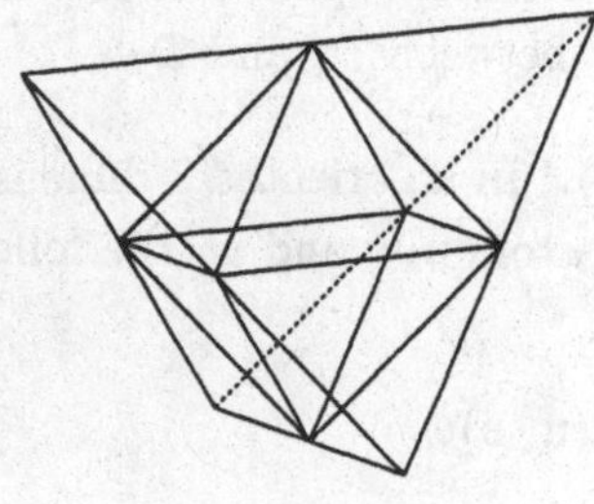

Fig. 14

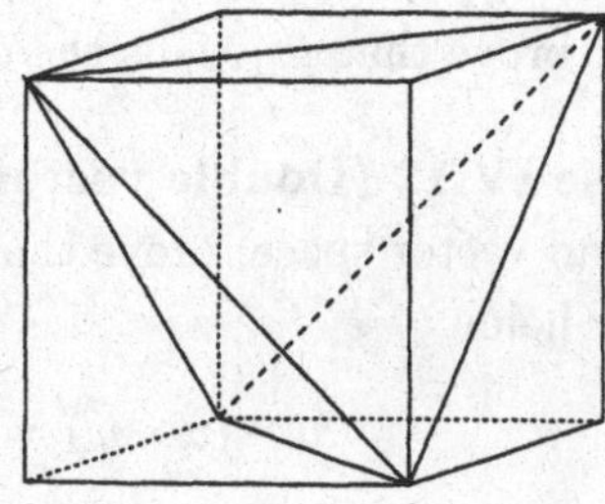

Fig. 15

[12]This can be used as a *definition* of D_{2n}.

Exercise IV.17 (Tetrahedron and cube). Prove that the diagonals of the faces of a cube form two regular tetrahedra (one of which is depicted in Figure 15). Given an isometry that preserves the cube, what is its effect on one of these tetrahedra?

Prove that the group of isometries (*resp.* of rigid motions) that preserve a regular tetrahedron is a subgroup of index 2 in the group of isometries (*resp.* of rigid motions) that preserve a cube.

Exercise IV.18 (Isometries preserving a tetrahedron). Let G be the group of all the isometries that preserve a regular tetrahedron $ABCD$. Prove that an isometry that preserves the tetrahedron $ABCD$ preserves the set of four points $\{A, B, C, D\}$. Deduce that there exists a morphism

$$G \longrightarrow \mathfrak{S}_4$$

from the group G into the symmetric group $\mathfrak{S}_4$. Find an isometry that fixes A and B and exchanges C and D. Deduce that the group of isometries that preserve a regular tetrahedron is isomorphic with $\mathfrak{S}_4$.

Exercise IV.19. Let $ABCD$ be a (scalene) tetrahedron. Determine the group of all the affine transformations that preserve it. What is the group of isometries that preserve it? Let G be a subgroup of order 3 in the alternating group $\mathfrak{A}_4$. Construct a tetrahedron whose group of rigid motions is isomorphic with G. What is its group of isometries[13]?

Exercise IV.20 (Rigid motions preserving a cube). Using Exercises IV.16 and IV.18, prove that the group of rigid motions preserving a cube has 24 elements. Make a list of all these rigid motions.

Prove that a rigid motion that preserves a cube transforms a main diagonal into a main diagonal. Deduce a morphism from the group of rigid motions of the cube into the symmetric group $\mathfrak{S}_4$. Prove that this is, in fact, an isomorphism.

Exercises

Exercise IV.21. Let $ABCD$ be a regular tetrahedron. What is the composition of the half-turns

$$s_{AD} \circ s_{AC} \circ s_{AB}?$$

Exercise IV.22. Let A', B' and C' be the (respective) midpoints of the three sides BC, CA and AB of a triangle and let G be its centroid. Prove that the six small triangles such that $AB'G$ have the same area (Figure 16).

(13) This proves that, as could be expected, "the more the tetrahedron is regular, the bigger its isometry group is".

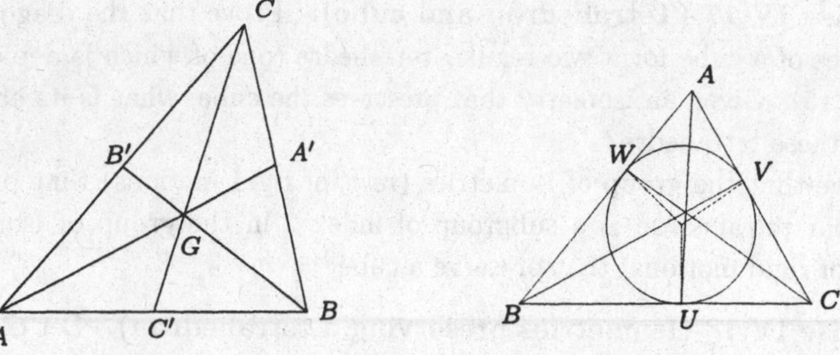

Fig. 16 Fig. 17. The Gergonne point

Exercise IV.23. Let ABC be a triangle, U, V and W be the projections of the center I of the incircle on the three sides BC, CA and AB, and a, b and c be the lengths of these three sides.

What are the barycentric coordinates of I in the triangle ABC?

Compute $\overrightarrow{CB} \cdot \overrightarrow{CU}$. Deduce the values of the ratios

$$\frac{\overrightarrow{UB}}{\overrightarrow{UC}}, \quad \frac{\overrightarrow{VC}}{\overrightarrow{VA}}, \quad \frac{\overrightarrow{WA}}{\overrightarrow{WB}}$$

and prove that the lines AU, BV and CW are concurrent (at a point called the *Gergonne point* of the triangle, see Figure 17).

Exercise IV.24 (Stereographic projection). Let $\mathcal{S}$ be a sphere of radius R and $\mathcal{P}$ a plane through the center O of $\mathcal{S}$. Let N (the north pole) be one of the two points of $\mathcal{S}$ such that $ON \perp \mathcal{P}$.

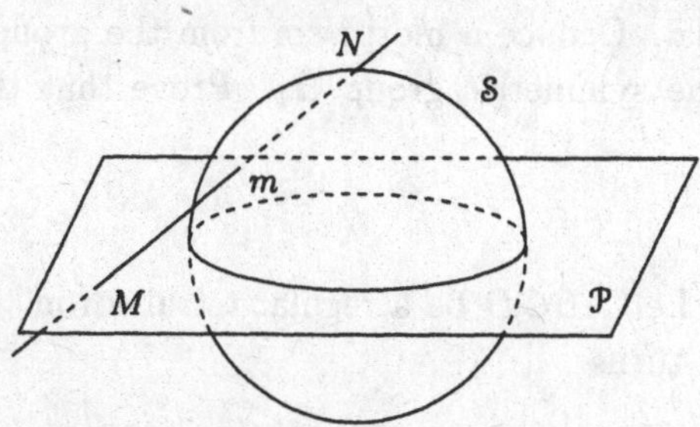

Fig. 18. Stereographic projection

The *stereographic projection* $\varphi : \mathcal{P} \to \mathcal{S} - \{N\}$ is the mapping that, with a point M of the plane, associates the point m at which the line NM intersects the sphere $\mathcal{S}$ (Figure 18). Prove that φ is the restriction to $\mathcal{P}$ of the inversion of pole N and power $2R^2$ and that φ is a homeomorphism.

Let m' be the symmetric of m with respect to $\mathcal{P}$ and $M' = \varphi^{-1}(m')$. Prove that M' is the image of M by the inversion of circle $\mathcal{C} = \mathcal{S} \cap \mathcal{P}$.

This exercise continues in Exercise IV.43.

Exercise IV.25 (Geography). One considers the Earth as a sphere $\mathcal{S}$ endowed with an axis or, what amounts to the same, with two antipodal points N and S, its poles. The great circles through N and S are the *meridians*, and the intersections of the planes orthogonal to the axis NS with $\mathcal{S}$ are the *parallels* (parallel circles).

What are the images of the meridians and parallels under the stereographic projection[14] of pole N? Prove that, on a map obtained by stereographic projection, angles are correct but ratios of distances are wrong[15].

Exercise IV.26 (Latitude and longitude). The latitude of a point M of the Earth is the measure in $[-\pi/2, \pi/2]$ of the angle $(\overrightarrow{Om}, \overrightarrow{OM})$, where O is the center of the Earth, assumed to be spherical, m is the projection of M on the equatorial plane and the plane defined by O, M and N is oriented[16] by the basis $(\overrightarrow{Om}, \overrightarrow{ON})$.

Are you able to define the *longitude* without making an extra choice? Do not hesitate to re-read [**Her45**] (for instance) to measure the consequences of a too quick answer to this question.

From the mysterious Island... In [**Ver74**], having arrived on an unknown island, Gédéon Spilett is very careful to keep his watch working, in order to be able later to know his longitude. Explain why.

... to Fur Country. What do you think of the way the heroes of [**Ver73**] plot their position?

> At two o'clock in the afternoon, Lieutenant Hobson and Thomas Black located with the sextant the elevation of the sun above the horizon. They intended to repeat this operation the next day at about ten in the morning, to deduce from the two heights the longitude of the point then occupied by the island on the polar ocean.

Exercise IV.27 (Separation).

Let C be a compact convex subset of a Euclidean affine space $\mathcal{E}$ and let A be a point out of C. Prove that the function that, with any point M of C,

---

(14) The reader should take the opportunity to contemplate a map of the Antarctic in a beautiful atlas, *e.g.*, [**Ran84**].

(15) Anyway, according to Exercise IV.39, it is impossible to have exact distances on a map.

(16) Like the other planes of the space, this plane does not have an orientation that is better than the others; there is a geopolitical choice here.

associates the distance from A to M has a minimum over C, reached at a point M_0. Let $d = AM_0$. Why is $d > 0$? Let Q be the hyperplane defined by

$$Q = \left\{ M \in \mathcal{E} \mid \overrightarrow{AM} \cdot \overrightarrow{AM_0} = \frac{d^2}{2} \right\}.$$

Prove that A lies in one of the two open half-spaces defined by Q and that C is in the other one[(17)].

Exercise IV.28 (Dual of a convex). We are in a Euclidean vector space E (endowed with its canonical affine structure). If $C \subset E$ is a convex subset, its *dual* $\check{C}$ is defined by

$$\check{C} = \{u \in E \mid u \cdot v \leqslant 1 \text{ for all } v \text{ of } C\}.$$

Check that $\check{C}$ is convex. Prove that, if the interior of C contains 0, then $\check{C}$ is bounded.

Assume that 0 is in the interior of C and that C is compact. Prove that C is the dual of $\check{C}$.

Let P be a convex polyhedron the interior of which contains 0. Prove that $\check{P}$ is a convex polyhedron.

Exercise IV.29. Prove that the compact convex set C containing 0 in its interior and its dual $\check{C}$ as defined in Exercise IV.28 have the same group of isometries.

Exercise IV.30. Let P be a regular polyhedron of center O. Let $A_1, \ldots, A_s$ be the vertices of a face of P and M be the center of the regular polygon $A_1, \ldots, A_s$. Prove that the line OM is perpendicular to the face $A_1, \ldots, A_s$.

Let P' be the convex hull of the centers of the faces of P. Prove that the convex subset dual of P' is the image of P by some dilatation.

Exercise IV.31. What can be said of the figure formed by the midpoints of the sides of a plane convex quadrilateral? Do you think that the result of Exercise IV.30 holds for any polyhedron?

Exercise IV.32 (Bathroom). Is it possible to tile a spherical bathroom with hexagonal floor-tiles (we do not assume that they are all the same, or that they are all regular, or that there are three of them at each vertex)?

---

(17) It is said that Q *separates* the point A and the convex C. Much stronger separation results, as the Hahn–Banach theorem, in [Ber77].

A variant: explain why a soccer ball is *not* made of only hexagonal leather pieces[18].

Exercise IV.33 (Soccer ball). Regular pentagons and regular hexagons with the same side length are glued together so that there are three polygons at each vertex and the result is a convex polyhedron. How many pentagons are needed?

More theoretical exercises

Exercise IV.34. A plane convex polygon is "triangulated", that is to say that it is covered by triangles that intersect along edges. Let V be the number of vertices of the triangles in the triangulation, E the number of edges and F the number of faces. Prove that

$$F - E + V = 1.$$

Exercise IV.35 (Simplicity of $O^+(3)$). We want to prove that $O^+(3)$ is simple, that is to say that its only normal subgroups are the trivial subgroups $O^+(3)$ and $\{\mathrm{Id}\}$. Let thus N be a normal subgroup of $O^+(3)$. Assuming that $N \neq \{\mathrm{Id}\}$, we want to prove that $N = O^+(3)$.

(1) Check that it is enough to prove that N contains *a* half-turn (see Exercise IV.7).
(2) Prove that it can be assumed that N contains a rotation f of angle $\theta \in]0, \pi[$. Let then a be a unit vector directing the axis of f, x a vector orthogonal to a and $y = f(x)$ its image by f. Let $d = \|x - y\|$ be the distance from x to y. Prove that, for any real number $m \in [0, d]$, there exists a unit vector x_1 such that $x_2 = f(x_1)$ is at distance m from x_1.
(3) Fix $m \in [0, d]$. Let y_1 and y_2 be two unit vectors such that $\|y_1 - y_2\| = m$. Prove that there exists a rotation f' in N such that $f'(y_1) = y_2$ (see Exercise IV.10).
(4) Let $n \in \mathbf{N}$ be an integer and let ρ_n be the rotation of axis a and angle π/n. Assume that n is large enough[19] so that $\|x - \rho_n(x)\| \leqslant d$. Let

$$x_0 = x, \quad x_1 = \rho_n(x), \ldots \quad x_{i+1} = \rho_n(x_i), \ldots$$

---

[18] Those of our readers who have never seen a soccer ball are invited to look at the one drawn by Leonardo da Vinci which is shown, *e.g.*, in [**Ber77**, Chapter 12]. Other polyhedra that seem to have only hexagonal faces also appear in nature; see the strange constructions shown in [**Wey52**].

[19] It is traditional to mention at this point that this proof uses the fact that the sequence $1/n$ tends to 0... in more pedantic terms, the fact that $\mathbf{R}$ is an Archimedean field. See [**Ber94**] or [**Per96**, Exercise VI.6.2] about this remark.

What is x_n? Prove that there exists a rotation u_i in N such that $u_i(x_i) = x_{i+1}$. Let $v = u_{n-1} \circ \cdots \circ u_1$. What is $v(x)$? Prove that v is a half-turn. Conclude[(20)].

Exercise IV.36 (Intrinsic distance on the sphere). Let x and y be two unit vectors in a Euclidean vector space E of dimension 3, considered as two points of the unit sphere S. Put

$$d(x, y) = \arccos(x \cdot y)$$

(this is the angle of the two vectors x and y).

(1) Prove that, if x, y and z are three points of S and if α is the angle at x of the triangle xyz, $a = d(y, z)$, $b = d(z, x)$ and $c = d(x, y)$ are the lengths of the three sides, then

$$\cos a = \cos b \cos c + \sin b \sin c \cos \alpha.$$

(2) Deduce that d is a distance on S.
(3) Check that d defines the same topology on S as that induced by the topology of E as a normed space.
(4) Prove that if $y \neq -x$, there is a unique shortest path by broken lines from x to y, which is the unique arc of great circle from x to y.

Exercise IV.37 (Isometries of the sphere). With the notation of Exercise IV.36, let $\varphi : S \to S$ be an isometry for the distance d. Prove that φ is the restriction to S of an isometry of the Euclidean space E. Deduce that the group of isometries of (S, d) is isomorphic to $O(E)$ (one can use Exercise II.5).

Exercise IV.38 (Spherical triangles, continuation). If a, b and c are three real numbers in $]0, \pi[$ that satisfy

$$|b - c| < a < b + c \quad \text{and} \quad a + b + c < 2\pi,$$

prove that there exists a spherical triangle of sides a, b and c, and that it is unique up to isometry.

For every real number α in $]\pi/3, \pi[$, prove that there exists an equilateral triangle all of whose angles have measure α.

Exercise IV.39. Let U be a nonempty open subset on the unit sphere S. Prove that there exists no mapping $f : U \to \mathbf{R}^2$ that preserves the

(20) When n is even, the group $O^+(n)$ has a nontrivial center $\{\pm \mathrm{Id}\}$. It can be shown that the quotient $PO^+(n)$ of $O^+(n)$ by its center is simple for $n \geqslant 5$. Using the quaternions (Exercise IV.45), it can be shown that, on the contrary, $PO^+(4)$ is not simple. See [**Per96**].

distances[21] in the following sense:

$$\|f(x) - f(y)\| = d(x, y).$$

Exercise IV.40 (Construction of the dodecahedron). Let F, F' and

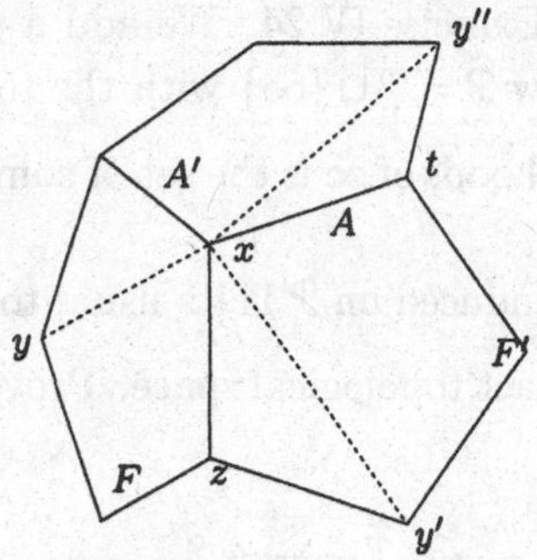

Fig. 19. Construction of the dodecahedron

F'' be three regular isometric pentagons. Prove that they can be glued along edges A (a side common to F' and F''), A' (common to F'' and F) and A'' (common to F and F') at a vertex x and that the figure obtained is unique up to isometry. One could begin by showing that there exists a spherical triangle whose sides have length $3\pi/5$ and that there is only one such triangle up to isometry and then prove the uniqueness of the gluing up to isometry by taking x', z and t to be the three vertices of an equilateral triangle of side length $3\pi/5$.

Using the fact that yz is parallel, while zt is orthogonal to A', prove that $yz \perp zt$. Use then the symmetry about the perpendicular bisector plane to A'' to prove that $xy \perp xy'$; with the notation of Figure 19, deduce that the lines xy, xy' and xy'' are three edges of a cube.

Exercise IV.41 (Another construction of the dodecahedron). Denote by τ the golden number, namely the number $\tau = \dfrac{\sqrt{5}-1}{2}$. Prove that the convex hull of the set of points whose coordinates in an orthonormal frame are $(\pm 1, \pm 1, \pm 1)$, $(0, \pm\tau, \pm\tau^{-1})$, $(\pm\tau^{-1}, 0, \pm\tau)$, $(\pm\tau, \pm\tau^{-1}, 0)$ is a regular dodecahedron.

Exercise IV.42 (Rigid motions of the dodecahedron). Contemplating Figure 9, discover five cubes in a dodecahedron. Prove that the group of rigid

(21) This is eventually because the sphere is curved, while the plane is flat. See Theorem VIII-3.10.

motions that preserve a dodecahedron is isomorphic with the alternating group $\mathfrak{A}_5$.

Why is the group of isometries that preserve the dodecahedron not isomorphic with $\mathfrak{S}_5$?

Exercise IV.43 (Stereographic projection, continuation). We use the same notation as in Exercise IV.24. We add a point, denoted by ∞, to the plane $\mathcal{P}$ and we endow $\widehat{\mathcal{P}} = \mathcal{P} \cup \{\infty\}$ with the topology such that:

- a basis of neighborhoods of ∞ is the set of complements of the compact subsets of $\mathcal{P}$
- and the topology induced on $\mathcal{P}$ is its usual topology[22].

Prove that $\widehat{\mathcal{P}}$ is a compact topological space. Prove that φ can be extended as a homeomorphism

$$\widehat{\varphi} : \widehat{\mathcal{P}} \longrightarrow \mathcal{S}.$$

What is the image of a circle by φ? the image of a line? What do you think of Theorem III-4.16 and the discussion[23] that follows it?

Exercise IV.44 (Finite subgroups of $O^+(3)$). Let G be an order-n subgroup of $O^+(3)$. Assume that G is not the trivial group (namely that $n > 1$). It acts on the unit sphere S of $\mathbf{R}^3$. Consider

$$\Gamma = \{(g, x) \in (G - \{\mathrm{Id}\}) \times S \mid g(x) = x\}$$

with its two projections on S and $G - \{\mathrm{Id}\}$. Let X be its image in S.

(1) Characterize geometrically the elements of X and check that G stabilizes X. Let $P_1, \ldots, P_k$ be the orbits of this action and e_i the order of the stabilizer of x when $x \in P_i$.

How many rotations in G fix the vector x of P_i? Computing the cardinal of Γ in two different ways, prove that

$$2(n-1) = n \sum_{i=1}^{k} \left(1 - \frac{1}{e_i}\right)$$

(a special case of the *class formula*).

(2) Deduce that $k = 2$ or 3. Prove that, if $k = 2$, G is also a cyclic group of order n.

---

(22) It is said that $\widehat{\mathcal{P}}$ is the *one point compactification* of $\mathcal{P}$. See, *e.g.*, [**Bou89**].

(23) See also Chapter V, in particular its § 7, together with § VI-7.

(3) Assume now that $k = 3$. Prove that, up to the order of the e_i's, (n, e_1, e_2, e_3) is of the form

$$\begin{aligned} &(2p, 2, 2, p)\\ \text{or}\quad &(12, 2, 3, 3)\\ \text{or}\quad &(24, 2, 3, 4)\\ \text{or}\quad &(60, 2, 3, 5). \end{aligned}$$

(4) Assume that $(n, e_1, e_2, e_3) = (2p, 2, 2, p)$. Prove that the stabilizer of the points of the orbit P_3 is a cyclic group of order p. Deduce that G is the group of all the rigid motions that preserve a plane regular p-gon and that this is a dihedral group D_{2p} of order $2p$.

(5) Assume that $(n, e_1, e_2, e_3) = (12, 2, 3, 3)$. How many elements has the orbit P_2? Prove that these points form a regular tetrahedron (one can use Exercise IV.19). Conclude that G is the group of rigid motions that preserve a regular tetrahedron. What can be said of the orbit P_3?

(6) Assume that $(n, e_1, e_2, e_3) = (24, 2, 3, 4)$. Prove similarly that the points of P_2 form a cube and that G is the group of rigid motions that preserve this cube. What is the figure formed by the points of the orbit P_3?

(7) Assume that $(n, e_1, e_2, e_3) = (60, 2, 3, 5)$. Prove that the points of the orbit P_2 are the vertices of a regular dodecahedron[24] and that G is the group of rigid motions that preserve this polyhedron. What can be said of the points of the orbit P_3?

Exercise IV.45 (The quaternions). Consider, on the Euclidean vector space $\mathbf{R}^4$ endowed with an orthonormal basis $(1, i, j, k)$, the algebra structure defined by the following:

- 1 is the neutral element
- $i^2 = j^2 = k^2 = -1$
- $ij = k$, $jk = i$, $ki = j$
- $ij + ji = 0$, $jk + kj = 0$, $ki + ik = 0$.

Let $\mathbf{H}$ be the algebra defined this way. If $q \in \mathbf{H}$,

$$q = a \cdot 1 + b \cdot i + c \cdot j + d \cdot k,$$

define its conjugate $\overline{q}$ by

$$\overline{q} = a \cdot 1 - b \cdot i - c \cdot j - d \cdot k$$

and its *norm* $N(q)$ par $N(q) = q\overline{q}$.

(1) Prove that $\overline{qq'} = \overline{q'}\,\overline{q}$.

(24) This is more complicated here.

(2) Prove that $\mathbf{H}$ is a (noncommutative) field, the field of *quaternions* and that the subspace generated by 1 is a subfield, isomorphic to $\mathbf{R}$ (if $a \in \mathbf{R}$, we thus write a for $a1$). What is the center of $\mathbf{H}$?

(3) The component of a quaternion on the vector 1 is called its *real part* (notation $\operatorname{Re}(q)$). Interpret $\operatorname{Re}(q\overline{q'})$. Prove that

$$S = \{q \in \mathbf{H} \mid N(q) = 1\}$$

is the unit sphere of $\mathbf{R}^4$ and a subgroup of the multiplicative group of $\mathbf{H}$. Prove that $\operatorname{Re}(sq\overline{s}) = N(s)\operatorname{Re}(q)$.

(4) Let $Q \subset \mathbf{H}$ be the subspace generated by i, j and k (*pure* quaternions). Prove that, if $s \in S$ and $q \in Q$, $sqs^{-1} \in Q$ and that this makes S act by *isometries* on Q.

(5) Deduce a homomorphism from S to $O^+(3)$. Prove that its kernel is the group $\{\pm \operatorname{Id}\}$. Let s be a quaternion both pure and unitary ($s \in S \cap Q$). Prove that s acts on Q as the half-turn of axis s. Deduce that $O^+(3)$ is isomorphic (as a group) and homeomorphic (as a topological space)[25] to $S/\{\pm 1\}$.

[25] We have seen in § II-3 that $O^+(2)$ is isomorphic, as a group and as a topological space, to a circle. This exercise gives an analogous property of $O^+(3)$.

Chapter V

Projective Geometry

Projective geometry was created as a completion of affine geometry in which there are no parallels, which is a nice property, as it gives neater statements.

It requires a small amount of abstraction to define the new framework, but it's worth the effort.

One can think of a projective plane as being an affine plane, to which a "line at infinity" has been added, all parallel lines meeting on this line. This is what I want to explain in this chapter. The idea comes, of course, of perspective in art: think of a vanishing point on the horizon line. One can understand that the idea is related to the plane representation of a 3-dimensional space. A wonderful reference is the geometry book of Dürer [**Dür95**].

In this chapter, all spaces, affine, linear and projective are defined over a field **K**. The reader is allowed to think that this field is **R** or **C**.

Good bibliographical references for this chapter (and for the projective aspects of the next one) are [**Sam88**], [**Ber77, Ber94**] and [**Sid93**].

1. Projective spaces

Let E be a (finite-dimensional) vector space. The *projective space* $P(E)$ deduced from E is the set of all lines (1-dimensional linear subspaces) in E, in other words, this is the set of equivalence classes of $E - \{0\}$ under the relation:

$$v \sim w \text{ if and only if } v = \lambda w \text{ for some nonzero } \lambda \text{ in } K.$$

The *dimension* of $P(E)$ is $\dim E - 1$ (this is a definition). If E consists of the point 0, it does not contain any line and $P(E)$ is empty. We shall thus implicitly assume that $\dim E \geqslant 1$. If $\dim E = 1$, E itself is a line, thus the set of its lines contains a unique element, $P(E)$ is a point. If E is a plane, $P(E)$ is a *projective line*, if $\dim E = 3$, $P(E)$ is a *projective plane*.

We shall denote by $\mathbf{P}_n(\mathbf{K}) = P(\mathbf{K}^{n+1})$ (or simply $\mathbf{P}_n$) the standard projective space of dimension n.

Topology. So far, we have only defined a set. If the field $\mathbf{K}$ over which the vector space E is defined is $\mathbf{R}$ or $\mathbf{C}$, the space $P(E)$ is also a *topological* space. The space E (and its subset $E - \{0\}$ as well) has a natural topology and $P(E)$ can be endowed with the *quotient topology* (1), which is defined so that the projection

$$p : E - \{0\} \longrightarrow P(E)$$

is a continuous mapping. Here, as the equivalence relation is defined by the action of the group $\mathbf{K}^\star$, the projection p is even an open mapping.

Proposition 1.1. *Real and complex projective spaces are compact and path-connected topological spaces.*

Proof in the real case. Endow the real projective space(2) E with a Euclidean structure. Denote $S(E)$ the unit sphere of E. Recall that, E being finite dimensional, the unit sphere is compact (as it is obviously closed and bounded).

Any real line contains two unit vectors, so that $P(E)$ is also the quotient of the unit sphere by the equivalence relation that identifies u and $-u$. Thus, $P(E)$ is the quotient of a compact space, or, if one prefers, the image of the compact topological space $S(E)$ by the quotient mapping... and this mapping is continuous by the very definition of the quotient topology. In short, $P(E)$ is the image of a compact space by a continuous mapping.

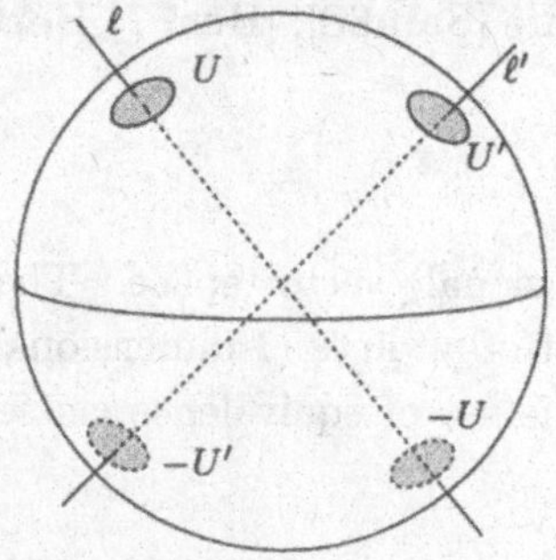

Fig. 1

(1) I understand that the notion of quotient topology is rather delicate, but everything here will be very simple, as the reader will see in the next proof.

(2) The proof for the complex case is the subject of Exercise V.37.

Therefore, to prove that it is compact, we only have to make sure that it is Hausdorff. Choose two (distinct) points in $P(E)$, namely two lines ℓ and ℓ' in E. Let v and v' be unit vectors in these lines. The sphere $S(E)$ being a Hausdorff space, there exists open sets U and U' such that:

– U contains v,
– U' contains v',
– U, $-U$, U', $-U'$ are disjoint (Figure 1).

Then the images of U and U' are open subsets (the projection is an open mapping, as I have already mentioned) of $P(E)$, neighborhoods of ℓ and ℓ' respectively, and they are disjoint.

The sphere $S(E)$ being path connected (as soon as the dimension of E is at least 2), projective spaces of positive dimension are path connected. The unit sphere of a real line consists of two points, but they are identified in a single point in the associated projective space, so that the latter is path connected as well. □

2. Projective subspaces

A subset V of $P(E)$ is a *projective subspace* if this is the image of a nonzero vector subspace F of E. This is then the set of all lines in E that are contained in F... that is to say, this is the set of lines of F; in short, one has

$$V = P(F) \subset P(E).$$

We thus have a one-to-one correspondence between the set of vector subspaces of dimension $(k+1)$ in E and the set of projective subspaces of dimension k in $P(E)$.

As I announced at the beginning of this chapter, we are now in a geometry without parallels. The next proposition tells us why:

Proposition 2.1. *Let V and W be two projective subspaces of $P(E)$.*

(1) *If their dimensions satisfy the inequality*

$$\dim V + \dim W \geqslant \dim P(E),$$

then their intersection $V \cap W$ is not empty. In particular, any two lines in a plane intersect.

(2) *Let H be a projective hyperplane of $P(E)$ and let m be a point not in H. Every line through m intersects H at a unique point.*

Remark 2.2. Beginners in projective geometry are often bewildered when the time comes to act and to *prove* this kind of result. The difficulty probably comes from the fact that the projective space is defined as a quotient, a not so straightforward procedure.

But a statement such as 2.1 only deals with incidence properties. It must thus be the translation, in the new (and perhaps disconcerting) language of projective geometry, of some very simple linear algebra statement. The only problem is to do the translation. The moral of the next proof (and, moreover, the methodological lesson to learn from it) is: translate into linear terms—linear algebra, that's easy!

Proof. To be able to translate the statement into linear terms, let F and G be the two subspaces of E of which V and W are deduced: $V = P(F)$, $W = P(G)$. The assumption on the dimensions is translated, at the level of vector subspaces, into:

$$(\dim F - 1) + (\dim G - 1) \geqslant (\dim E - 1),$$

namely

$$\dim F + \dim G \geqslant \dim E + 1.$$

Once we have done the translation, let us use the *ad hoc* linear algebra property:

$$\dim F + \dim G = \dim(F + G) + \dim(F \cap G) \leqslant \dim E + \dim(F \cap G).$$

We deduce that

$$\dim(F \cap G) \geqslant 1.$$

Thus there is a line in $F \cap G$. This is the conclusion in the framework of vector spaces. We still have to translate it back to projective geometry: $V \cap W$ is not empty.

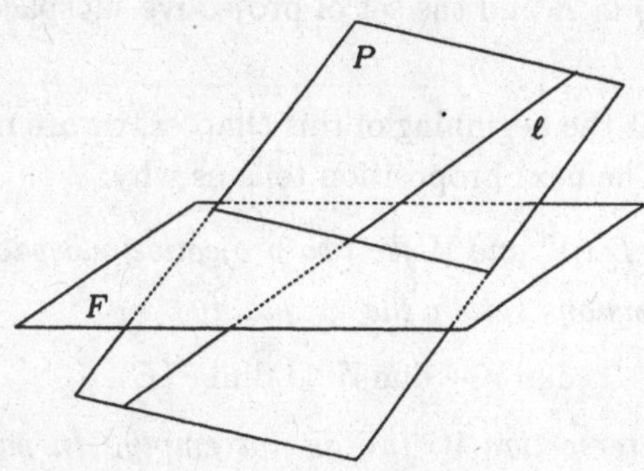

Fig. 2

Let us pass now to the second property. The projective hyperplane H is the image of a (linear) hyperplane F of E. The point m is the image of a line ℓ of E. Let us translate the assumption that m is not in H: the line ℓ is not contained in the hyperplane F.

Let us look now at the projective lines through the point m, namely, at the (linear) planes containing the line ℓ. The assertion that we have to prove is, in the linear algebra language, the fact that, if ℓ is not contained in F, any plane containing ℓ meets F along a unique line (Figure 2). And, as ℓ is not contained in F, for any plane P containing ℓ, we have indeed $P+F=E$. Hence

$$\begin{aligned}\dim(P\cap F) &= \dim P + \dim F - \dim(P+F)\\ &= 2+\dim E - 1 - \dim E = 1,\end{aligned}$$

which is exactly what we wanted to prove. □

One can define, as in the linear or affine cases, the *projective subspace generated* or *spanned* by a subset of $P(E)$. For instance, to say that two lines generate a (unique) plane can be translated into saying that two points generate a (unique) projective line (in other words, any two distinct points lie on a unique line).

3. Affine *vs* projective

I show now that this geometry without parallels is completely analogous to affine geometry, in a very precise sense: a projective space, this is nothing other than an affine space to which something has been added, projective lines are affine lines with an additional point and the incidence relations are, apart from that, the same relations as in affine geometry.

This is still quite easy, we are doing elementary linear algebra, but we play now on three levels. There is a *vector* space E of dimension $n+1$, the *projective* space $P(E)$, space of its lines and also an *affine* space (actually a bunch of affine spaces) of dimension n that we will be able to consider, both as a subset of $P(E)$ and as an affine hyperplane of E.

Introduction: the projective line. To begin with[(3)], consider the case of 1-dimensional projective spaces. Consider a (vector) plane E. The projective line $P(E)$ is the set of lines in E.

Choose a basis (e_1, e_2) of E. All lines in E have a unique directing vector of coordinates $(x, 1)$, except for the x-axis. In other words, all lines in E intersect the affine line of equation $y=1$ at a unique point, except for the x-axis, which is parallel to it (Figure 3). We have just set a one-to-one correspondence between the projective line $P(E)$ minus a point (the x-axis)

[(3)] I will not hesitate to repeat myself in this part.

and the affine line $y = 1$. We can thus identify the affine line in question with the complement of a point in the projective line[(4)].

Conversely, we can consider that the projective line was obtained by the addition of a point to the affine line. The tradition is to call this point the *point at infinity:* in the real and complex cases, the mapping

$$\begin{array}{rcl}(\text{affine line } y = 1) \cup \{\infty\} & \longrightarrow & P(E)\\ (x, 1) & \longmapsto & (\text{vector}) \text{ line generated by } (x, 1)\\ \infty & \longmapsto & (\text{vector}) \text{ line generated by } (1, 0)\end{array}$$

is a homeomorphism if the left-hand space is endowed with the topology of the *one point compactification* of the line[(5)].

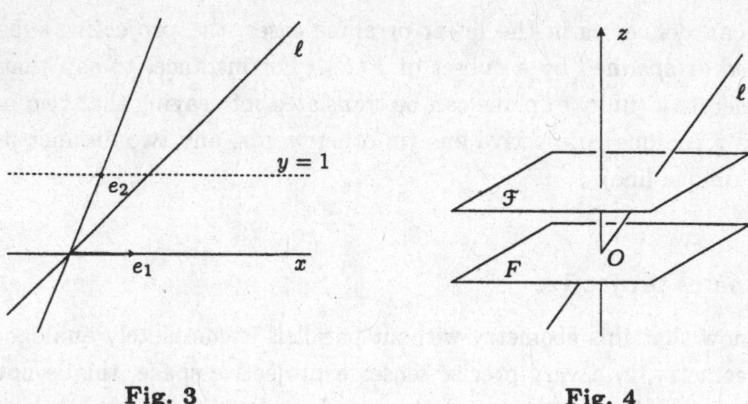

Fig. 3 Fig. 4

The projective plane. We proceed in the same way with a vector space E of dimension 3: we choose a basis in which the coordinates of the vectors we consider are denoted (x, y, z). Let F be the plane of equation $z = 0$ and $\mathcal{F}$ the affine plane (directed by F) of equation $z = 1$. A line ℓ of E intersects $\mathcal{F}$ at a unique point... except if it is contained in F, in which case it does not intersect it at all (Figure 4).

We thus have a one-to-one correspondence between $P(E) - P(F)$ and $\mathcal{F}$. Conversely, we can consider that the projective plane $P(E)$ is obtained by adding to the affine plane $\mathcal{F}$ the projective line $P(F)$. The latter is said to be the *line at infinity* of $\mathcal{F}$.

In coordinates, the points $(x, y, 1)$ are those of $\mathcal{F}$, and the points at infinity are $(x, y, 0)$.

(4) I will come back to this, but we can already notice that the affine line $y = 1$, the one-to-one correspondence and the missing point *depend* on the choice of the basis.

(5) This topology is such that the family of the complements of the compacts of the affine line is a basis of neighborhoods of ∞ (see [**Bou89**]).

The general case: projective completion of an affine space. Let $\mathcal{F}$ be an affine space of dimension n. To begin with, we embed it as an affine hyperplane in a vector space E of dimension $n+1$. This can be done in a quite intrinsic way, but this is too abstract for this book. We choose an origin in $\mathcal{F}$... and we consider the vector space $E = \mathcal{F} \times \mathbf{K}$. Let F be the hyperplane of equation $x_{n+1} = 0$ (x_{n+1} is the coordinate on the $\mathbf{K}$ summand) and identify $\mathcal{F}$ with the affine hyperplane of equation $x_{n+1} = 1$ (this way, we consider that $\mathcal{F} = \mathcal{F} \times \{1\}$).

Every line of E that is not contained in F intersects $\mathcal{F}$ at a unique point. We get a one-to-one correspondence between $\mathcal{F}$ and $P(E) - P(F)$. The projective hyperplane $P(F)$ is called the *hyperplane at infinity* of $\mathcal{F}$. Notice that it is not in $\mathcal{F}$ and consists of the directions of the affine lines of $\mathcal{F}$.

Affine lines, projective lines. We keep the same notation and consider now the affine lines of the affine space $\mathcal{F}$. Let $\mathcal{D}$ be such a line. There is a unique plane that contains this line, the (affine) subspace of E generated by 0 and $\mathcal{D}$ (Figure 5). Call it $P_{\mathcal{D}}$. Like all the vector subspaces of E, it defines a projective subspace (here a line) in $P(E)$, the set $P(P_{\mathcal{D}})$ of its (vector) lines.

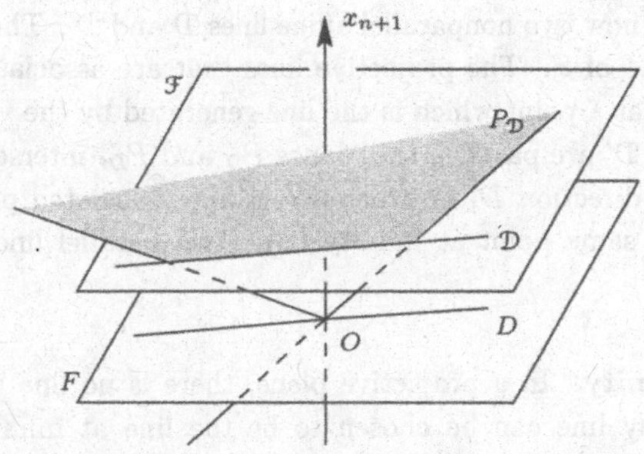

Fig. 5

Under the embedding of $\mathcal{F}$ in $P(E)$, the line $\mathcal{D}$ of $\mathcal{F}$ is mapped to the projective line associated with $P_{\mathcal{D}}$, which can be considered as its own projective completion.

We thus have a bijection between the set of affine lines of $\mathcal{F}$ and the set of projective lines in $P(E)$ that are not contained in the hyperplane at infinity $P(F)$.

Intersections of lines in the plane. Let $\mathcal{F}$ be an affine plane directed by F. We have just constructed a projective plane P, which is the union of $\mathcal{F}$ and of the projective line $P(F)$, the line at infinity of $\mathcal{F}$.

Let us make the list of all the projective lines of P. These are the images of all the planes in the 3-dimensional vector space. We thus have the line at infinity, in other words $P(F)$, and all the other lines come from the affine lines of $\mathcal{F}$ to which we have added a point (which, by the way, is in $P(F)$). If $\mathcal{D}$ is an affine line in $\mathcal{F}$, we denote this point $\infty_{\mathcal{D}}$.

Remark 3.1. The line at infinity is the set of all lines D of F, namely the set of affine lines of $\mathcal{F}$. Every point of the line at infinity is a direction of parallel lines in $\mathcal{F}$.

We have seen (in Proposition 2.1) that two projective lines of a projective plane always intersect. Let us check this statement on our list of lines.

- The line at infinity and the line coming from the affine line $\mathcal{D}$ intersect at $\infty_{\mathcal{D}}$: in vector terms, we are considering the intersection of the plane $P_{\mathcal{D}}$ with the plane F, this is the line D, which is both the direction of $\mathcal{D}$... and the point at infinity $\infty_{\mathcal{D}}$ of $\mathcal{D}$ (I have used the notation of Figure 5).
- Consider now two nonparallel affine lines $\mathcal{D}$ and $\mathcal{D}'$. They intersect at a point M of $\mathcal{F}$. The projective lines that are associated with them intersect at a point which is the line generated by the vector $\overrightarrow{OM}$.
- If $\mathcal{D}$ and $\mathcal{D}'$ are parallel, the planes $P_{\mathcal{D}}$ and $P_{\mathcal{D}'}$ intersect along their common direction D, in other words, the associated projective lines have the same point at infinity $\infty_{\mathcal{D}}$: two parallel lines intersect at infinity.

Choice of infinity. In a projective plane, there is no line at infinity. In other words, any line can be chosen to be the line at infinity. Consider more generally the projective space $P(E)$. Choose any hyperplane F in E and remove from $P(E)$ the corresponding projective hyperplane $P(F)$. The result is an affine space $\mathcal{F}$ directed by F and whose hyperplane at infinity is $P(F)$.

Any hyperplane may be considered as the hyperplane at infinity. This remark has numerous applications. Rather than presenting a lot of theory, let us give two applications, proofs of the most general theorems of Pappus and Desargues as consequences of the simple special cases encountered in Chapter I. I will come back to the question of the choice of infinity after the proofs of these two theorems.

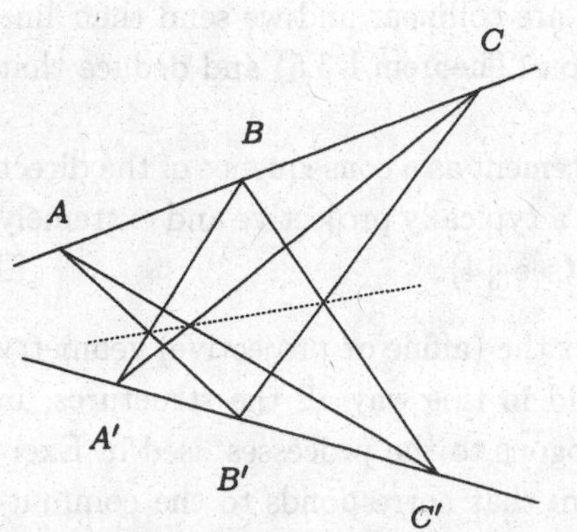

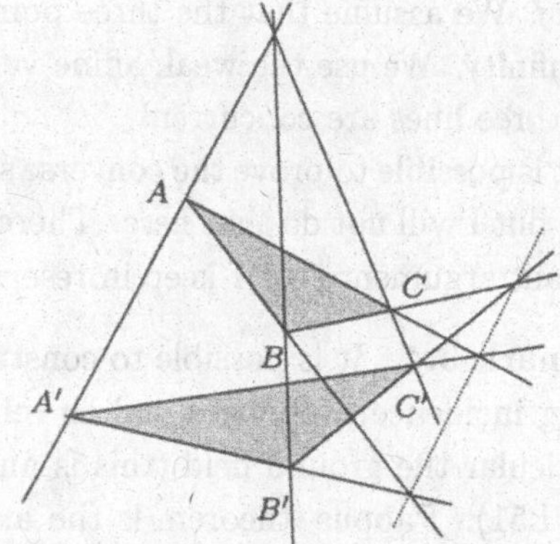

Fig. 6. Pappus' theorem

Fig. 7. Desargues' theorem

Pappus' theorem. Here is the general statement of the theorem of Pappus. Notice that it is a bit simpler than that given in Exercise I.39: we do not need to assume that the lines intersect. They do intersect because we are in a projective plane... or because we are in an affine plane and we consider that two parallel lines intersect on the line at infinity.

Theorem 3.2. *Let $\mathcal{D}$ and $\mathcal{D}'$ be two lines; let A, B and C be three points of $\mathcal{D}$, and A', B' and C' three points of $\mathcal{D}'$. Let α, β and γ be the intersection points of $B'C$ and $C'B$, $C'A$ and $A'C$, and $A'B$ and $B'A$ respectively. Then α, β and γ are collinear.*

Proof. We consider the projective line $\alpha\gamma$ as the line at infinity. In other words, we investigate the problem in the affine plane obtained by removing this line from the projective plane. To say that $B'C$ and $C'B$ intersect on the line at infinity, this is to say that they are parallel. Similarly, $A'B$ and $B'A$ are parallel and we just need to apply the weak affine version[6] (that is, Theorem I-3.4) to conclude: $C'A$ and $A'C$ are parallel and thus intersect on the line at infinity... that is, the point β is indeed on the line $\alpha\gamma$. □

Desargues' theorem. We are going to proceed in the same way for the theorem of Desargues (same remark about the statement of Exercise I.40):

Theorem 3.3. *Let ABC and $A'B'C'$ be two triangles. Let α, β and γ be the intersection points of BC and $B'C'$, CA and $C'A'$, AB and $A'B'$. Then the points α, β and γ are collinear if and only if the lines AA', BB' and CC' are concurrent.*

(6) There is a "quite projective" proof in Exercise V.13.

Proof. We assume that the three points are collinear and we send their line to infinity. We use the weak affine version (Theorem I-3.5) and deduce that the three lines are concurrent.

It is possible to prove the converse statement as a consequence of the direct one, but I will not do that here. There is a typically projective and extremely elegant argument that I keep in reserve (see § 4). □

Remark 3.4. It is possible to construct the (affine or projective) geometry using incidence properties and to rebuild in this way all the structures, in particular the ground field (this is analogous to the processes used in Exercise I.51). Pappus' theorem is the axiom that corresponds to the commutativity of the field, while Desargues' gives its associativity (see [**Art57**]).

Choice of infinity (continuation). The reader may wonder why, in the statements 3.2 and 3.3 above, I have not made precise whether we were in an affine or in a projective plane. Let us look again at both Figure 6 here and Figure 6 of Chapter I.

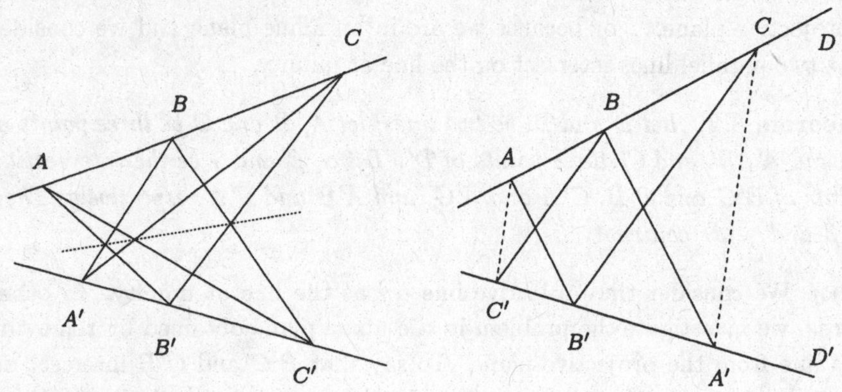

Fig. 8. Pappus in Chapter V **Fig. 9.** Pappus in in Chapter I

Figure 8 is drawn in an affine plane $\mathcal{E}$. One completes this plane in a projective plane P by the addition of a line at infinity. Now, remove the dotted line from P. The result is an affine plane $\mathcal{E}'$, this is the plane in which Figure 9 is drawn. One understands that the two pictures are drawn in two *different* affine planes $\mathcal{E}$ and $\mathcal{E}'$ in P.

But... the two pictures are drawn on the same sheet of paper. It is not absurd to try to consider them in the same affine plane (after all, this is what the drawer has done). In this case, the philosophy would rather be to consider that one is obtained from the other by a suitable transformation. I will come back to this in § 5.

4. Projective duality

Let F be a vector subspace in a vector space E. Consider the subspace F' of the dual $E^\star$ defined by

$$F' = \{\varphi \in E^\star \mid \varphi|_F = 0\}.$$

This is the set of all linear forms that vanish on F. This is indeed a vector subspace of $E^\star$ (is this clear?) of dimension

$$\dim F' = \dim E - \dim F$$

(if necessary, see Exercise V.1).

For example, if E is of dimension 3, a line F in E defines a plane in $E^\star$, a plane in E defines a line in $E^\star$. The next two tables show all the profit we can expect to get from this remark in plane projective geometry. The first one is a summary of the points of view on these spaces. The second one is about the incidence relations, consequences of the following more or less obvious fact:

$$F \subset G \Longleftrightarrow F' \supset G'.$$

It uses the same notation as the first table, except that we have suppressed the useless $'$'s and E's.

$P(E)$	E	$E^\star$	$P(E^\star)$
$a \in P(E)$ point	$a \subset E$ (vector) line	$a' \subset E^\star$ (vector) plane	$P(a') = A' \subset P(E^\star)$ projective line
$D = P(d) \subset P(E)$ projective line	$d \subset E$ (vector) plane	$d' \subset E^\star$ (vector) line	$d' \in P(E^\star)$ point

a point in P	A line in $P^\star$
D line in P	d point in $P^\star$
$a \in D$	$A \ni d$
three collinear points in P	three concurrent lines in $P^\star$
line ab	point $A \cap B$

Figure 10 shows what happens[7] to three concurrent lines P, Q, R through the three vertices a, b, c of a triangle (plane on the left) when you apply duality: three collinear points p, q and r on the three sides A, B and C of a triangle (plane on the right).

---

[7] This translation is called "metamorphosis" in **[Sid93]**.

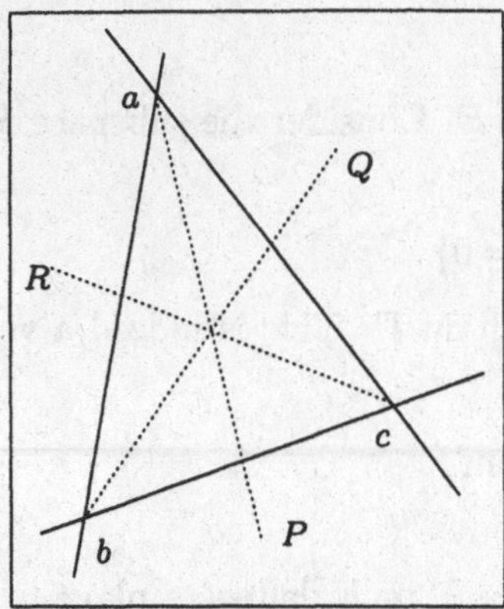

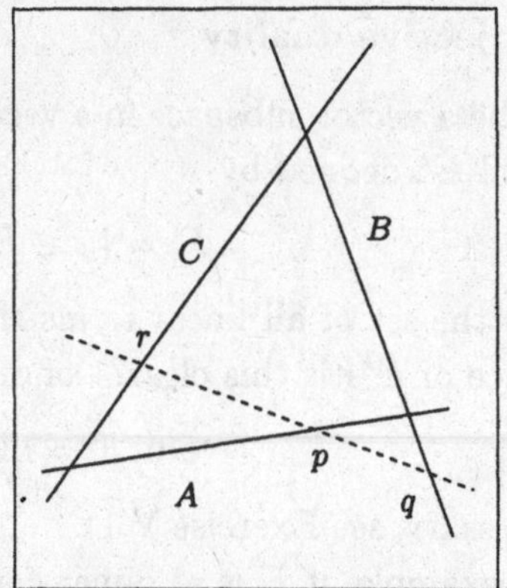

Fig. 10

These properties are a multiplying-theorems-by-two-machine: every statement in the plane P gives, for free, another statement in the plane $P^\star$... which is as an ordinary projective plane as P is. Let us examine, as an example, Desargues' theorem.

Proof of the converse in Desargues' theorem. We use the same notation as in the statement of Theorem 3.3.

- The triangles ABC and $A'B'C'$ in the plane P become triangles with sides $A^\star$, $B^\star$, $C^\star$ and $A'^\star$, $B'^\star$, $C'^\star$ in the dual plane $P^\star$.
- Let abc, $a'b'c'$ be these triangles, denoting their vertices in such a way that $a = B^\star \cap C^\star$, *etc.* Thus, the point a of $P^\star$ is dual to the line BC in P.
- The points α, β and γ of P become lines in $P^\star$. As $\alpha = BC \cap B'C'$, the line $\alpha^\star$ contains the points a and a', thus $\alpha^\star = aa'$, *etc.*
- The lines AA', BB' and CC' of P become points $\mathfrak{a}$, $\mathfrak{b}$ and $\mathfrak{c}$ in $P^\star$. By definition, $\mathfrak{a} = A^\star \cap A'^\star = bc \cap b'c'$, *etc.*

We have shown that, given two triangles ABC and $A'B'C'$ in a plane P, if the three points α, β and γ are collinear, then the three lines AA', BB' and CC' are concurrent. Let us apply this result to the triangles abc and $a'b'c'$ in the plane $P^\star$: we know that if the three points $\mathfrak{a}$, $\mathfrak{b}$ and $\mathfrak{c}$ are collinear, then the three lines aa', bb' and cc' are concurrent.

Let us now translate this statement from $P^\star$ to P.

- Firstly the hypothesis. To say that the three points $\mathfrak{a}$, $\mathfrak{b}$ and $\mathfrak{c}$ are collinear is to say that the three lines AA', BB' and CC' are concurrent.
- Then the conclusion. To say that the three lines aa', bb' and cc' are concurrent is to say that the three points α, β and γ are collinear.

Let us sum up all these translations in a table.

In P	In $P^\star$
triangle with vertices A, B and C triangle with vertices A', B' and C'	triangle with sides $A^\star$, $B^\star$ and $C^\star$ triangle with sides $A'^\star$, $B'^\star$ and $C'^\star$
side BC of the triangle ABC *etc.*	vertex a of the above triangle $a = B^\star \cap C^\star$ *etc.**
the point $\alpha = BC \cap B'C'$ idem	the line $\alpha^\star = aa'$ idem*
the line AA' ...	the point $A^\star \cap A'^\star$ namely $\mathfrak{a} = bc \cap b'c'$ ...
points α, β and γ are collinear	lines aa', bb' and cc' are concurrent
lines AA', BB' and CC' are concurrent	points $\mathfrak{a}$, $\mathfrak{b}$ and $\mathfrak{c}$ are collinear

Using this trick, we have shown the converse of Desargues' theorem. More precisely, we have simply remarked that the converse statement is the statement dual to the direct one (and conversely!). □

5. Projective transformations

Let E and E' be two vector spaces, $p : E - \{0\} \to P(E)$ and $p' : E' - \{0\} \to P(E')$ be the two projections.

A *projective transformation* $g : P(E) \to P(E')$ is a mapping such that there exists a linear isomorphism $f : E \to E'$ with $p' \circ f = g \circ p$, in other words such that the diagram

$$\begin{array}{ccc} E - \{0\} & \xrightarrow{\ f\ } & E' - \{0\} \\ p \downarrow & & \downarrow p' \\ P(E) & \xrightarrow{\ g\ } & P(E') \end{array}$$

commutes.

Very concretely, the image under g of a point ℓ of $P(E)$ (line of E) is the line $\ell' = f(\ell)$ of E' (point of $P(E')$).

A projective transformation of a projective line is called a *homography*.

Remark 5.1. Be careful, a linear map $f : E \to E'$ does not always define a mapping from $P(E)$ into $P(E')$: in general, the image of $E - \{0\}$ is not contained in $E' - \{0\}$. Said in another way, the image of a line by a linear mapping may very well not be a line, it may be the set 0. A linear mapping f defines a mapping $P(E) - P(\operatorname{Ker} f) \to P(E')$ and nothing more. We shall thus restrict ourselves to isomorphisms.

The image of a plane by a linear isomorphism is again a plane. In projective translation: the image of a projective line by a projective transformation is a projective line.

The projective group. This is the group of projective transformations of $P(E)$.

Proposition 5.2. *The set of projective transformations from $P(E)$ to itself is a group for the composition of mappings.*

The mapping which, with a linear isomorphism $E \to E$, associates a projective transformation of $P(E)$ is a surjective group homomorphism, the kernel of which is the group of dilatations of E.

Remark 5.3. In particular, $GP(E)$ is isomorphic to $GL(E)/\{\text{dilatations}\}$. This group is also often denoted $PGL(E)$, and even $PGL(n, \mathbf{K})$ when E is $\mathbf{K}^n$ and $P(E)$ is $\mathbf{P}_{n-1}(\mathbf{K})$.

Proof. By definition, it is clear that, if g and g' are projective transformations, coming from vector space isomorphisms f and f', then $g \circ g'$ is the projective transformation coming from $f \circ f'$, that the identity of $P(E)$ is a projective transformation coming from the identity of E and that g^{-1} is a projective transformation coming from the isomorphism f^{-1}. We thus get both the fact that $GP(E)$ is a group and the fact that the mapping $f \mapsto g$ is a group morphism. The latter is onto by definition. Let us look for his kernel. Let $f : E \to E$ be an isomorphism such that the associated projective transformation is the identity of $P(E)$. This is to say that f preserves all the lines in E, or that for any x in E, there exists some element $\lambda \in \mathbf{K}$ (depending *a priori* on x) such that $f(x) = \lambda x$. The fact that f is a dilatation is deduced as usually (see, *e.g.*, Exercise I.23). □

Homogeneous coordinates and projective frames. Given a basis $(e_1, \ldots, e_{n+1})$ of the vector space E, the vectors of E can be described by their coordinates $(x_1, \ldots, x_{n+1})$ with respect to this basis. A point m in $P(E)$ can also be described by the coordinates of a nonzero vector x that generates the line m. The $(n+1)$-tuples $(x_1, \ldots, x_{n+1})$ and $(x'_1, \ldots, x'_{n+1})$

represent the same point m of $P(E)$ if and only if there exists a nonzero scalar λ such that

$$x_i' = \lambda x_i \text{ for all } i.$$

The equivalence class of $(x_1, \ldots, x_{n+1})$ is called a set of *homogeneous coordinates* for m. It is often denoted $[x_1, \ldots, x_{n+1}]$.

To give the points $p(e_1), \ldots, p(e_{n+1})$ is not enough to determine the basis $(e_1, \ldots, e_{n+1})$, even up to a multiplicative coefficient: they obviously determine the lines generated by the vectors $e_1, \ldots, e_{n+1}$, namely the $\lambda_1 e_1, \ldots, \lambda_{n+1} e_{n+1}$, but nothing more.

However, if we add to them the $(n+2)$-th point $p(e_1 + \cdots + e_{n+1})$, we can reconstitute the vectors $e_1, \ldots, e_{n+1}$ up to a *unique* coefficient. This is the contents of Lemma 5.5 below and this is what justifies the next definition.

Definition 5.4. If E is a vector space of dimension $n+1$, a *projective frame* of $P(E)$ is a system of $n+2$ points $(m_0, m_1, \ldots, m_{n+1})$ of $P(E)$ such that $m_1, \ldots, m_{n+1}$ are the images of the vectors $e_1, \ldots, e_{n+1}$ in a basis of E and m_0 is the image of $e_1 + \cdots + e_{n+1}$.

A projective frame is thus a set of $n+2$ points such that none of them belongs to the projective subspace determined by n of the others (Exercise V.6).

The homogeneous coordinates of m in the basis $(e_1, \ldots, e_{n+1})$ only depend on the projective frame $(p(e_1), \ldots, p(e_{n+1}))$. It is thus possible to speak of homogeneous coordinates in a projective frame.

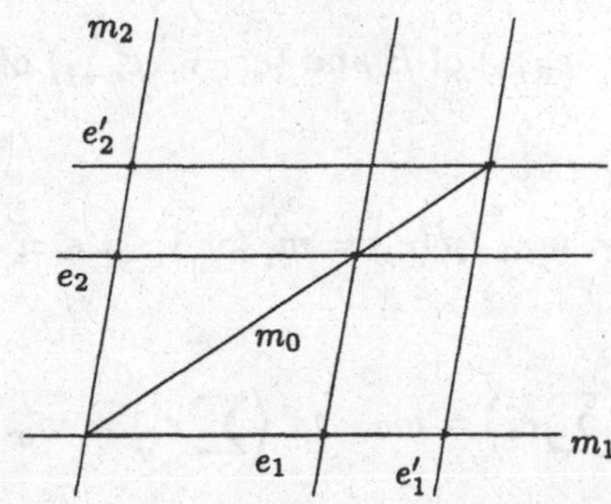

Fig. 11

Lemma 5.5. *Let $(m_0, m_1, \ldots, m_{n+1})$ be a projective frame of $P(E)$. If the two bases $(e_1, \ldots, e_{n+1})$ and $(e_1', \ldots, e_{n+1}')$ of E are such that $p(e_i) = p(e_i') = m_i$ and $p(e_1 + \cdots + e_{n+1}) = p(e_1' + \cdots + e_{n+1}') = m_0$, then they are proportional.*

Proof. Consider the points m_i of $P(E)$ as lines of E. The vectors e_i and e'_i generate the line m_i thus $e'_i = \lambda_i e_i$ for some nonzero scalar λ_i (this for $1 \leqslant i \leqslant n+1$).

Use the $(n+2)$-th point to conclude: $e'_1 + \cdots + e'_{n+1}$ is proportional to $e_1 + \cdots + e_{n+1}$ thus there exists a nonzero λ such that

$$\lambda_1 e_1 + \cdots + \lambda_{n+1} e_{n+1} = \lambda(e_1 + \cdots + e_{n+1}).$$

As we are dealing with a basis, we get that $\lambda_i = \lambda$ for all i, thus all the λ_i are equal and the two bases are proportional. □

In particular, a projective frame of a line consists of *three* distinct points. One should notice, however, that to give two points is enough to determine a projective line: two independent vectors determine a plane.

Proposition 5.6. *Let $P(E)$ and $P(E')$ be two projective spaces of dimension n. Any projective mapping from $P(E)$ to $P(E')$ maps a projective frame of $P(E)$ onto a projective frame of $P(E')$. If $(m_0, \ldots, m_{n+1})$ and $(m'_0, \ldots, m'_{n+1})$ are projective frames of $P(E)$ and $P(E')$ respectively, there exists a unique projective transformation $g : P(E) \to P(E')$ such that $m'_i = g(m_i)$ for all i.*

Proof. The first assertion is clear. To prove the second, choose two bases

$$(e_1, \ldots, e_{n+1}) \text{ of } E \text{ and } (e'_1, \ldots, e'_{n+1}) \text{ of } E'$$

such that

$$p(e_i) = m_i, \quad p'(e'_i) = m'_i \text{ for } 1 \leqslant i \leqslant n+1$$

and

$$p\left(\sum e_i\right) = m_0, \quad p'\left(\sum e'_i\right) = m'_0.$$

There exists a unique isomorphism f from E to E' such that $f(e_i) = e'_i$. It must satisfy $f(\sum e_i) = \sum e'_i$ and the projective transformation g induced by f maps the first frame on the second.

Let us now prove the uniqueness of the projective transformation g determined this way. If g and g' are two such projective transformations, the projective transformation $g'^{-1} \circ g$ from $P(E)$ into itself keeps the frame $(m_0, \ldots, m_{n+1})$ invariant. Due to the previous lemma, it thus comes from a dilatation $\lambda \operatorname{Id}_E$, and is the identity transformation. □

Homographies. For instance, to know a homography from a projective line into itself, it is enough to know the images of three points.

What are the homographies, in this case? Let us use the notation of § 3. A homography comes from a linear isomorphism $f : E \to E$ where E is a plane endowed with a basis (e_1, e_2). In other words,

$$f(x,y) = (ax + by, cx + dy)$$

with $ad - bc \neq 0$. Write now $z = x/y$, so that $(x,y) \sim (z,1)$. Then

$$f(z,1) = (az + b, cz + d) \sim \left(\frac{az+b}{cz+d}, 1\right)$$

... as long as $y \neq 0$ so that we have the right to divide. The image of the point at infinity $\infty = (1,0)$ is (a,c) (in general, $(a/c, 1)$) and that of $(-d, c)$ is $(-ad + bc, 0) \sim (1,0) \sim \infty$.

It is as simple to write that the homography is

$$z \longmapsto \frac{az+b}{cz+d}$$

in this case, having implicitly in mind the usual conventions on the image and pre-image of ∞.

Choice of infinity (continuation). Let us come back to the question of the "expedition to infinity" that we have evoked around the proofs of Pappus' and Desargues' theorems in § 3.

Consider again the two configurations drawn for Pappus' theorem (Figures 8 and 9), now trying to see them in the *same* affine plane $\mathcal{E}$ (the point of view of the drawer).

Figure 9 is "simpler" than Figure 8. We want to consider that it is obtained from Figure 9 by a transformation of the plane. This cannot be an affine transformation, since some intersecting lines are mapped on parallel lines... this is not even a transformation of the affine plane (for the same reason). This is of course a projective transformation of the projective completion P of the affine plane $\mathcal{E}$. In the specific case of the two pictures that illustrate Pappus' theorem, this is a projective transformation that maps the dotted line onto the line at infinity.

The philosophy of the "expedition to infinity" is here the projective version of the very general principle that consists in transforming a problem into a simpler problem, or into an already solved problem... by an *ad hoc* transformation.

Projective and affine transformations. ... and this leads us to consider the affine transformations of the affine space $\mathcal{E}$ among the projective transformations of the projective completion P: they are the projective transformations that (globally) preserve the hyperplane at infinity.

Proposition 5.7. *Let $\mathcal{E}$ be an affine space directed by E and $P(E \oplus \mathbf{K})$ be its projective completion. Any projective transformation of $P(E \oplus \mathbf{K})$ that preserves the hyperplane at infinity defines, by restriction, an affine transformation of $\mathcal{E}$. Conversely, any affine transformation of $\mathcal{E}$ can be extended in a unique projective transformation of $P(E \oplus \mathbf{K})$.*

Proof. Let $g : P(E \oplus \mathbf{K}) \to P(E \oplus \mathbf{K})$ be a projective transformation. Since g preserves the hyperplane at infinity $P(E)$, it preserves its complement $\mathcal{E}$. Thus, it induces a transformation

$$\varphi : \mathcal{E} \longrightarrow \mathcal{E},$$

which we want to prove is an affine transformation.

If f is a linear automorphism of $E \oplus \mathbf{K}$ that defines the projective transformation g, we have

$$\begin{cases} f(u,0) &= (h(u),0) \\ f(0,1) &= (v,a) \end{cases}$$

where h is a linear automorphism of E and a is nonzero. One can choose f (and thus h) so that a is equal to 1, namely, so that $f(0,1)$ is in the affine subspace $\mathcal{E}$ of $E \oplus \mathbf{K}$. Let O be the point $(0,1)$ of $\mathcal{E}$. For any point $M = (u,1)$ in $\mathcal{E}$, we have

$$\begin{aligned} \overrightarrow{\varphi(O)\varphi(M)} &= f(u,1) - f(0,1) \\ &= (h(u),0) \end{aligned}$$

and thus φ is indeed an affine transformation: the associated linear mapping is h.

Conversely, if φ is an affine transformation of $\mathcal{E}$, let h be the associated linear isomorphism E ($h = \overrightarrow{\varphi}$) and define a mapping

$$\begin{aligned} f : \ E \oplus \mathbf{K} &\longrightarrow E \oplus \mathbf{K} \\ (u,a) &\longmapsto a\varphi(0,1) + (h(u),0) \end{aligned}$$

in such a way that f coincides with h on E and with φ on $\mathcal{E}$. It is clear that f is linear and bijective. The projective transformation g deduced from f has thus all the expected properties.

The uniqueness of a projective transformation g extending φ is automatic. With any point of $\mathcal{E}$, g associates the same point as φ does. As for the points of the hyperplane at infinity $P(E)$, they are the directions of the affine lines in $\mathcal{E}$, on which φ acts, obviously by the projective transformation of $P(E)$ defined by the linear map h associated with φ. □

6. The cross-ratio

If D is a projective line, three (distinct) points a, b and c form a projective frame. There exists a unique projective mapping from the line to $\mathbf{K} \cup \{\infty\}$ that maps a to ∞, b to 0 and c to 1. If d is another point of D, the image of d under this projective mapping is a point of $\mathbf{K} \cup \{\infty\}$ that is called the *cross-ratio* of (a, b, c, d) and that is denoted $[a, b, c, d]$.

One can speak of the cross-ratio of four collinear points in any projective space.

Remark 6.1. By definition, the cross-ratio $[a, b, c, d]$ is equal to

- ∞ when $d = a$
- 0 when $d = b$
- 1 when $d = c$.

The cross-ratio is an element of $\mathbf{K} - \{0, 1\}$ if and only if the four points are distinct. Moreover for all $k \in \mathbf{K} \cup \{\infty\}$, there exists a unique point d on the line D such that $[a, b, c, d] = k$.

The most important property of the cross-ratio is its preservation under projective transformations, which is expressed in the next proposition.

Proposition 6.2. *Let a_1, a_2, a_3 and a_4 be four points in D (the first three being distinct) and a'_1, a'_2, a'_3 and a'_4 be four points of a line D' (satisfying the same assumption). There exists a homography $f : D \to D'$ such that $f(a_i) = a'_i$ if and only if $[a_1, a_2, a_3, a_4] = [a'_1, a'_2, a'_3, a'_4]$.*

Proof. This is a straightforward consequence of the definition. Assume, to begin with, that $f : D \to D'$ is a projective mapping that sends a_i to a'_i. Let $g' : D' \to \mathbf{K} \cup \{\infty\}$ be the unique homography that maps a'_1 to ∞, a'_2 to 0 and a'_3 to 1. Then $[a'_1, a'_2, a'_3, a'_4] = g'(a'_4)\ldots$ but then $g' \circ f$ is a (and thus the unique) homography that maps a_1 to ∞, a_2 to 0 and a_3 to 1. Hence

$$[a_1, a_2, a_3, a_4] = g' \circ f(a_4) = g'(a'_4) = [a'_1, a'_2, a'_3, a'_4]$$

and the cross-ratio is conserved.

Conversely, there is a unique homography f that maps a_i to a'_i (for $i \leqslant 3$). We still have to check that $f(a_4) = a'_4$, which is true due to the direct part of the proposition. □

Formulas for the cross-ratio. Although they are not very often useful (and although I prefer to avoid using them), there are formulas to express the cross-ratio. To begin with, the most classical one.

Proposition 6.3. *Let a, b, c and d be four points on an affine line, the first three being distinct. Then*

$$[a,b,c,d] = \frac{d-b}{d-a}\Big/\frac{c-b}{c-a}.$$

Proof. The unique homography that maps a to ∞, b to 0 and c to 1 is

$$z \longmapsto \frac{z-b}{z-a}\Big/\frac{c-b}{c-a}$$

as it has a pole at a and a zero at b. Thus the formula gives, indeed, the image of d. □

There are more general (?) and more complicated formulas. They are based on the next remark.

Lemma 6.4. *Let a, b and c be three distinct points on a projective line $D = P(E)$. Let x and y be vectors in E such that $a = p(x)$, $b = p(y)$, $c = p(x+y)$. Then d is the image of $hx+ky$ if and only if the cross-ratio $[a,b,c,d]$ is the image of (h,k) in $\mathbf{P}_1(\mathbf{K}) = \mathbf{K} \cup \{\infty\}$.*

Proof. Since a and b are distinct, the vectors x and y constitute a basis of E. Let f be the isomorphism $E \to \mathbf{K}^2$ that maps the basis (x,y) to the canonical basis. The homography g defined by f maps a to ∞, b to 0 and c to 1. It thus sends d to the cross-ratio $[a,b,c,d]$. But f maps $hx+ky$ to $(h,k) \in \mathbf{K}^2$, hence g maps d to the image of (h,k) in $\mathbf{P}_1(\mathbf{K})$. □

Let now E be a (vector) plane with basis (e_1,e_2). Consider four points a_i (for $1 \leqslant i \leqslant 4$) in $P(E)$ and assume that the first three are distinct. Assume that a_i is the image of a vector with coordinates (λ_i,μ_i). We shall prove that

$$[a_1,a_2,a_3,a_4] = \frac{\begin{vmatrix}\lambda_3 & \lambda_1\\ \mu_3 & \mu_1\end{vmatrix}}{\begin{vmatrix}\lambda_3 & \lambda_2\\ \mu_3 & \mu_2\end{vmatrix}}\Bigg/\frac{\begin{vmatrix}\lambda_4 & \lambda_1\\ \mu_4 & \mu_1\end{vmatrix}}{\begin{vmatrix}\lambda_4 & \lambda_2\\ \mu_4 & \mu_2\end{vmatrix}}.$$

For each i, choose one of the vectors x_i generating the line a_i. The points a_1 and a_2 being distinct, x_1 and x_2 are independent and one can write x_3 and x_4 in this basis

$$\begin{aligned} x_3 &= \alpha x_1 + \beta x_2 \\ x_4 &= \gamma x_1 + \delta x_2 \quad = \frac{\gamma}{\alpha}(\alpha x_1) + \frac{\delta}{\beta}(\beta x_2). \end{aligned}$$

Applying the previous lemma to $x = \alpha x_1$, $y = \beta x_2$, one gets

$$[a_1,a_2,a_3,a_4] = \frac{\gamma}{\alpha}\Big/\frac{\delta}{\beta}.$$

Expressing everything in the basis (e_1, e_2), one has

$$\begin{cases} \lambda_1\alpha + \lambda_2\beta &= \lambda_3 \\ \mu_1\alpha + \mu_2\beta &= \mu_3 \end{cases}$$

thus

$$\alpha = \frac{\begin{vmatrix} \lambda_3 & \lambda_2 \\ \mu_3 & \mu_2 \end{vmatrix}}{\begin{vmatrix} \lambda_1 & \lambda_2 \\ \mu_1 & \mu_2 \end{vmatrix}}, \qquad \beta = \frac{\begin{vmatrix} \lambda_1 & \lambda_3 \\ \mu_1 & \mu_3 \end{vmatrix}}{\begin{vmatrix} \lambda_1 & \lambda_2 \\ \mu_1 & \mu_2 \end{vmatrix}}.$$

Similarly $x_4 = \gamma x_1 + \delta x_2$ gives analogous formulas for γ and δ (replace 3 by 4 in the formula giving α and β).

The announced formula follows. □

Here are some useful and concrete examples.

- When the projective line is the completion of an affine line $\mathcal{D}$ and when all the points are on $\mathcal{D}$, we find the classical formula of Proposition 6.3: since $a_i \neq \infty_{\mathcal{D}}$, we can choose each μ_i equal to 1.
- If a, b and c are three points on an affine line $\mathcal{D}$, one can compute $[a, b, c, \infty_{\mathcal{D}}]$: here, one can choose $\mu_i = 1$ for $i \leqslant 3$ and $\mu_4 = 0$. One gets:

$$[a, b, c, \infty_{\mathcal{D}}] = \frac{\lambda_3 - \lambda_1}{\lambda_3 - \lambda_2} = \frac{\overrightarrow{ac}}{\overrightarrow{bc}}.$$

From the same formula, the first series of equalities of the next proposition is immediately deduced[8].

Proposition 6.5. *If a, b, c and d are four collinear distinct points, the following equalities hold:*

$$[a, b, c, d] = [b, a, c, d]^{-1} = [a, b, d, c]^{-1}$$

$$[a, b, c, d] + [a, c, b, d] = 1.$$

Proof. Let us prove the second one. Write $a = p(x)$, $b = p(y)$, $c = p(x + y)$ and $d = p(hx + ky)$ (as in Lemma 6.4).

Let $x' = -x$ (so that $p(x') = a$) and let $y' = x + y$ (so that $p(y') = c$). Then

$$y = x' + y' \text{ and } hx + ky = (k - h)x' + ky'.$$

Thus

$$[a, c, b, d] = \frac{k - h}{k} = 1 - \frac{h}{k} = 1 - [a, b, c, d],$$

which is what we wanted to prove. □

---

[8] But this is not the best proof; see Exercise V.26.

Remark 6.6. If $[a, b, c, d] = k$, the twenty-four cross-ratios obtained by permutations of the four points take in general six values. These values are

$$k, \quad \frac{1}{k}, \quad 1 - k, \quad 1 - \frac{1}{k}, \quad \frac{1}{1-k}, \quad \frac{k}{k-1}.$$

Harmonic range. It is said that four distinct points form a *harmonic range* when their cross-ratio is -1. For instance, on an affine line $\mathcal{D}$, the points a, b, c and $\infty_{\mathcal{D}}$ form a harmonic range when $\overrightarrow{ac} = -\overrightarrow{bc}$, that is, when c is the midpoint of the segment ab.

7. The complex projective line and the circular group

We now assume that $\mathbf{K} = \mathbf{C}$ and we investigate the projective line $\mathbf{P}_1(\mathbf{C})$.

Fourteen ways to describe the line[9]**.** We will consider $\mathbf{P}_1(\mathbf{C})$ as:

- the (quotient) space of lines in $\mathbf{C}^2$,
- that of unit vectors in the Hermitian space $\mathbf{C}^2$—the 3-sphere S^3—modulo the action of the group $\mathbf{U}$ of complex numbers of module 1 by coordinate multiplication,
- the projective completion of the complex affine line: $\mathbf{P}_1(\mathbf{C}) = \mathbf{C} \cup \{\infty\}$ (this avatar of $\mathbf{P}_1(\mathbf{C})$ is often called the *Riemann sphere*),
- the unit sphere S^2 in the space $\mathbf{C} \oplus \mathbf{R}$, *via* the stereographic projection (see Exercises IV.24 and IV.43),
- ...

We will also keep in mind the fact that $\mathbf{C}$ contains $\mathbf{R}$ and thus that $\mathbf{P}_1(\mathbf{C})$ contains $\mathbf{P}_1(\mathbf{R})$. We can keep track of the subspace $\mathbf{P}_1(\mathbf{R})$ in all the previous descriptions of $\mathbf{P}_1(\mathbf{C})$ (Exercise V.45). I will just say here that:

$$\mathbf{P}_1(\mathbf{R}) = \mathbf{R} \cup \{\infty\} \subset \mathbf{C} \cup \{\infty\}$$

(the point at infinity of $\mathbf{P}_1(\mathbf{C})$ is real). In order to better understand this inclusion, one should look at Figure 12 (in which the real axis and the unit circle in $\mathbf{C}$ are shown) and its image under the stereographic projection, Figure 13.

Now that we have the framework, we are going to look mainly at $\mathbf{C}$ as being the (real) Euclidean affine plane.

---

(9) See § 32 in [**HCV52**] or listen to [**Eis41**].

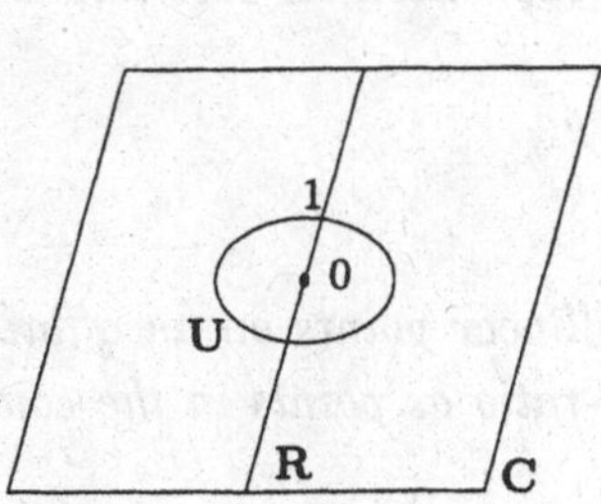

Fig. 12. The line $\mathbf{R}$ in $\mathbf{C}$

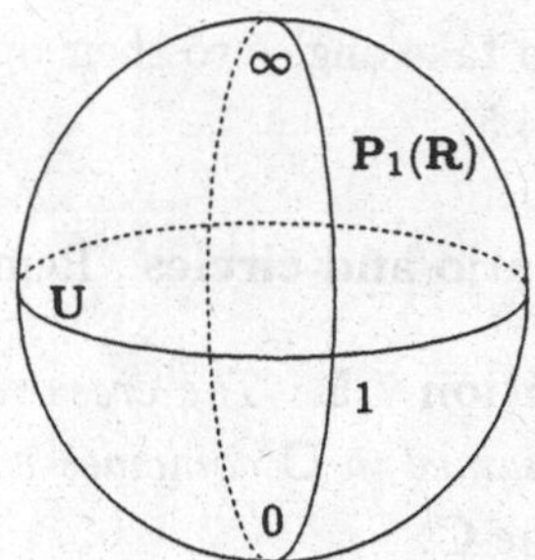

Fig. 13. The line $\mathbf{P}_1(\mathbf{R})$ in $\mathbf{P}_1(\mathbf{C})$

Important remark. The objects we want to investigate are the real affine lines and the circles, but we look at them in the projective completion $\mathbf{P}_1(\mathbf{C}) = \mathbf{C} \cup \{\infty\}$ in which a point has been added to the plane, *not* in the completion $\mathbf{P}_2(\mathbf{R})$ in which a whole projective line has been added. See Remark 7.9 below.

The homographies. The group $PGL(2, \mathbf{C})$ of projective transformations consists of the transformations of the form

$$z \longmapsto \frac{az+b}{cz+d} \qquad ad - bc \neq 0.$$

Proposition 7.1. *The group $PGL(2, \mathbf{C})$ is generated by the direct similarities (of the form $z \mapsto az + b$ with $a \neq 0$) and the mapping $z \mapsto 1/z$.*

Proof. We simply perform the Euclidean division of $az + b$ by $cz + d$:

$$\frac{az+b}{cz+d} \begin{cases} = \dfrac{a}{c} + \dfrac{bc-ad}{c}\dfrac{1}{cz+d} & \text{if } c \neq 0 \\ \text{is a similarity} & \text{if } c = 0, \end{cases}$$

which gives the result. □

Corollary 7.2. *The elements of $PGL(2, \mathbf{C})$ preserve oriented angles.*

Remark 7.3. This has to be understood in the same sense as Corollary III-4.5: elements of $PGL(2, \mathbf{C})$ do not preserve lines.

Remark 7.4. One could simply apply the fact that the holomorphic transformations preserve the angles (Exercise III.65). For the reader who does not appreciate holomorphic functions the way she should, this is also a consequence of the previous proposition: direct similarities preserve angles, so that it suffices to check that the same is true of the mapping $z \mapsto 1/z$. But this is the composition of the reflection $z \mapsto \overline{z}$ and the inversion $z \mapsto 1/\overline{z}$

and both take angles to their opposites (see Proposition III-1.10 and Corollary III-4.5). □

Cross-ratio and circles. Remark firstly:

Proposition 7.5. *The cross-ratio of four collinear points on an affine real line contained in* $\mathbf{C}$ *coincides with their cross-ratio as points in the complex affine line* $\mathbf{C}$.

Proof. Let f be the unique homography (in $PGL(2,\mathbf{R})$) that maps ∞ to a, 0 to b and 1 to c. Let $k = [a,b,c,d]_{\mathbf{R}}$. Then $f(k) = d$.

But $PGL(2,\mathbf{R})$ is contained in $PGL(2,\mathbf{C})$, hence, as $f(k) = d$,

$$[a,b,c,d]_{\mathbf{C}} = k = [a,b,c,d]_{\mathbf{R}}.$$

□

Notice in particular that this cross-ratio is real. The next proposition and corollary are very useful.

Proposition 7.6. *Four points of* $\mathbf{C}$ *are collinear (on an affine real line) or cyclic if and only if their cross-ratio is real.*

Proof. As we are speaking of the cross-ratio of the four points a, b, c and d, the first three must be distinct. If d coincides with a, b or c, there are only three points, thus they are cyclic (or collinear) and their cross-ratio is indeed real, being equal to ∞, 0 or 1. We can thus assume that the four points are distinct.

In this case, an argument of the ratio

$$\frac{c-a}{c-b} \Big/ \frac{d-a}{d-b}$$

is a measure of the angle

$$(\overrightarrow{CA}, \overrightarrow{CB}) - (\overrightarrow{DA}, \overrightarrow{DB})$$

thus the cross-ratio of the four points is real if and only if the four points are cyclic (according to Corollary III-1.19). □

Remark 7.7. This statement simply translates an angle equality in terms of a condition on the cross-ratio. See some spectacular applications in Exercise V.38.

Corollary 7.8. *Any homography of the complex projective line transforms a circle or a line of* $\mathbf{C}$ *into a circle or a line of* $\mathbf{C}$. □

This result should remind the reader of the section on plane inversions (§ III-4). Notice that the lines, and the circles as well, are the images, *via* the stereographic projection, of the circles drawn on the sphere. This is why it is natural to call them *circles* (of $\mathbf{P}_1(\mathbf{C})$).

Remark 7.9. This is the place where we can appreciate the role played by *this* specific completion of the plane: to make a line look like a circle, you have to add *a single* point to it—but two lines intersect at a single point, unlike two circles, which can intersect at two points. Thus you have to add *the same point* to all the lines.

Thus, Corollary 7.8 tells us that projective transformations preserve circles. We know that the same is true of inversions. I presume that the reader has understood that inversions are *not* homographies[10].

The circular group. Let thus G be the group of transformations of $\mathbf{P}_1(\mathbf{C})$ generated by:

– the homographies

$$z \longmapsto \frac{az+b}{cz+d} \quad ad-bc \neq 0$$

– and the symmetry $z \mapsto \overline{z}$.

Thus G contains the projective group $PGL(2, \mathbf{C})$ and all the inversions. This group is called the *circular group*. This terminology will be justified by Theorem 7.12. Let us exhibit first a system of generators for this group.

Proposition 7.10. *The circular group G is generated by the inversions and the reflections.*

Proof. Let φ be an element of G. If $\varphi(\infty) = \infty$, φ is a (direct or opposite) similarity, it is thus the composition of a dilatation h and an isometry u (Proposition III-3.2). We know that u is a composition of reflections (Theorem II-2.2) and that h is a composition of inversions (Proposition III-4.7).

If $\varphi(\infty) = A$ is in $\mathbf{C}$, compose φ with an inversion I of pole A so that $I \circ \varphi(\infty) = \infty$. Then apply what we have just done. □

We get immediately:

Corollary 7.11. *Elements of G preserve nonoriented angles.* □

Proof. Reflections reverse oriented angles (Proposition III-1.10) as do inversions (Corollary III-4.5). □

---

[10] In the unlikely case where this would not be true, have a look at Exercise III.14: if you write an inversion in complex numbers, you have to use complex conjugates.

Let us now characterize the elements of the circular group among all the punctual transformations of $\mathbf{C} \cup \{\infty\}$.

Theorem 7.12. *The elements of G are the transformations that preserve the set of circles and lines.*

Proof. From Proposition 7.10 and the properties of reflections and inversions (§ III-4), one deduces that the elements of G transform circles and lines into circles and lines.

Conversely, let $\varphi : \mathbf{P}_1(\mathbf{C}) \to \mathbf{P}_1(\mathbf{C})$ be a transformation that preserves the set of circles and lines. Up to the composition of φ with a homography, we can assume that $\varphi(\infty) = \infty$ and thus that φ is a transformation of $\mathbf{C}$ which maps any circle to a circle and any line to a line.

We could now use the so-called "fundamental theorem of affine geometry" (Exercise I.51) to say that φ is (real) affine and to check that an affine transformation that preserves circles is a similarity (direct or opposite). If the second step is easy, the first one is harder than necessary. Let us use another method (in which, eventually, one proves the only part of the fundamental theorem that we really need).

Let us show first that φ preserves harmonic ranges. The claim is that Figure 14 shows a ruler construction of the unique point d such that $[a, b, c, d]$ is equal to -1 in the cases where the three points a, b and c are three noncollinear points. Figure 15 shows a ruler construction of the fourth point when the first three are collinear points.

– Case where a, b, c are not collinear (Figure 14). Send d to ∞, so that

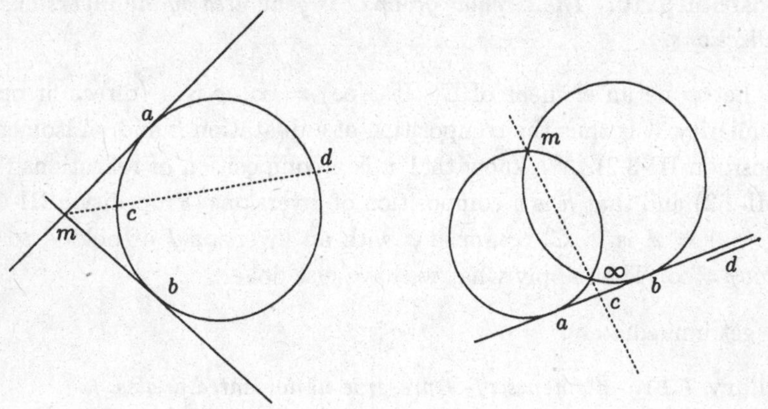

Fig. 14

ma and mb become two circles intersecting at (the image of) m and at another point (image of ∞) and the circle $abcd$ becomes a projective line tangent to both circles. The point c is now the midpoint of (the image of) ab, so that the cross-ratio $[a, b, c, d]$ is indeed -1.

– If a, b and c are collinear (Figure 15). Transform the line mc to the line

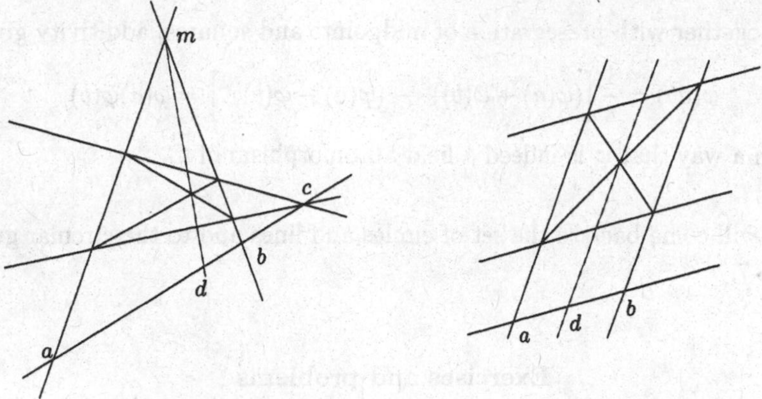

Fig. 15

at infinity (we are in $\mathbf{P}^2(\mathbf{R})$ for this proof), we are reduced to proving that the diagonals of the parallelogram intersect at their midpoints.

From this possibility of constructing the "fourth" point in a harmonic range, we deduce the assertion.

We conclude the proof of the theorem with the next lemma.

Lemma 7.13. *Let $\varphi : \mathbf{C} \to \mathbf{C}$ be a transformation that fixes* 0 *and* 1 *and preserves harmonic ranges. Then φ is a field automorphism of $\mathbf{C}$.*

Once the lemma is proved, compose φ with a similarity to make it preserve the real axis. Then, this is a field automorphism of $\mathbf{C}$ that preserves $\mathbf{R}$, thus this is either the identity or the complex conjugation. □

Proof of the lemma. Recall that $[a, b, c, \infty] = -1$ if and only if c is the midpoint of ab (as we have said at the end of § 6) and use it above. Thus φ conserves the midpoints:

$$\varphi\left(\frac{a+b}{2}\right) = \frac{\varphi(a)+\varphi(b)}{2}.$$

In particular (with $b = 0$), we have

$$\varphi(a) = 2\varphi\left(\frac{a}{2}\right)$$

and thus

$$\varphi(a+b) = \varphi(a) + \varphi(b)$$

hence φ is additive.

Notice now that $[a, -a, a^2, 1] = -1$ (direct verification) so that φ also preserves squares: $\varphi(a^2) = \varphi(a)^2$ for any complex number a. Eventually:

$$ab = \frac{1}{4}\left[(a+b)^2 - (a-b)^2\right],$$

thus, together with preservation of midpoints and squares, additivity gives:

$$\varphi(ab) = \frac{1}{4}\left[(\varphi(a)+\varphi(b))^2 - (\varphi(a)-\varphi(b))^2\right] = \varphi(a)\varphi(b)$$

in such a way that φ is indeed a field automorphism of $\mathbf{C}$. □

We will come back to the set of circles and lines and to the circular group in § VI-7.

Exercises and problems

Direct applications and complements

Exercise V.1. Let E be a vector space and $F \subset E$ be a subspace. Using a basis of F completed in a basis of E, find a basis of $F' \subset E^\star$. Deduce the dimension of F'.

Exercise V.2. Let E be a vector space and let $\mathcal{H}(E)$ be the space of hyperplanes in E. Prove that there exists a one-to-one correspondence between $\mathcal{H}(E)$ and the projective space $P(E^\star)$ associated with the dual of E.

Exercise V.3. Using a stereographic projection, prove that $\mathbf{P}_1(\mathbf{R})$ is homeomorphic to a circle.

Exercise V.4. The complement of a line in a real affine plane has two connected components (see Exercise I.30). Prove that the complement of a line in a real projective plane is connected. What happens with the complement of the real projective line $\mathbf{P}_1(\mathbf{R})$ in $\mathbf{P}_1(\mathbf{C})$?

Exercise V.5. Let E be a vector space and let $\mathcal{F} \subset E$ be an *affine* subspace not containing 0. Prove that the projection

$$p : E - \{0\} \longrightarrow P(E)$$

restricts to an injective mapping from $\mathcal{F}$ to $P(E)$.

Exercise V.6. Prove that $(m_0, \ldots, m_{n+1})$ is a projective frame of the projective space $P(E)$ of dimension n if and only if, for all i and k, m_i is not in the projective subspace generated by the m_j's for $j \neq i, k$.

Exercise V.7. Let $g : P(E) \to P(E)$ be a projective transformation. Prove that if $\mathbf{K} = \mathbf{C}$ (or if $\mathbf{K} = \mathbf{R}$ and the dimension of $P(E)$ is even), g always has a fixed point. Find a projective transformation $\mathbf{P}_1(\mathbf{R}) \to \mathbf{P}_1(\mathbf{R})$ without a fixed point.

Exercise V.8. Let f and g be two homographies of a line, each having two distinct fixed points. Prove that f and g commute if and only if they have the same fixed points.

Exercise V.9. What are the homographies of $\mathbf{P}_1(\mathbf{K})$ that preserve ∞? that preserve 0 and ∞? Prove that the subgroup of $PGL(2, \mathbf{K})$ consisting of all projective transformations that preserve two (distinct) points a and b is isomorphic with the multiplicative group $\mathbf{K}^\star$.

Exercise V.10 (Pencils of lines). A *pencil of lines* in a projective plane P is the family, denoted by $m^\star$, of all the lines through a point m. Prove that a pencil of lines of P is... a line of $P^\star$.

Exercise V.11 (Incidences). Let P be a projective plane. Let D be a line and m be a point in P not in D. Let $m^\star \subset P^\star$ be the dual line, that is, the set of lines through m (see § 4 and Exercise V.10). One defines the *incidence mapping*

$$i : m^\star \longrightarrow D$$

associating with any line through m its intersection point with D. Prove that i is a projective transformation.

Exercise V.12 (Perspectivities). Let H and H' be two hyperplanes in the projective space $P(E)$, m a point which is neither in H nor in H'. Let x be a point in H. Prove that the line mx intersects H' at a unique point, that we denote by $g(x)$. Prove that g is a projective transformation. The mapping g is called the *perspectivity* of center m from H to H' (or projection from H to H').

Prove that, if D and D' are two lines in the projective plane P and if m is a point of P (neither in D nor in D'), the perspectivity of center m from D to D' is the composition of two incidences.

Exercise V.13 (Pappus, again (the most projective proof)). We use the notation of the statement of Pappus' theorem (Theorem 3.2). Let $\mathcal{O}$ be the intersection point of $\mathcal{D}$ and $\mathcal{D}'$, M that of BC' and $A'C$, N that of AC' and $A'B$. Consider the composition

$$f : BC' \longrightarrow \mathcal{D}' \longrightarrow A'B$$

of the perspectivities of center C from BC' to $\mathcal{D}'$ and of center A from $\mathcal{D}'$ to $A'B$. Determine $f(B)$, $f(M)$, $f(C')$ and $f(\alpha)$. Consider then the perspectivity of center β

$$g : BC' \longrightarrow A'B.$$

Determine $g(B)$, $g(M)$ and $g(C')$. Identify $g(\alpha)$ and conclude.

Exercise V.14 (Pencil of lines (continuation)). Let d_1, d_2, d_3 and d_4 be four lines in a pencil and let D be a secant line (a line not containing m). Let a_i be the intersection point of d_i and D. Prove that their cross-ratio $[a_1, a_2, a_3, a_4]$ does not depend on the chosen secant D. It is called the cross-ratio of the four lines and it is denoted by $[d_1, d_2, d_3, d_4]$.

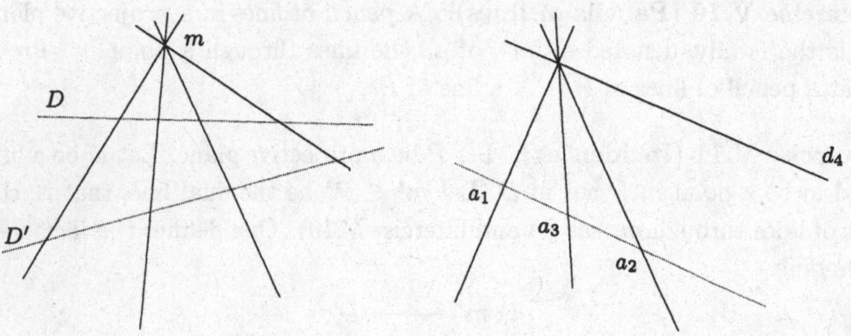

Fig. 16. The cross-ratio of four lines

Fig. 17. A harmonic pencil

Exercise V.15 (Cross-ratio and duality). We use the notation of Exercise V.14. Prove that the cross-ratio $[a_1, a_2, a_3, a_4]$ is actually that of the four points d_1, d_2, d_3, and d_4 on the projective line $m^\star$.

Exercise V.16 (Harmonic pencil). It is said that four concurrent lines d_1, d_2, d_3 and d_4 in an affine plane $\mathcal{P}$ constitute a *harmonic pencil* if their cross-ratio equals -1. Prove that, in order that the four concurrent lines lines d_1, d_2, d_3 and d_4 form a harmonic pencil, it is necessary and sufficient that a parallel to d_4 intersect d_1, d_2 and d_3 at points a_1, a_2, a_3 such that a_3 is the midpoint of a_1a_2 (Figure 17).

Exercise V.17. On an affine real line, prove that if four collinear points A, B, C, D form a harmonic range, one and only one of the points C and D is in the interior of the segment $[AB]$.

Exercise V.18. In an affine plane, let A', B' and C' be three points on the sides BC, CA and AB of a triangle. The line $B'C'$ intersects BC at D. Prove the equality

$$\frac{\overrightarrow{A'C}}{\overrightarrow{A'B}} \cdot \frac{\overrightarrow{B'A}}{\overrightarrow{B'C}} \cdot \frac{\overrightarrow{C'B}}{\overrightarrow{C'A}} = [C, B, A', D].$$

Prove that the product

$$\frac{\overrightarrow{A'C}}{\overrightarrow{A'B}} \cdot \frac{\overrightarrow{B'A}}{\overrightarrow{B'C}} \cdot \frac{\overrightarrow{C'B}}{\overrightarrow{C'A}}$$

is invariant under projective transformations.

Exercise V.19 (The return of Thales). Four lines d, d', d'' and d_0 pass through a point m, two lines $\mathcal{D}_1$ and $\mathcal{D}_2$ intersect d_0 at m_1, m_2. Let $A_i = \mathcal{D}_i \cap d$, $A'_i = \mathcal{D}_i \cap d'$, $A''_i = \mathcal{D}_i \cap d''$. What is the significance, in an affine plane, of the equality of cross-ratios

$$[A_1, A'_1, A''_1, m_1] = [A_2, A'_2, A''_2, m_2]$$

(proved in Exercise V.14) when d_0 is the line at infinity?

In Chapter I, we have shown that Thales' theorem is a translation of the fact that the projections are affine mappings. What property have you used here?

Exercise V.20. Two lines $\mathcal{D}$ and $\mathcal{D}'$ meet at A. Three points B, C and D are given on $\mathcal{D}$, and three points B', C' and D' are given on $\mathcal{D}'$. Prove that the lines BB', CC' and DD' are concurrent if and only if

$$[A, B, C, D] = [A, B', C', D'].$$

Exercise V.21. In a Euclidean affine plane, let $\mathcal{D}$ and $\mathcal{D}'$ be two secant lines, Δ and Δ' their bisectors. Prove that $\mathcal{D}$, $\mathcal{D}'$, Δ and Δ' form a harmonic pencil.

Conversely, if $\mathcal{D}$, $\mathcal{D}'$, Δ and Δ' form a harmonic pencil and if Δ and Δ' are orthogonal, prove that Δ and Δ' are the bisectors of $\mathcal{D}$ and $\mathcal{D}'$.

Find again in this way the result of Exercise III.26.

Exercise V.22. In a Euclidean affine plane, let $\mathcal{C}$ and $\mathcal{C}'$ be two circles of centers O and O'. The line OO' intersects $\mathcal{C}$ at A and B, $\mathcal{C}'$ at M and N. Prove that $\mathcal{C}$ and $\mathcal{C}'$ are orthogonal if and only if $[A, B, M, N] = -1$.

Exercise V.23. In a projective plane, consider two lines D and D' intersecting at a point O and a point A not in $D \cup D'$. Consider Figure 18 and construct the unique line d such that (D, D', d, OA) is a harmonic pencil.

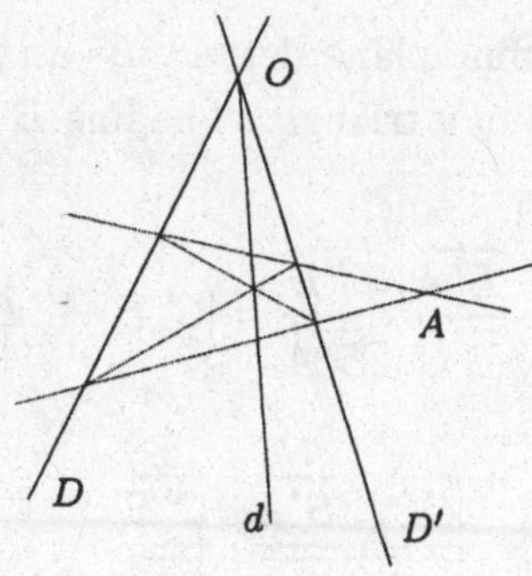

Fig. 18

Exercise V.24. Let (a, b, c, d) be a projective frame of a projective plane. Let α be the intersection of ab and cd, β that of ad and bc, γ that of ac and bd and eventually δ that of ac and $\alpha\beta$. This way, the four points a, c, γ and δ are collinear. Prove that $[a, c, \gamma, \delta] = -1$.

Exercise V.25. Let F be a figure in a projective plane. Remove a line from this plane, thus getting a figure F' in an affine plane. It is said that F is F' viewed in perspective. For instance, one can consider a part of Figure 18 as a parallelogram and its diagonals viewed in perspective (is this clear?). Draw

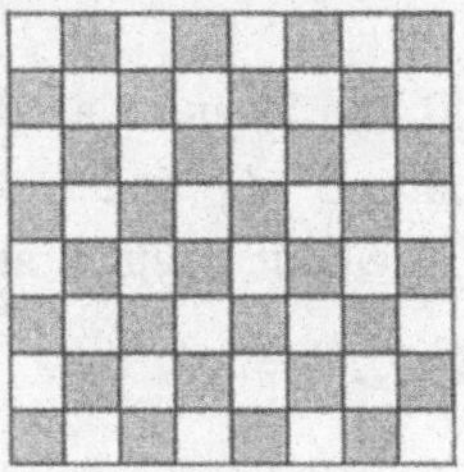

Fig. 19

a chessboard (Figure 19) in perspective.

Exercise V.26. What are the homographies of $\mathbf{P}_1(\mathbf{K})$ that fix 1 and exchange 0 and ∞? Deduce a proof (without computation) of the equality $[b, a, c, d] = [a, b, c, d]^{-1}$. Prove in the same way that the equalities $[a, b, d, c] = [a, b, c, d]^{-1}$ and $[a, c, b, d] = 1 - [a, b, c, d]$ hold.

Exercise V.27. Let a, b, m, n and p be five distinct points on a projective line. Prove that the equality

$$[a, b, m, n][a, b, n, p][a, b, p, m] = 1$$

holds.

Exercise V.28 (Involutions). An *involution* is a projective transformation g that is not the identity and is such that $g^2 = \mathrm{Id}$. Prove that a homography g of a projective line is an involution if and only if there exist two (distinct) points p and p' on this line such that $g(p) = p'$ and $g(p') = p$.

Two series of three distinct points a_i and a'_i ($1 \leqslant i \leqslant 3$) are given on a projective line. Let f be the unique homography of this line that maps a_i to a'_i. Prove that f is an involution if and only if

$$j = 1, 2, 3 \implies [a_1, a_2, a_3, a'_j] = [a'_1, a'_2, a'_3, a_j].$$

Let g be an involutive homography of a real projective line. Prove that if g has a fixed point, then it has exactly two fixed points. Call them a and b. Prove that, for any point m on the line, one has $[a, b, m, g(m)] = -1$.

Exercise V.29. For each of the following parts of $\mathbf{C}$, describe its image under the indicated projective transformations[11]:

- The first quadrant ($x > 0$ and $y > 0$) under
$$z \longmapsto \frac{z-i}{z+i}.$$
- The half-disk $|z| < 1$, $\mathrm{Im}(z) > 0$ under
$$z \longmapsto \frac{2z-i}{2+iz}.$$
- The sector $x > 0$, $y > 0$ and $y < x$ under
$$z \longmapsto \frac{z}{z-1}.$$
- The strip $0 < x < 1$ under
$$z \longmapsto \frac{z-1}{z} \text{ and under } z \longmapsto \frac{z-1}{z-2}.$$

Exercises

Exercise V.30. Let H be a fixed hyperplane of a vector space E. Let E_H be the set of lines of E that are not contained in H. What is the complement of E_H in $P(E)$? Let $\ell \in E_H$ be a fixed line. Prove that there exists a bijective mapping

$$E_H \longrightarrow \mathcal{L}(\ell, H)$$

from E_H on the space of linear mappings from ℓ to H.

(11) This exercise has been copied from [Sil72].

Exercise V.31 (Aerial photography). Consider three hyperplanes H, H_1 and H_2 and two points m_1 and m_2 in a projective space $P(E)$. Assume that m_i is neither in H, nor in H_i and let g_i be the perspectivity of center m_i from H to H_i (see Exercise V.12). What can be said of $g_2 \circ g_1^{-1} : H_1 \to H_2$? Imagine now (in dimension 3) that H is a part (that we assume to be plane) of the Earth, that m_1 and m_2 are the objectives of two cameras placed in two airplanes and that H_1 and H_2 are the planes of the films in these cameras. Using four points of H (trees, monuments, *etc.*) explain how it is possible to glue the pictures taken by the two cameras(12).

Exercise V.32. What is the statement dual to Pappus' theorem (Theorem 3.2)?

Exercise V.33. Two points are given on a sheet of paper. You have a very short ruler. Use it to construct the line through the two points.

Exercise V.34 (Desargues' second theorem). A triangle ABC, a line d not passing through its vertices and three distinct points P, Q and R of d are given. Let P', Q' and R' be the intersection points of d with the lines BC, CA and AB. Prove that the lines AP, BQ and CR are concurrent if and only if there exists an involution of d that maps P to P', Q to Q' and R to R'.

Exercise V.35 (Orthogonality in projective geometry). The real affine plane is now endowed with a Euclidean structure. Complete it to get a projective plane. Prove that the orthogonality of line directions defines an involutive homography on the line at infinity.

Exercise V.36 (The orthocenter, projective proof). Let ABC be a triangle in an affine Euclidean plane. Using the second theorem of Desargues (Exercise V.34) and the involution defined on the line at infinity by the orthogonality relation (Exercise V.35), prove that the three altitudes are concurrent.

More theoretical exercises

Exercise V.37. Prove Proposition 1.1 in the complex case (be careful in proving that the quotient is Hausdorff!).

Exercise V.38 (The six cross-ratio theorem, with applications). Let A, B, C, D, A', B', C' and D' be eight distinct points on a projective

(12) See a "practical" application in [**Ber94**].

line. Prove that the equality

$$[A,B,C',D'][B,C,A',D'][C,A,B',D'] \cdot [A',B',C,D][B',C',A,D][C',A',B,D]=1$$

holds.

Cubic systems. Let a *cubic system* denote the data of an eight element set $\mathcal{X}$ and three subsets of four elements, $\mathcal{A}$, $\mathcal{B}$ and $\mathcal{C}$ given in such a way that:

- the subsets $\mathcal{A}\cap\mathcal{B}$, $\mathcal{B}\cap\mathcal{C}$ and $\mathcal{C}\cap\mathcal{A}$ are distinct and all of them consist of two elements,
- the intersection $\mathcal{A}\cap\mathcal{B}\cap\mathcal{C}$ contains a unique element.

The six subsets $\mathcal{A}$, $\mathcal{B}$, $\mathcal{C}$, $\mathcal{X}-\mathcal{A}$, $\mathcal{X}-\mathcal{B}$, $\mathcal{X}-\mathcal{C}$ are called the *faces* of the system.

Prove that, if $\mathcal{X}$ is the set of vertices of a cube and if $\mathcal{A}$, $\mathcal{B}$ and $\mathcal{C}$ are the (sets of vertices of) three faces through a given vertex, $(\mathcal{X},\mathcal{A},\mathcal{B},\mathcal{C})$ is a cubic system. Prove that it is possible to denote the elements of $\mathcal{X}$ by

$$\mathcal{X}=\{a,b,c,d,a',b',c',d'\}$$

in such a way that

$$\mathcal{A}=\{b,c,a',d'\},\quad \mathcal{B}=\{c,a,b',d'\},\quad \mathcal{C}=\{a,b,c',d'\}.$$

Conversely, prove that any cubic system is in one-to-one correspondence with a cubic system of this form (hence the terminology).

Let $(\mathcal{X},\mathcal{A},\mathcal{B},\mathcal{C})$ be a cubic system of points of $\widehat{\mathbf{C}}=\mathbf{P}_1(\mathbf{C})$. Prove that if five faces of $\mathcal{X}$ consist of cyclic or collinear points, the same is true of the sixth face.

Applications. Find again the results of Exercises III.32 (Miquel's theorem), III.33 (the Simson line) and III.35 (the pivot)(13).

Exercise V.39 (Grassmannians). In this exercise, $\mathbf{K}=\mathbf{R}$ or $\mathbf{C}$. Consider the set (the *Grassmannian*(14)) $G_k(\mathbf{K}^n)$ of all the linear subspaces of dimension k in $\mathbf{K}^n$. Prove that $G_k(\mathbf{K}^n)$ can be considered as a quotient of the set of the k-tuples of independent vectors in $\mathbf{K}^n$. Use this to endow it with a topology(15). Prove that $G_k(\mathbf{K}^n)$ is a compact topological space.

Prove that the projective lines of $\mathbf{P}_n(\mathbf{K})$ constitute a topological space.

(13)The text of this exercise is due to Daniel Perrin.

(14)From the name of Grassmann.

(15)The Grassmannian may also be endowed with a topology when being considered as a subset of the projective space $P(\Lambda^k\mathbf{K}^n)$, where Λ^k denotes the k-th exterior power: associate with the subspace E of $\mathbf{K}^n$ the line generated by $x_1\wedge\cdots\wedge x_k$, for any basis $(x_1,\ldots,x_k)$ of E... this gives the same topology; this is both another exercise and the *Plücker embedding*.

Exercise V.40 (Affine lines). The aim of this exercise is to investigate the set of affine lines in a real affine space.

(1) To what is the space of oriented lines through zero in $\mathbf{R}^{n+1}$ homeomorphic?

(2) Let D_n be the set of all affine lines in $\mathbf{R}^n$. Embed $\mathbf{R}^n$ in $\mathbf{R}^{n+1}$ as the affine hyperplane $x_{n+1} = 1$ (and as usual). Prove that D_n may be considered as a subset of the set of projective lines in $\mathbf{P}_n(\mathbf{R})$ (as in Exercise V.39). This endows it with a topology. Is this a compact space? What is its complement?

(3) Let $\widetilde{D}_n$ be the set of oriented affine lines in $\mathbf{R}^n$. Prove that there is a bijection

$$\widetilde{D}_n \longrightarrow \{(p,u) \in \mathbf{R}^n \times \mathbf{R}^n \mid \|p\| = 1 \text{ and } p \cdot u = 0\}.$$

This gives $\widetilde{D}_n$ a topology.

(4) Prove that, with all these spaces, we get a commutative diagram (in which all the mappings are continuous):

$$\begin{array}{ccc} \widetilde{D}_n & \longrightarrow & D_n \quad \subset G_2(\mathbf{R}^{n+1}) \\ \downarrow & & \downarrow \\ S^{n-1} & \longrightarrow & \mathbf{P}_n(\mathbf{R}) \end{array}$$

Exercise V.41. Prove that the set of all affine lines in the affine plane $\mathbf{K}^2$ may be identified with the complement of a point in the projective plane $\mathbf{P}_2$.

Exercise V.42 (Barycentric coordinates and projective frames). Three noncollinear points A, B and C are given in an affine plane $\mathcal{P}$. Any point M of $\mathcal{P}$ has barycentric coordinates in the affine frame (A, B, C). Under which condition is a triple (x, y, z) a system of barycentric coordinates for some point M? Under which condition do two systems (x, y, z) and (x', y', z') determine the same point of $\mathcal{P}$?

Let O be a point of barycentric coordinates (x_0, y_0, z_0). Under which conditions do the four points (O, A, B, C) form a projective frame in the projective completion of $\mathcal{P}$?

Exercise V.43 (Barycentric coordinates and projective frames, continuation). Let E be a vector space of dimension 3 and let (u, v, w) be a basis of E. Let $\mathcal{P}$ be the affine plane of E generated by these three vectors. The projective plane $P(E)$ is equipped with the projective frame

$$(p(u + v + w), p(u), p(v), p(w)) = (O, A, B, C).$$

Consider $\mathcal{P}$ as a subset of $P(E)$ (if necessary, see Exercise V.5). Prove that (A, B, C) is an affine frame of $\mathcal{P}$ and that the barycenter of the system $((A, \alpha), (B, \beta), (C, \gamma))$ (with $\alpha + \beta + \gamma \neq 0$) is the point of $\mathcal{P}$ that has homogeneous coordinates $[\alpha, \beta, \gamma]$ in the projective frame (O, A, B, C).

Exercise V.44 (The so-called "fundamental" theorem of projective geometry). Let $P(E)$ and $P(E')$ be two real projective spaces of the same dimension $n \geqslant 2$. Let $f : P(E) \to P(E')$ be a bijective mapping such that, if A, B and C are collinear, then $f(A)$, $f(B)$ and $f(C)$ are collinear as well. Prove that f is a projective transformation.

Exercise V.45. In each of the descriptions of $\mathbf{P}_1(\mathbf{C})$ given in § 7, find the real projective line $\mathbf{P}_1(\mathbf{R})$.

Exercise V.46. We want to prove that there is not continuous way to choose an orientation of the lines in an oriented real vector space of dimension $n \geqslant 2$. Prove that such a choice would give a continuous mapping from $\mathbf{P}_{n-1}(\mathbf{R})$ to the unit sphere such that the composition

$$\mathbf{P}_{n-1} \longrightarrow S^{n-1} \xrightarrow{\ p\ } \mathbf{P}_{n-1}$$

is the identity and a homeomorphism from S^{n-1} to $\mathbf{P}_{n-1} \times \{\pm 1\}$. Conclude.

Exercise V.47 (Topology of the real projective plane). We have seen (Exercise V.3) that $\mathbf{P}_1(\mathbf{R})$ is homeomorphic with a circle and (see § 7) $\mathbf{P}_1(\mathbf{C})$ with a sphere of dimension 2. The goal of this exercise is to study $\mathbf{P}_2(\mathbf{R})$—which is much more intricate.

(1) Prove that

$$\mathbf{P}_2(\mathbf{R}) = \left\{(x, y, z) \in \mathbf{R}^3 \mid x^2 + y^2 + z^2 = 1\right\} / (x, y, z) \sim (-x, -y, -z)$$

and that $\mathbf{P}_2(\mathbf{R})$ is the quotient of a hemisphere by the equivalence relation that identifies two antipodal points on the equator (Figure 20). If one tries to imagine[16] this gluing embedded in our 3-dimensional

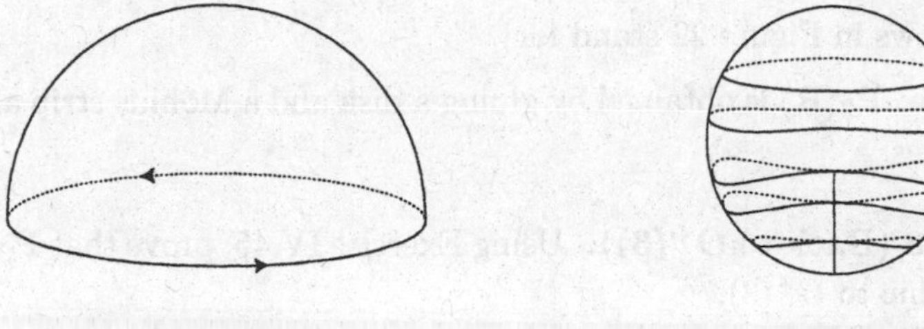

Fig. 20 Fig. 21

space, one can think of Figure 21. Criticize this figure.

[16] See [HCV52] and [Apé87] for beautiful illustrations of the various representations of $\mathbf{P}_2(\mathbf{R})$.

(2) One removes from $\mathbf{P}_2(\mathbf{R})$ the spherical calotte image D in $\mathbf{P}_2(\mathbf{R})$ of

$$\left\{(x,y,z) \;\middle|\; |z| \geqslant \frac{1}{2}\right\}.$$

Check that D is homeomorphic with a disk and prove that $\mathbf{P}_2 - D$ is homeomorphic with

$$M = \mathbf{U} \times \left]-\tfrac{1}{2},\tfrac{1}{2}\right[/ (u,t) \sim (-u,-t).$$

(3) About M. Prove that the first projection

$$\mathbf{U} \times \left]-\tfrac{1}{2},\tfrac{1}{2}\right[\longrightarrow \mathbf{U}$$

defines a continuous mapping $p : M \to \mathbf{U}$, all the "fibers" $p^{-1}(u)$ of which are the open intervals $\left]-\frac{1}{2},\frac{1}{2}\right[$. Hence M is a kind of cylinder. Prove that the image of the circle by

$$\mathbf{U} \times \{0\} \subset \mathbf{U} \times \left]-\tfrac{1}{2},\tfrac{1}{2}\right[\xrightarrow{\;p\;} M$$

is homeomorphic to a circle (this is the image of a $\mathbf{P}_1 \subset \mathbf{P}_2$) but that the complement of this circle in M is connected! One can deduce that M is not homeomorphic to a cylinder; this is a *Möbius strip* (see also Exercise VIII.20).

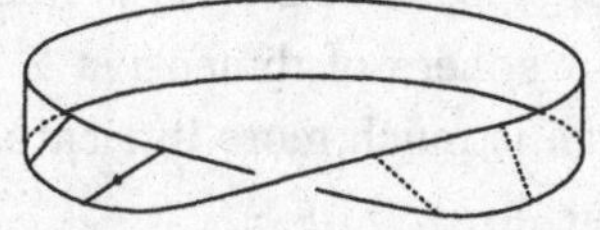

Fig. 22

(4) Prove that M is homeomorphic to the quotient of $[0,\pi] \times \left]-\frac{1}{2},\frac{1}{2}\right[$ by the equivalence relation that identifies $(0,t)$ and $(\pi,-t)$ (this is what the arrows in Figure 22 stand for).

As a conclusion: $\mathbf{P}_2(\mathbf{R})$ is obtained by gluing a disk and a Möbius strip along their boundaries.

Exercise V.48 (Back to $O^+(3)$). Using Exercise IV.45, prove that $\mathbf{P}_3(\mathbf{R})$ is homeomorphic to $O^+(3)$.

Exercise V.49 (Automorphisms of the disk). In this exercise, D denotes the open unit disk

$$D = \{z \in \mathbf{C} \mid |z| < 1\}$$

in C and the word "automorphism" denotes a biholomorphic bijective map. Using Schwarz's lemma(17), prove that any automorphism of D that fixes 0 is a rotation.

If a and b are complex numbers where $|a|^2 - |b|^2 = 1$, prove that the mapping defined by:

$$z \longmapsto \frac{az+b}{\overline{b}z+\overline{a}}$$

is an automorphism of D. If $w \neq 0$ is a point of D, find a projective transformation that preserves D and maps w to 0. Deduce that any automorphism of D is a homography and describe its form.

Find a homography that maps D onto the upper half-plane H. Deduce that any automorphism of H is a projective transformation of the form

$$z \longmapsto \frac{az+b}{cz+d} \quad \text{with } a, b, c \text{ and } d \text{ real and } ad - bc > 0.$$

Exercise V.50 (The Poincaré half-plane). Let $H \subset \mathbf{C}$ be the upper half-plane ($\operatorname{Im} z > 0$). Let "lines" of H denote the half-circles centered on the real axis and the half-lines that are orthogonal to this axis (Figure 23). It is said that two "lines" are parallel if they do not intersect.

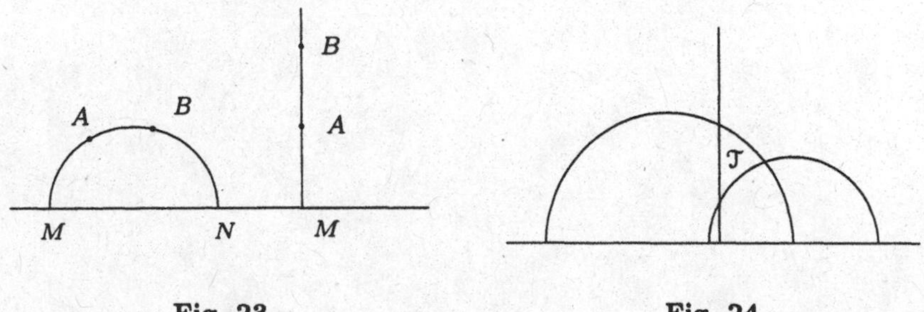

Fig. 23 **Fig. 24**

(1) Check that any two points lie on a unique "line" and that a point not on a "line" lies on an infinite number of parallels to this "line" pass(18).
(2) If $A, B \in H$, the "line" AB is a half-circle of diameter MN or a half-line orthogonal to the x-axis at M (notice that M and N are not points of H). Put

$$d(A,B) = \begin{cases} |\log[A,B,M,\infty]| & \text{in the first case} \\ |\log[A,B,M,N]| & \text{in the second one} \end{cases}$$

(17) See, if necessary [**Car95**] or [**Sil72**].

(18) This model of non-Euclidean geometry, embedded in Euclidean geometry, was invented by Poincaré to prove the independence of the parallel axiom from the other axioms of Euclidean geometry.

(where log denotes the Napierian logarithm). Prove that $d(A, B)$ is a well-defined nonnegative real number (it depends only on A and B) and that d defines a distance on H.

(3) The *angle* of two "lines" denotes their geometric angle in the Euclidean sense. What can be said of the sum of the angles of a triangle $\mathcal{T}$ (Figure 24)?

(4) Prove that the subgroup $PSL(2, \mathbf{R})$ of the group $PGL(2, \mathbf{C})$ of homographies of $\mathbf{C}$ acts on H preserving the distance and angles[19].

(5) Construct a regular hexagon in H all the angles of which are right angles.

---

[19] In Exercise V.49, it is shown that the real homographies are the only homographies that preserve H, it is also possible to prove that these are the only direct isometries of (H, d).

Chapter VI

Conics and Quadrics

This chapter is devoted to quadrics and especially to conics. I have tried to keep a balance between:

- the algebraic aspects: a quadric is defined by an equation of degree 2 and this has consequences...
- the geometric aspects: a conic of the Euclidean plane can be given a strictly metric definition... and this also has consequences.

The two first sections of this chapter are elementary and devoted respectively to these two aspects of affine conics.

For the benefit of the more advanced reader, I then come to projective conics and quadrics.

- I investigate the properties that are related to the algebraic definition, mainly polarity and duality.
- I sketch the relation to homographies, around Pascal's theorem.
- I also show that the affine *and* metric properties of affine conics may be deduced from the projective approach.

In the last section, I show how the results of §§ III-4 and V-7 about pencils of circles and inversions can be expressed very simply by looking at the lines intersecting a suitable quadric in a (suitable) space of dimension 3.

For the convenience of the reader, I include an appendix on quadratic forms and a few "stretching" exercises.

Notations. I will use simultaneously a quadratic form q, its polar form φ and the linear mapping $\widetilde{\varphi} : E \to E^\star$ that it defines (see, if necessary, § 8).

The ground field is $\mathbf{K} = \mathbf{R}$ (almost always) or $\mathbf{C}$.

1. Affine quadrics and conics, generalities

Definition of affine quadrics. Let us start with a (rather fundamental) problem: should a quadric be defined as the set of points satisfying an equation or as this equation itself?

Assume we are in an affine space $\mathcal{E}$. We would like to *define* an affine quadric as the set of points satisfying a degree-2 equation

$$\mathcal{C} = \{M \in \mathcal{E} \mid f(M) = 0\}$$

where f is a *polynomial of degree* 2, which means that there exists a point O, a nonzero quadratic form q, a linear form L_O and a constant c_O such that

$$f(M) = q(\overrightarrow{OM}) + L_O(\overrightarrow{OM}) + c_O.$$

Notice that the form of the expression does not depend on the point O we have chosen: if O' is another point,

$$\begin{aligned} f(M) &= q(\overrightarrow{OO'} + \overrightarrow{O'M}) + L_O(\overrightarrow{OO'} + \overrightarrow{O'M}) + c_O \\ &= q(\overrightarrow{O'M}) + L_{O'}(\overrightarrow{O'M}) + c_{O'}, \end{aligned}$$

where $L_{O'}(\overrightarrow{O'M}) = 2\varphi(\overrightarrow{OO'}, \overrightarrow{O'M}) + L_O(\overrightarrow{O'M})$ and $c_{O'} = q(\overrightarrow{OO'}) + L_O(\overrightarrow{OO'}) + c_O$. This computation also shows that the quadratic part of the polynomial, namely the form q, does not either depend on the point O. The linear form and the constant do depend on O.

Remark 1.1. We have just defined the notion of a degree-2 polynomial, without using any monomial! If, instead of choosing just an origin O, we choose a whole affine frame, the polynomial f will be written, in the associated coordinates $(x_1, \ldots, x_n)$

$$\underbrace{f(x_1, \ldots, x_n)}_{M} = \underbrace{\sum a_{i,j} x_i x_j}_{q(\overrightarrow{OM})} + \underbrace{\sum b_i x_i}_{L(\overrightarrow{OM})} + c.$$

I guess this is what the reader will agree to call a polynomial of degree 2.

The definition we have given is not completely satisfactory with respect to the questions of classification. Consider, for instance, in an affine real plane endowed with an affine frame, the equations:

- $x^2 + y^2 + 1 = 0$, on the one hand,
- $x^2 + 1 = 0$, on the other.

They clearly define the same set of points (the empty set!). However, it is not reasonable to consider that they are equivalent: if one looks at them in a *complex* plane, the first one describes a genuine conic while the other one describes two parallel lines.

This is why it is better[1] to consider the polynomial f rather than the points it defines. Of course, two proportional equations f and λf $(\lambda \neq 0)$ should be considered as equivalent.

Definition 1.2. The equivalence class of a degree-2 polynomial $f : \mathcal{E} \to \mathbf{K}$ for the relation "$f \sim g$ if and only if g is a scalar multiple of f" is called an *affine quadric*. A plane quadric is called a *conic*. The set of points of $\mathcal{E}$ satisfying the equation $f(M) = 0$ is the *image* of the quadric.

Remark 1.3. I insist: the degree of the polynomial f is indeed 2—this means that its quadratic part q is assumed to be nonzero.

Still working in an affine plane endowed with an affine frame, we want to make a difference between equations such as $xy = 0$ (which describes two intersecting lines) or $x^2 = 0$ (which describes a line, said to be a double line) and $x^2 + y^2 - 1 = 0$ or $x^2 - y = 0$ (which describe "genuine" conics).

One could be tempted to require that the quadratic form q that shows up in the equation f be nondegenerate, but this is the case for xy (which we want to avoid) and not for $x^2 - y$ (which we want to allow). The solution[2] is to use an extra variable.

Definition 1.4. The quadric defined by the polynomial

$$f(M) = q(\overrightarrow{OM}) + L(\overrightarrow{OM}) + c$$

in the affine space $\mathcal{E}$ is said to be *proper* if the quadratic form defined on $E \times \mathbf{K}$ by

$$Q(u, z) = q(u) + L(u)z + cz^2$$

is nondegenerate. This quadratic form is called the *homogeneous form* of f.

Remark 1.5. To change the point O amounts to replacing $Q(u, z)$ by

$$Q'(u, z) = Q(u + z\overrightarrow{OO'}, z).$$

The form Q' is thus $Q' = Q \circ \varphi$, where φ is the isomorphism from $E \times \mathbf{K}$ to itself defined by

$$\varphi(u, z) = (u + z\overrightarrow{OO'}, z).$$

The nondegeneracy of Q' is thus equivalent to that of Q (see Remark 8.3); the notion of proper quadric is thus well defined.

---

(1) This kind of problem, although rather elementary here, lies at the foundations of modern algebraic geometry. On an algebraically closed field, as $\mathbf{C}$ is, it is actually equivalent to give the equation or the set of points that satisfy it; this is a simple case of the *Nullstellensatz* (see Exercise VI.39).

(2) Which should not look artificial to the "projective" reader.

Examples 1.6. Let us come back to the above-mentioned examples. For $f = xy$, the homogeneous form is $Q = xy$; for $f = x^2$, this is $Q = x^2$, and both are *degenerate* quadratic forms in the three variables (x, y, z); for $x^2 + y^2 - 1$ we find $x^2 + y^2 - z^2$ and for $x^2 - y$, $x^2 - yz$, and both are nondegenerate.

Intersection of a quadric and a line. Since the equation of a quadric has degree 2, to look for the intersection points of the image of a quadric with a line leads us to solve a degree-2 equation (Figure 1). Consider a line D directed by the vector u and passing through the point A. As the previous computation shows, the equation of the quadric $\mathcal{C}$ may be written (the point A is fixed):

$$q(\overrightarrow{AM}) + L_A(\overrightarrow{AM}) + c_A = 0$$

and the point M lies on the line D if and only if $\overrightarrow{AM} = \lambda u$ for some scalar λ. The intersection consists thus of the points M on the line that are defined by the roots of the equation in λ,

$$\lambda^2 q(u) + \lambda L_A(u) + c_A = 0.$$

The line is contained in the image of the quadric when

$$q(u) = L_A(u) = c_A = 0.$$

They intersect at a unique (simple[(3)]) point when $q(u) = 0$ and $L_A(u) \neq 0$. When $q(u) \neq 0$, the line D intersects $\mathcal{C}$:

- at two points (it is said to be *secant*) if the degree-2 equation has two distinct roots,
- at a unique (double) point if it has a double root, namely if

$$L_A(u)^2 - 4c_A q(u) = 0,$$

- (only for $\mathbf{K} = \mathbf{R}$) at no point at all if this equation has no real root.

Tangent lines to a quadric. In the case where there is a double root, we want to say that the line is *tangent* to the quadric. Let us check that it is indeed tangent, in the sense of differential geometry.

Notice first that, if $A \in \mathcal{C}$, namely if $c_A = 0$, the line through A and directed by u intersects $\mathcal{C}$ at the points M such that

$$\begin{cases} \overrightarrow{AM} = \lambda u \\ \lambda(\lambda q(u) + L_A(u)) = 0 \end{cases}$$

---

(3) There is another intersection point at infinity, as the reader will discover in § 3 and/or in Exercise VI.37.

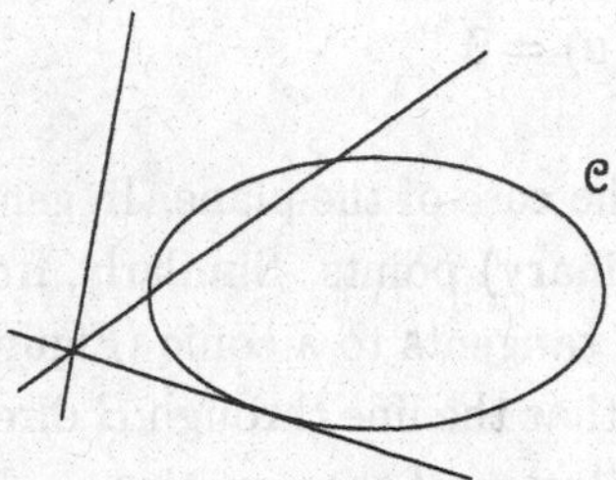

Fig. 1. The intersection of a line and a conic

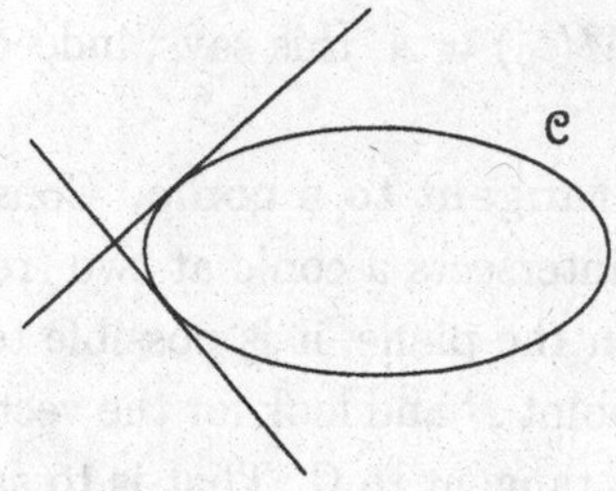

Fig. 2. Tangents to a conic

that is, at A (for $\lambda = 0$) and, if $q(u) \neq 0$, at another point defined by

$$\lambda = -\frac{L_A(u)}{q(u)}.$$

The latter coincides with A if and only if $L_A(u) = 0$.

Proposition 1.7. *A line intersects a quadric $\mathcal{C}$ at a unique double point A or is contained in $\mathcal{C}$ if and only if it is tangent to $\mathcal{C}$ at the point A, in the sense of differential geometry.*

Proof. Let D be the line passing through A and directed by u. Notice first that, for any point M,

$$(df)_M(u) = 2\varphi(\overrightarrow{AM}, u) + L_A(u)$$

using the specific form[4] of f. At $M = A$, we thus have

$$(df)_A(u) = L_A(u).$$

This gives the equivalence $L_A(u) = 0$ if and only if $(df)_A(u) = 0$. The first equality expresses the fact that D intersects $\mathcal{C}$ at the unique double point A. The second one expresses the fact that D is contained in the hyperplane tangent to the quadric $\mathcal{C}$ (see if necessary Proposition VIII-2.10). □

Remark 1.8. For the reader who is not quite familiar with differential geometry in large dimensions, let us, in the case of a conic, identify the kernel of of $(df)_A$ with the set of vectors directing the tangent line at A. Parametrize the conic (in a neighborhood of A) by a differentiable mapping $t \mapsto M(t)$ defined on an interval of $\mathbf{R}$ and such that $M(t)$ is in $\mathcal{C}$ for all t and $M(t_0) = A$. The point $M(t)$ being on $\mathcal{C}$ for all t, $f(M(t))$ is identically zero. Differentiating this relation with respect to t, we find, at t_0,

$$0 \equiv \frac{d}{dt}(f(M(t))) = df_A\left(\frac{dM}{dt}\right) = 2\varphi\left(\overrightarrow{AM(t_0)}, \frac{dM}{dt}\right) + L_A\left(\frac{dM}{dt}\right).$$

(4) Recall—see § 8 if necessary—that $dq_x(y) = 2\varphi(x, y)$.

Since $M(t_0) = A$, this says, indeed, that $L_A(u) = 0$.

Lines tangent to a conic. Consider now the case of the plane. In general, a line intersects a conic at two (real or imaginary) points. Similarly, from a point in the plane, it is possible to draw two tangents to a conic (Figure 2). Fix a point B and look for the vectors u such that the line through B directed by u is tangent to $\mathcal{C}$. That is to say, u is a solution of the equation

$$L_B(u)^2 - 4c_B q(u) = 0.$$

As L_B is a linear form, q a quadratic form and c_B is a constant, the solutions are the vectors u that are isotropic for the *quadratic* form $L_B^2 - 4c_B q$. We are looking for vectors spanning lines, in particular, for nonzero vectors (and two collinear vectors define the same line). In general, there are thus solution lines that may be imaginary (see some comments in Exercises VI.16 and, more intrinsically, VI.65).

Central quadrics. The computation above expressing $f(M)$ in terms of the various points O, O' gives the idea of investigating whether there is a point for which the expression of f contains no linear term: does there exist a point Ω in $\mathcal{E}$ for which

$$f(M) = q(\overrightarrow{\Omega M}) + c'?$$

Remark 1.9. If this was the case, one would have

$$c' = f(\Omega) \text{ and } f(M) - f(\Omega) = q(\overrightarrow{\Omega M}).$$

If M' is the symmetric of M with respect to Ω, one has

$$f(M') - f(\Omega) = q(\overrightarrow{\Omega M'}) = q(-\overrightarrow{\Omega M}) = q(\overrightarrow{\Omega M}) = f(M) - f(\Omega)$$

hence $M \in \mathcal{C}$ if and only if $M' \in \mathcal{C}$: Ω is a center of symmetry for $\mathcal{C}$.

Definition 1.10. A point Ω such that $L_\Omega = 0$ is said to be a *center* of the quadric. When the quadric has a *unique* center, it is said to be a *central quadric*.

Theorem 1.11. *An affine quadric is a central quadric if and only if the quadratic part q of a polynomial that defines it is nondegenerate.*

Remark 1.12. We are indeed speaking of the initial form q, not of the homogeneous form Q.

Proof. Assume a point O is given, so that an equation of $\mathcal{C}$ has the form

$$f(M) = q(\overrightarrow{OM}) + L(\overrightarrow{OM}) + c$$

and (taking the calculation at the beginning of this section into account) we are looking for a point Ω such that

$$2\varphi(\overrightarrow{O\Omega}, \overrightarrow{\Omega M}) + L(\overrightarrow{\Omega M}) = 0.$$

In other words, we are looking for a point Ω and a vector $u = \overrightarrow{O\Omega}$ such that

$$(\forall x \in E) \quad \varphi(u, x) = -\frac{1}{2}L(x)$$

or, in a more abstract, but equivalent way, such that

$$\widetilde{\varphi}(u) = -\frac{1}{2}L \in E^{\star}.$$

This equation has a unique solution u if and only if $\widetilde{\varphi} : E \to E^{\star}$ is injective, that is, if and only q is nondegenerate. □

Remarks 1.13. Notice firstly that this is a simple (although smart) way of using the (abstract) definition of nondegeneracy.

Moreover, this proof gives slightly more precise results than the bare statement of the theorem. It may happen that q is degenerate and L is not in the image of $\widetilde{\varphi}$, in which case there is no center. We will see that this is the case for the parabola, for instance. But it may also happen that L is in the image of $\widetilde{\varphi}$ although q is degenerate, in which case the set of centers is an affine subspace. This is what happens in the case of conics consisting of two parallel lines.

2. Classification and properties of affine conics

In this section, the conics of a real affine plane $\mathcal{E}$ are investigated.

Euclidean classification of affine conics. Assume first[5] that $\mathcal{E}$ is Euclidean. The Euclidean classification of conics relies on the simultaneous orthogonalization of the quadratic form q and the Euclidean scalar product, that is, on the "diagonalization" of q in an orthonormal basis (see if necessary Theorem 8.8).

The first classification result is usually stated as follows:

Proposition 2.1. *A proper central conic (with nonempty image) can be written, in an orthonormal frame with origin at the center:*

– *either* $\dfrac{x^2}{a^2} + \dfrac{y^2}{b^2} = 1$ *(the conic is then an ellipse),*

[5]Once in a while does no harm.

– *or* $\frac{x^2}{a^2} - \frac{y^2}{b^2} = 1$ *(the conic is a hyperbola)*

for two nonnegative real numbers a and b (such that $0 < b \leqslant a$ *if the conic is an ellipse).*

Let us explain why this is indeed a *classification* result: the proposition describes the orbits, in the set of proper central conics, of the action of the group of affine isometries. To say that there exists an orthonormal basis in which the conic has this or that equation is to say that there exists an isometry mapping our conic to the conic defined by this equation. Any proper central conic is thus isometric to an ellipse or a hyperbola of the above form.

Moreover, two proper central conics are isometric if and only if they have the same type with the same corresponding numbers a and b.

Proof of the proposition. We have said that to have a unique center means to have an equation with a nondegenerate quadratic form q. In an orthonormal frame with origin at the center that is orthogonal for the quadratic form, the equation of the conic has the form

$$\alpha x^2 + \beta y^2 + c = 0,$$

with α and β nonzero. The conic being proper, the homogeneous quadratic form Q, which is written

$$\alpha x^2 + \beta y^2 + cz^2,$$

is nondegenerate, hence c is nonzero. It is thus possible to write the equation in the form

$$-\frac{\alpha}{c}x^2 - \frac{\beta}{c}y^2 = 1.$$

If the two coefficients are negative, the image of the conic is empty, if they are nonnegative, we have indeed the equation of an ellipse (Figure 3); if they have opposite signs, this is a hyperbola (Figure 4). □

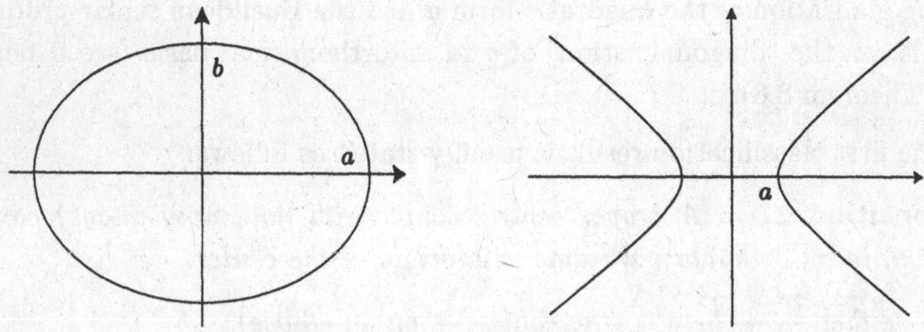

Fig. 3. An ellipse **Fig. 4.** A hyperbola

Remarks 2.2

– If $c = 0$, the conic is not proper, it contains its own center, and the equation is $\alpha x^2 + \beta y^2 = 0$, which is easily put in the form:
  - either $\dfrac{x^2}{a^2} + \dfrac{y^2}{b^2} = 0$, in the case where the image is a point,
  - or $\dfrac{x^2}{a^2} - \dfrac{y^2}{b^2} = 0$, in the case where it consists of two secant lines.

– It should be noticed that an ellipse is compact, while a hyperbola has infinite branches. The asymptotes of a hyperbola of equation $\dfrac{x^2}{a^2} - \dfrac{y^2}{b^2} = 1$ are the two secant lines defined by the equation

$$\frac{x^2}{a^2} - \frac{y^2}{b^2} = 0$$

(see Exercise VI.38... another opportunity to advertise the projective section of the present chapter!).

– An ellipse is connected (for instance because this is the image of the connected space $\mathbf{R}$ under the continuous mapping $t \mapsto (a\cos t, b\sin t)$) but a hyperbola has two connected components (why?).

The complement of a point in an ellipse is still connected (why?). From this, we deduce that an ellipse is completely contained in one of the closed half-planes defined by any of its tangent lines: a tangent line intersects the conic only at its contact point, as was noticed in § 1. Let us write the plane as a disjoint union

$$\mathcal{P} = \mathcal{P}_1 \cup \mathcal{D} \cup \mathcal{P}_2$$

where $\mathcal{D}$ is the tangent to the ellipse at a point A and $\mathcal{P}_1$, $\mathcal{P}_2$ are the two open half-planes it determines. Then the complement of A in $\mathcal{C}$ is the disjoint union

$$\mathcal{C} - \{A\} = (\mathcal{C} \cap \mathcal{P}_1) \cup (\mathcal{C} \cap \mathcal{P}_2).$$

The complement $\mathcal{C} - \{A\}$ being connected, it is contained in one of the open half-planes.

Consider now the case where the quadratic form q is degenerate. Its rank must be equal to 1. There is an orthonormal frame in which the equation of the conic is

$$aY^2 + \alpha X + \beta Y + \gamma = 0 \text{ with } a \neq 0.$$

Putting $x = X$, $y = Y + \dfrac{\beta}{2a}$, $c = \gamma - \dfrac{\beta^2}{4a}$ (namely, changing the origin), this equation becomes

$$ay^2 + \alpha x + c = 0.$$

Let us identify the plane E with its dual using the orthonormal basis. Then the linear mapping

$$\widetilde{\varphi} : \mathbf{R}^2 \longrightarrow \mathbf{R}^2$$

associated with the quadratic form $q(x, y) = ay^2$ is

$$\widetilde{\varphi}(x, y) = (0, ay).$$

As for the linear form $L(x, y) = \alpha x$, this is $L = (\alpha, 0) \in \mathbf{R}^2$.

Thus L is in the image of $\widetilde{\varphi}$ if and only if $\alpha = 0$.

- In this case ($\alpha = 0$), the equation is $ay^2 + c = 0$, all the points of the x-axis are centers of symmetry and the conic consists of:
  - two parallel lines if $-c/a > 0$,
  - a (double) line if $c = 0$
  - and it is empty if $-c/a < 0$.

  The form Q can be written $ay^2 + cz^2$ and, of course, the conic is *not* proper.
- In the opposite case (namely, when $\alpha \neq 0$), the conic is proper, it has no center of symmetry at all and, up to a change of origin, it has an equation of the form

  $$y^2 = bx \text{ (where it can be assumed that } b > 0).$$

  This is a *parabola* (Figure 5).

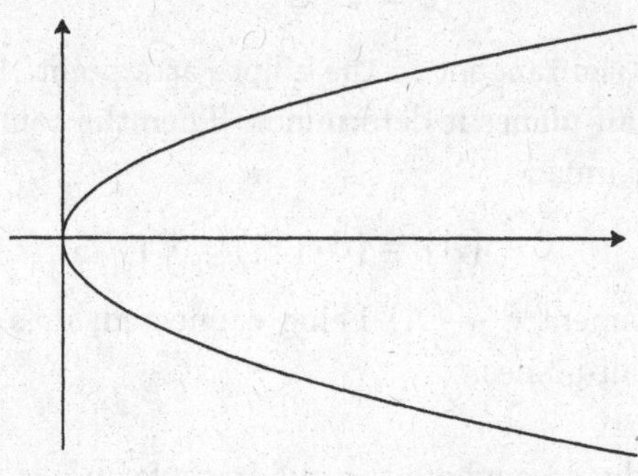

Fig. 5. A parabola

Proposition 2.3. *An equation in an orthonormal frame of a proper noncentral conic with nonempty image is* $y^2 = 2px$ *for some nonnegative real number* p. □

A little bit of terminology. The equations

$$\frac{x^2}{a^2}+\frac{y^2}{b^2}=1, \qquad \frac{x^2}{a^2}-\frac{y^2}{b^2}=1, \qquad y^2=2px$$

are said to be *reduced*. The nonnegative numbers $2a$, $2b$, p are respectively the *axes*[(6)] (case of central conics) and more precisely the *major* and the *minor* axis (ellipse) and the *parameter* (parabola). The origin of the frame in which the reduced equation of a parabola is written is the *vertex* of this parabola.

When $a = b$, an ellipse is a circle, and a hyperbola is said to be *rectangular* (this is the case where the asymptotes are orthogonal).

Any proper conic with nonempty image is in one (and only one) of the families "ellipses", "hyperbolas", or "parabolas", called its *type*.

Affine classification. An immediate consequence of Propositions 2.1 and 2.3 is:

Corollary 2.4. *Let $\mathcal{C}$ be a proper conic with nonempty image in an affine plane. There exists an affine frame in which an equation of $\mathcal{C}$ has one (and only one) of the forms*

$$x^2+y^2=1 \text{ (ellipse)}, \quad x^2-y^2=1 \text{ (hyperbola) or } \quad y^2=x \text{ (parabola)}.$$

The reader now should have understood that this is indeed an affine classification:

Corollary 2.5. *Let $\mathcal{C}$ and $\mathcal{C}'$ be two proper conics with nonempty images. There exists an affine transformation mapping $\mathcal{C}$ to $\mathcal{C}'$ if and only if $\mathcal{C}$ and $\mathcal{C}'$ have the same type.* □

Remarks 2.6

- As this classification result shows, for the proper real conics, to give the image is equivalent to giving an equation. We shall thus feel free to use "ellipse" *etc.* to denote the corresponding subsets of the plane.
- The type of the conic describes its topological type. In other words, the expression "affine transformation" can be replaced by "homeomorphism" in this statement: the ellipses are the compact conics, the parabolas are the connected, but noncompact conics, and the hyperbolas are the neither connected nor compact conics.

---

(6) This terminology is classical. It is not excellent: this is rather the length of a segment on an axis than an axis.

– From the affine viewpoint, in a Euclidean plane, there is no difference between an ellipse and a circle. This remark is often used to prove properties of an ellipse starting from the case where the ellipse is a circle (see, *e.g.*, Exercises VI.19 and VI.20).

Focus-directrix definition. In the Euclidean plane, conics can be given strictly metric descriptions.

Proposition 2.7. *For any proper conic with nonempty image that is not a circle, there exists a point F called focus, a line D not containing F, called directrix and a nonnegative real number e called the eccentricity such that the conic is the set of points M satisfying $FM = ed(M, D)$. Conversely, given F, D and e, the set of points M in the plane such that $FM = ed(M, D)$ is a proper conic, an ellipse if $e < 1$, a parabola if $e = 1$, or a hyperbola if $e > 1$.*

Proof. Let a point F and a line D not through F be given. Choose an orthonormal frame such that D is parallel to the y-axis and F is on the x-axis. The coordinates of the point F are $(c, 0)$, and an equation of the line D is $x = h$. To say that $FM = ed(M, D)$ is to say that the coordinates (x, y) of M satisfy

$$(x-c)^2 + y^2 = e^2(x-h)^2,$$

which is indeed the equation of a conic. Notice that the x^2 term vanishes if $e = 1$. This case must thus be considered separately.

– If $e = 1$, the origin O is chosen so that $h = -c$; we get $y^2 = 4cx$, and the conic is a parabola of parameter $2|c|$ and vertex O.
– If $e \neq 1$, we get similarly $h = c/e^2$, which puts our equation in the form

$$\frac{x^2}{(c^2/e^2)} + \frac{y^2}{((1-e^2)c^2/e^2)} = 1.$$

This is an ellipse if $e < 1$, a hyperbola if $e > 1$.

This shows that the data F, D and e determine, indeed, a proper conic. Conversely, it is easy to put the reduced equation of a conic (when the conic is not a circle) in one of the forms obtained above, thus allowing us to determine F, D and e satisfying the expected property. □

Notice also that F belongs to a symmetry axis of the conic (thus this axis is called the *focal axis*), and that D is orthogonal to this axis.

Remark 2.8. By symmetry, proper central conics that are not circles have two foci (located on the major axis in the case of an ellipse) and two parallel directrices.

Bifocal properties of central conics. It is in fact possible to describe ellipses and hyperbolas in terms of the two foci. Notice first that an ellipse is contained in the strip determined by its two directrices, while a hyperbola lies outside this strip: the previous calculation shows that the focus and the center are on the same side of the directrix in the ellipse case and on the two different sides in the hyperbola case.

Proposition 2.9. *An ellipse with foci F and F' is the set of points M such that $MF + MF' = 2a$ for some nonnegative real number a such that $2a > FF'$.*

A hyperbola with foci F and F' is the set of points M such that $|MF - MF'| = 2a$ for some nonnegative real number a such that $2a < FF'$.

Proof. I give the proof in the case of ellipses only, leaving the case of hyperbolas to the reader. If $\mathcal{C}$ is an ellipse with foci F and F' and if D and D' are the corresponding two directrices,

$$MF + MF' = e(d(M,D) + d(M,D')) = ed(D,D')$$

the ellipse being located between the two directrices. Put $a = ed(D,D')/2$. Up to now, we have proved that $\mathcal{C}$ is contained in

$$\mathcal{C}' = \{M \mid MF + MF' = 2a\}$$

with $2a > FF'$.

Consider, conversely, the set $\mathcal{C}'$. Let us prove that this is indeed the ellipse $\mathcal{C}$. For this, it is enough to prove that $\mathcal{C}'$ is a conic. Choose a frame whose origin is the midpoint of FF' and whose x-axis is the line FF', so that the points F and F' have coordinates $(c,0)$ and $(-c,0)$ (with $c > 0$). We have

$$MF^2 + MF'^2 = 2(x^2 + y^2 + c^2) \text{ and } MF'^2 - MF^2 = 4cx.$$

If the point M is such that $MF + MF' = 2a > 0$, it must satisfy the equality

$$MF' - MF = \frac{2c}{a}x.$$

This is solved in

$$MF = a - \frac{c}{a}x \text{ and } MF' = a + \frac{c}{a}x,$$

so that we have

$$\left(a - \frac{c}{a}x\right)^2 + \left(a + \frac{c}{a}x\right)^2 = 2(x^2 + y^2 + c^2),$$

an equation that can also be written

$$(a^2 - c^2)\left(\frac{x^2}{a^2} - 1\right) + y^2 = 0.$$

This is indeed the equation of a conic (and the conic must be an ellipse, due to the inequality $a > c$). □

Remark 2.10. This definition of the ellipse gives a practical method of drawing it using a pencil and a string between the two foci (Figure 6).

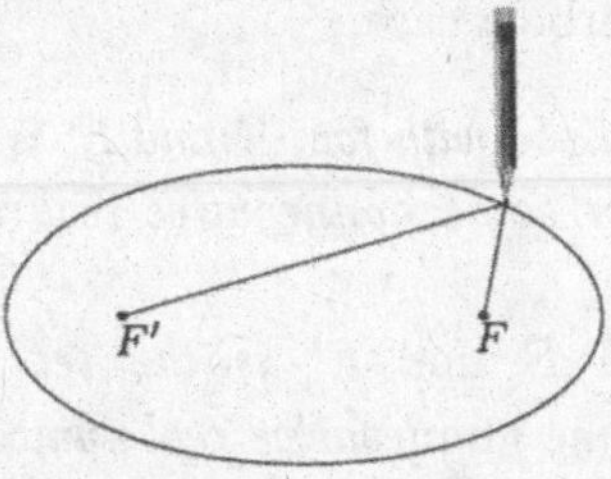

Fig. 6

Bisectors and tangents. If F is a focus of $\mathcal{C}$ and if M and N are two points of $\mathcal{C}$, the bisectors of the angle (FM, FN) are obtained very simply, as is expressed in the next proposition (see Figure 7).

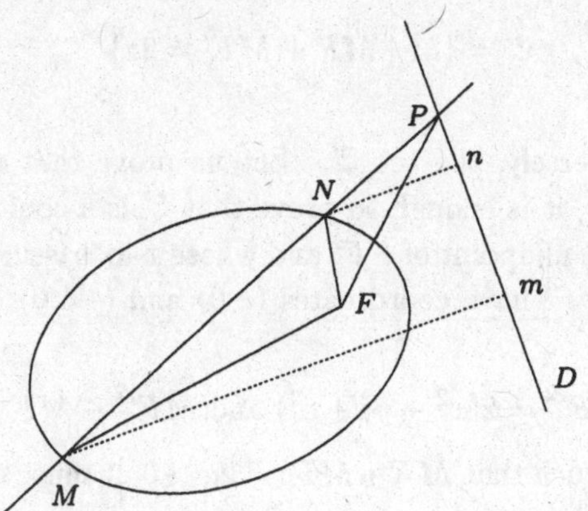

Fig. 7

Proposition 2.11. *Let $\mathcal{C}$ be a nonempty proper conic of focus F and directrix D. Let M and N be two points of $\mathcal{C}$. Assume the line MN intersects D at P. Then the line PF is a bisector of the angle (of lines) (FM, FN).*

Proof. Let m and n be the orthogonal projections of M and N onto D. The lines Mm and Nn are parallel and Thales' theorem gives

$$\frac{PM}{PN} = \frac{Mm}{Nn}.$$

By the definition of focus and directrix,

$$\frac{Mm}{Nn} = \frac{MF}{NF},$$

so that

$$\frac{PM}{PN} = \frac{FM}{FN}.$$

We deduce[7] that PF is one of the bisectors of the angle (FM, FN). □

Remark 2.12. If MN is parallel to D, the line PF can be replaced by the parallel to D through F. The result still holds (by a symmetry argument).

If N tends to M, the secant MN tends to the tangent to $\mathcal{C}$ at M. If the latter intersects D at a point P, the line PF is a bisector of the angle (FM, FM), meaning that it is orthogonal to FM. This is what the next proposition asserts.

Proposition 2.13. *Let $\mathcal{C}$ be a proper conic that is not a circle, let F be a focus of $\mathcal{C}$ and let D be the corresponding directrix. If M is a point of $\mathcal{C}$, let P be the intersection point of the perpendicular to MF at F with the line D. Then the line PM is tangent to $\mathcal{C}$ at M.* □

For the reader who does not appreciate the limit process as much as she should, let us sketch another proof[8].

Another proof of Proposition 2.13. The line PM already intersects $\mathcal{C}$ at M. To prove that it is tangent to $\mathcal{C}$, let us check that it contains no other point of $\mathcal{C}$. Assume N is a point of PM distinct from M. Let m and n be the orthogonal projections of M and N on D, N' the orthogonal projection of N on PF. If e is the eccentricity of $\mathcal{C}$, we have

$$e = \frac{MF}{Mm} = \frac{NN'}{Nn} \text{ thanks to Thales' theorem}$$

and $NN' < NF$ (Figure 8). We deduce the inequality

$$\frac{NF}{Nn} > e,$$

therefore the point N cannot lie on $\mathcal{C}$.

We still have to check that this unique point M is not a simple intersection point of the line PM with $\mathcal{C}$.

- In the case where $\mathcal{C}$ is an ellipse, there are no such secant lines, and the proof is thus complete.

---

[7] See if necessary Exercises III.26 or V.21.

[8] I am indebted to Daniel Perrin for this proof. See also [DC51] for a more geometric argument.

- In the case where $\mathcal{C}$ is a parabola, these secants are the lines parallel to the symmetry axis, that is, the lines orthogonal to D. But, if PM is orthogonal to D, the triangle PMF is right-angled at F *and* isosceles with vertex M (since $MF = MP$), which is impossible.
- The case where $\mathcal{C}$ is a hyperbola is a little more intricate, so I leave it to the reader (Exercise VI.32).

□

Remark 2.14. Here again, the line MF may be orthogonal to D (the point P may be at infinity). The line PM is replaced by the parallel to D passing through M.

Corollary 2.15. *If P is a point of the directrix D, the contact points with $\mathcal{C}$ of the tangents through P are collinear with F.* □

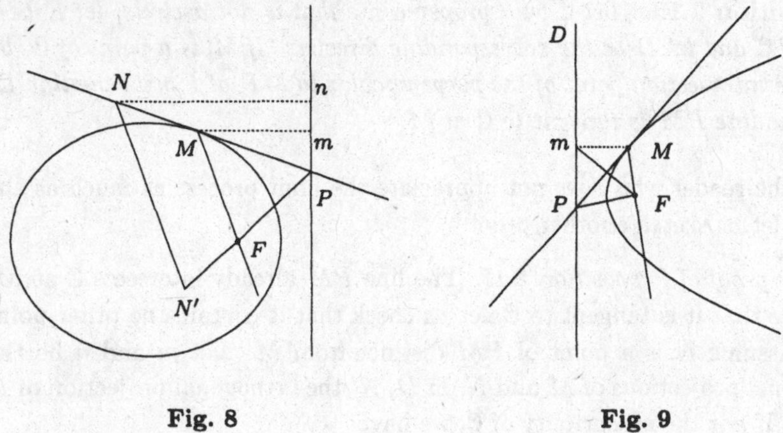

Fig. 8 **Fig. 9**

A useful property of the parabola and its tangents is derived. It gives a method to construct the tangent to a parabola at a given point (Figure 9).

Corollary 2.16. *Let $\mathcal{P}$ be a parabola with focus F and directrix D. The tangent to $\mathcal{P}$ at one of its points M is the perpendicular bisector of the segment Fm, where m is the orthogonal projection of M on D.*

Proof. The tangent to $\mathcal{P}$ at M intersects D at P; one has $PF \perp FM$ according to what we have just done. Thus F is the symmetric of m with respect to MP, which is indeed the perpendicular bisector of Fm. □

Corollary 2.17. *The orthogonal projection of the focus of a parabola on any tangent also lies on the tangent to the parabola at the vertex.*

Proof. With the same notation as above, the projection of F on the tangent at M is the image of m by the dilatation of center F and ratio $1/2$. But the image of the directrix by this dilatation is the tangent at the vertex. □

Here is now an application to central conics: the tangent at M is a bisector of the angle (MF, MF') (Figure 10).

Corollary 2.18. *Let $\mathcal{C}$ be a proper central conic with foci F and F'. The tangent to $\mathcal{C}$ at M is the internal (resp. external) bisector of the angle at M of the triangle MFF' if $\mathcal{C}$ is a hyperbola (resp. an ellipse).*

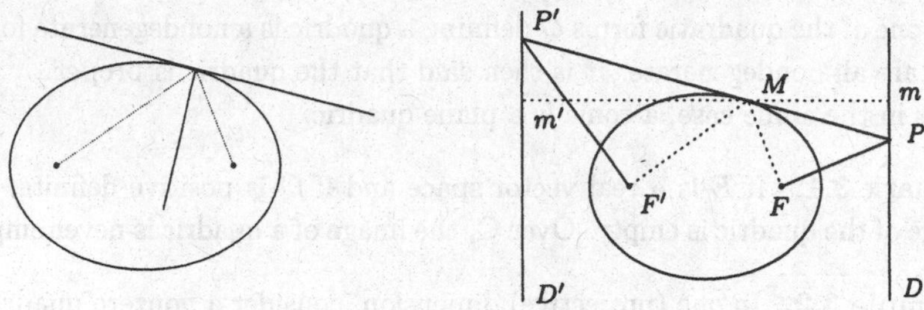

Fig. 10 **Fig. 11**

Proof. Let D and D' be the directrices associated with the foci F and F' and let P and P' be the intersection points of the tangent at M with D and D', m and m' the projections of M on D and D' (Figure 11).

We know that $PF \perp MF$ and similarly that $P'F' \perp MF'$. Moreover,

$$\frac{MF'}{MF} = \frac{Mm'}{Mm} = \frac{MP'}{MP}$$

so that the right-angled triangles MFP and $MF'P'$ are similar[(9)], whence the angle inequality.

As we have said above, the ellipse is contained in one of the closed half-planes defined by the tangent. The segment FF' is contained in the interior of the ellipse and the tangent cannot intersect it. This is thus the external bisector, due to Proposition III-1.13. The reader is kindly requested to check, similarly, that we get the internal bisector in the case of a hyperbola. □

Remark 2.19. These tangential properties are the origin of numerous envelope problems (see Chapter VII and its exercises).

---

(9) See the criteria for similarity of triangles in Exercise III.44.

3. Projective quadrics and conics

Consider now a real or complex projective space $P(E)$ of dimension n. As for Chapter V, one could use [**Sam88**], [**Ber77**] and [**Sid93**].

Definition of projective quadrics. A *quadric* in $P(E)$ is a nonzero quadratic form on E, up to multiplication by a nonzero scalar. In other words, a quadric is an element of the projective space $PQ(E)$ associated with the vector space $Q(E)$ of quadratic forms on E.

If Q is a quadratic form and $Q^{-1}(0) \subset E$ its isotropy cone, the *image* of the quadric is $p(Q^{-1}(0) - \{0\})$. It does not change when Q is replaced by λQ ($\lambda \neq 0$), in other words it depends only on the quadric defined by Q.

If one of the quadratic forms Q defining a quadric is a nondegenerate form, they are all nondegenerate. It is then said that the quadric is *proper*.

As in the affine case, a conic is a plane quadric.

Remark 3.1. If E is a real vector space and if Q is positive definite, the image of the quadric is empty. Over $\mathbf{C}$, the image of a quadric is never empty.

Example 3.2. In one (projective) dimension, consider a nonzero quadratic form on a 2-dimensional vector space. Its rank can only be 1 or 2. Thus it is equivalent either to x^2, or to $x^2 \pm y^2$. The image of the quadric is a point when the rank is 1, empty (only in the real case) or consists of two points when the rank is 2.

Affine vs projective. If $P(E)$ is the projective completion of an affine space $\mathcal{F}$ and if $\mathcal{F}$ is described by the equation $z = 1$ in $P(E)$ (the vector space E is $F \times \mathbf{K}$, as in § V-3), any affine quadric of equation

$$q(\overrightarrow{OM}) + L(\overrightarrow{OM}) + C = 0$$

defines a projective quadric, of equation

$$Q(u, z) = q(u) + L(u)z + Cz^2 = 0$$

(with the notation of § 1): we have simply *homogenized* the equation of the affine quadric. The affine quadric is proper if and only if the projective quadric is. The points (u, z) of the projective quadric are the points $(u, 1)$ such that u is a point of the affine quadric and the points (at infinity) $(u, 0)$ such that u is an isotropic vector for q.

Figure 12 shows the image of the affine quadric $\mathcal{C}$ in $\mathcal{F}$ and the isotropy cone $Q^{-1}(0)$ in E.

Conversely, if the quadratic form Q is the equation of a projective quadric in the projective completion of an affine space $\mathcal{F}$ and if it is not the product

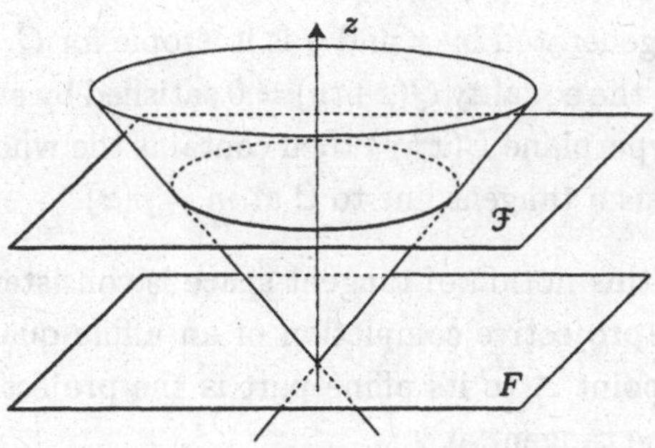

Fig. 12

of z by a linear form[10], it defines an affine quadric in $\mathcal{F}$, thanks to the reverse operation: one *dehomogenizes* the form Q by "making $z = 1$".

We will come back to these manipulations to deduce the properties of affine conics (and even of affine Euclidean conics) from those of projective conics in § 5.

Intersection of a line and a quadric. For the same reason as an affine line (see § 1), either a projective line intersects a projective quadric at zero, one or two points, or it is contained in this quadric.

Proof. Let x and y be two vectors generating a plane in the vector space E. The intersection of the projective line they determine with the quadric is obtained by solving the equation

$$Q(x+ty) = Q(x) + 2t\Phi(x,y) + t^2Q(y) = 0$$

(where Φ is, of course, the polar form of Q), which is an equation of degree 2 in t. □

When this equation is satisfied for all t, the plane generated by x and y is isotropic for Q and the projective line is contained in the quadric. Over $\mathbf{C}$, all the quadrics of dimension $\geqslant 2$ contain lines (see Exercise VI.41).

Tangent spaces. If $\mathcal{C}$ is a proper quadric defined by a quadratic form Q, then for all x in E, the orthogonal subspace $x^{\perp_Q}$ is a hyperplane. The projective hyperplane $P(x^{\perp_Q})$ is called the *tangent hyperplane* to $\mathcal{C}$ at the point $m = p(x)$.

It is said that a projective subspace is tangent to $\mathcal{C}$ at m if it is contained in the tangent hyperplane.

A line contained in a quadric is, in particular, tangent to this quadric.

[10] If Q is the product of z by a linear form, the projective quadric is degenerate (see Exercise VI.3) and contains the hyperplane at infinity.

Proof. If the plane generated by x and y is isotropic for Q, it is contained in the orthogonal of x: the equality $Q(x+ty)=0$ satisfied by every t implies that $\Phi(x,y)=0$. The hyperplane $P(x^{\perp_Q})$ then contains the whole line defined by x and y which is thus a tangent line to $\mathcal{C}$ at $m=p(x)$. □

Notice also that this notion of tangent space is consistent with the affine notion. If $\mathcal{C}$ is the projective completion of an affine quadric, the tangent hyperplane at the point A to its affine part is the projective completion of the affine hyperplane tangent at A.

Proof. An equation of the affine quadric is $f(M)=0$; we choose the origin of the affine plane at the point A of the quadric to write

$$f(M)=q(\overrightarrow{AM})+L(\overrightarrow{AM}).$$

The projective completion of the quadric is defined by the quadratic form

$$Q(u,z)=q(u)+zL(u).$$

The polar form of Q is the bilinear form Φ, defined by

$$\Phi((u,z),(u',z'))=\varphi(u,u')+\frac{1}{2}(zL(u')+z'L(u)).$$

The orthogonal for Q of the vector $(0,1)$ corresponding to A is thus the subspace

$$H=\left\{(v,z)\mid\frac{1}{2}L(v)=0\right\}$$

and this is the projective hyperplane obtained by completion of the affine hyperplane tangent to the affine quadric at A (the kernel of L, see the proof of Proposition 1.7) as we wanted to prove. □

Remark 3.3. Coming back to the intersection of a line and a quadric considered above, notice that a line is tangent to a quadric if and only if its intersection with this quadric consists of either one or more than three points.

Duality and polarity. We are going now to understand geometrically the notion of orthogonality with respect to a quadratic form. In this part, we will consider only nondegenerate quadratic forms (alias proper quadrics).

A nondegenerate quadratic form Q defines an isomorphism

$$\widetilde{\varphi}:E\longrightarrow E^\star$$

and thus a projective transformation

$$\psi:P(E)\longrightarrow P(E^\star)$$

that depends only on the class of Q in $PQ(E)$, that is, on the quadric defined by Q.

The mapping ψ is called the *polarity* with respect to the quadric defined by Q. For example, if m is a point of $P(E)$, $\psi(m)$ is a point of $P(E^\star)$, that is, a hyperplane of E, which is called the polar hyperplane of m (the polar (line) in the case of conic) and we denote it by $m^\perp$.

This is indeed orthogonality with respect to Q (it is cumbersome and useless to make the reference to Q appear in the notation as long as a single quadric is in question): if m and n are points of $P(E)$ with $m = p(x)$, $n = p(y)$, the equation $\varphi(x, y) = 0$ describes the orthogonality of x and y with respect to Q. It is said that m is orthogonal to n (notation $m \perp n$) if $\varphi(x, y) = 0$. The hyperplane $m^\perp$ consists of the (vectorial) lines orthogonal to m.

Similarly, if H is a hyperplane, $H^\perp$ is a point, the *pole* of H with respect to the quadric. The reader should already understand that the pole of the polar hyperplane of m is m, *etc.*

Example 3.4. Assume that E is a Euclidean vector space, or, if one prefers, that Q is a (real) quadratic form which is positive definite. The image of the quadric defined by Q is the set of the nonzero vectors x satisfying the equation

$$Q(x) = 0.$$

This image is empty. This quadric is nevertheless a very interesting object to consider, since the polarity with respect to it is everyday orthogonality... This example is important enough, as it is the source of all the metric properties of the Euclidean conics, as will be seen later.

Remarks 3.5

(1) If m is a point of the conic $\mathcal{C}$, we have already said that $m^\perp$ is the tangent hyperplane to $\mathcal{C}$ at m.

(2) In dimension 1. If D is a projective line and $\mathcal{C}$ a proper quadric with nonempty image in D, it consists of two points a and b. Then m is orthogonal to n if and only if the cross-ratio $[a, b, m, n]$ equals -1.

Proof. This is a simple calculation in the (linear) plane from which our projective line D comes. We choose a basis so that the points a and b are the images of the basis vectors. The quadric is defined by a nondegenerate quadratic form in two variables x and y, for which the two vectors $(1, 0)$ and $(0, 1)$ are isotropic. This is thus a multiple of the form $Q(x, y) = xy$.

The points m and n are the images of the vectors of coordinates (x, y) and (x', y') and the cross-ratio is

$$[a, b, m, n] = \frac{y}{x} \Big/ \frac{y'}{x'}.$$

Its value is -1 if and only if the vectors (x, y) and (x', y') satisfy the equality

$$xy' + yx' = 0,$$

that is, if and only if they are orthogonal with respect to Q. □

(3) The previous remarks can be used to construct the polar hyperplane of the point m. Let us concentrate on the case of conics.

Assume that two tangents to the conic $\mathcal{C}$ can be drawn through the point m. Let p_1 and p_2 be the two contact points. The point m lies on the tangents at p_1 and p_2: $m \in p_1^\perp \cap p_2^\perp$, hence p_1 lies in $m^\perp$, and similarly $p_2 \in m^\perp$. The line p_1p_2 is thus the polar of m with respect to $\mathcal{C}$ (Figure 13).

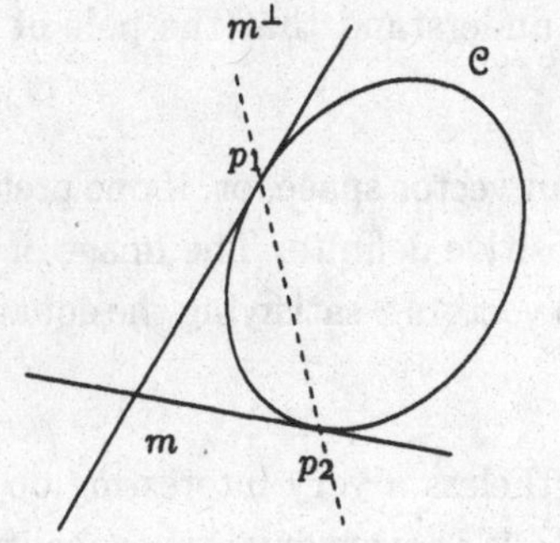

Fig. 13

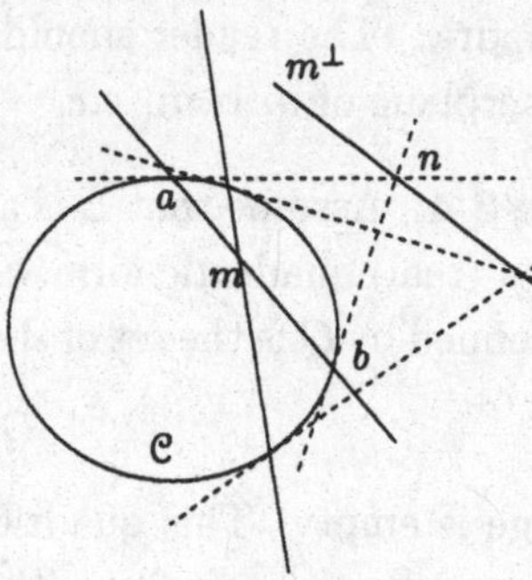

Fig. 14

(4) More generally[11], let us draw a line through m and intersecting the conic at two points a and b. The tangents at a and b intersect at a point n. According to what we have just seen, the line ab is the polar of the point n. Hence n is the pole of the line ab and the polar of m passes through n (Figure 14).

(5) It is also possible to use secants in place of tangents to construct the polar of the point m. See Exercise VI.46.

Projective classification. The projective group acts on the set of quadrics. Let $f : P(E) \to P(E)$ be a projective transformation, coming from an isomorphism $g : E \to E$ and let $\mathcal{C}$ be a quadric of equation Q in $P(E)$. Then $Q \circ g^{-1}$ is also a nonzero quadratic form. Denote $f(\mathcal{C})$ its class in the space of quadrics (it is clear that it depends only on $\mathcal{C}$ and not on the choice of the quadratic form Q and similarly that it depends only on the projective transformation f and not on the choice of the linear mapping g). Notice that "the image of $f(\mathcal{C})$ is the image by f of the image of $\mathcal{C}$".

---

(11) Because, in the real case, the tangents through m might not be real.

Proof. The point n is in the image of $f(\mathcal{C})$ if and only if it comes from a (nonzero) vector v such that $Q \circ g^{-1}(v) = 0$. Let m be the image in $P(E)$ of the unique nonzero vector u of E such that $g(u) = v$. Hence $n = f(m)$. The equality

$$Q \circ g^{-1}(v) = 0$$

is equivalent to $Q(u) = 0$, that is, to the fact that m is in the image of $\mathcal{C}$. We have proved that n is in the image of $f(\mathcal{C})$ if and only if it can be written $f(m)$ for some point m in the image of $\mathcal{C}$, that is, we have proved the equality

$$\operatorname{Im}(f(\mathcal{C})) = f(\operatorname{Im}(\mathcal{C})).$$

□

Proposition 3.6. *The orbits of the quadrics of the n-dimensional projective space $P(E)$ under the action of $PGL(E)$ are indexed:*

- *by the rank of an equation if* $\mathbf{K} = \mathbf{C}$,
- *by the ordered pairs (s, r) with $0 \leqslant r \leqslant s \leqslant n+1$ and $1 \leqslant r+s \leqslant n+1$ if* $\mathbf{K} = \mathbf{R}$.

Proof. This is an immediate consequence of the classification of the quadratic forms (see, if necessary, Corollary 8.2) together with the fact that, over $\mathbf{R}$ and for $\lambda < 0$, the signature of λQ is (r, s) when that of Q is (s, r). □

Consider for instance the case of real projective *proper* quadrics. The rank of the quadratic form is $n+1$ thus $s+r = n+1$. We find $\left[\dfrac{n+1}{2}\right]+1$ types of quadrics. For instance:

- for $n = 2$, the pairs (s, r) under consideration are $(0, 3)$ (in which case the image of the quadric is empty) and $(1, 2)$, where an equation may be chosen of the form $x^2 + y^2 - z^2 = 0$ (proper conic). The simplicity of this classification, compared to the affine classification (ellipses, parabolas, hyperbolas), should be noticed.
- for $n = 3$, we find similarly $(0, 4)$ (empty image), $(1, 3)$ (equation $x^2 + y^2 + z^2 - t^2 = 0$) and $(2, 2)$ (equation $x^2 + y^2 - z^2 - t^2 = 0$).

Remark 3.7. The total number of orbits is:

- in the complex case, $n+1$,
- in the real case, $(n^2 + 6n + 4)/4$ if n is even, $(n^2 + 6n + 5)/4$ if n is odd.

Pencils of quadrics. We shall now consider some projective subspaces of the space $PQ(E)$ of quadrics.

Proposition 3.8. *Let m be a point of $P(E)$. The set of all quadrics whose image contains m is a hyperplane of $PQ(E)$.*

Proof. Denote by $H(m)$ the set of quadrics whose image contains m, namely

$$H(m) = \{\mathcal{C} \in PQ(E) \mid m \in \operatorname{Im}(\mathcal{C})\}.$$

If $m = p(x)$ and $\mathcal{C} = p(Q)$, to say that $m \in \operatorname{Im}(\mathcal{C})$, this is to say that $Q(x) = 0$. The vector x being given, this equation is indeed linear in Q (and nonzero because $x \neq 0$). □

Corollary 3.9. *Let $m_1, \ldots, m_k$ be points of $P(E)$. The set of quadrics through the points $m_1, \ldots, m_k$ is a projective subspace of dimension at least equal to $\dfrac{n(n+3)}{2} - k$. In particular, $\dfrac{n(n+3)}{2}$ points always lie on at least one quadric.* □

For example, five points of the plane always lie on a conic, nine points of the space of dimension 3 always lie on a quadric.

A line in the space $PQ(E)$ is called a *pencil* of quadrics. A pencil may be given by two of its points: given two distinct quadrics $\mathcal{C}$ and $\mathcal{C}'$ of equations Q and Q' (assumed to be nonproportional), the set of quadrics of equations $\lambda Q + \lambda' Q'$ is a pencil.

The *base locus* of the pencil is the set of points that lie on all the quadrics of the pencil.

Remark 3.10. Given a pencil of quadrics $\mathcal{F}$ and a point m in $P(E)$, the point m always lies on a quadric of $\mathcal{F}$: we are actually considering the intersection in the projective space $PQ(E)$ of the line $\mathcal{F}$ with the hyperplane $H(m)$... an intersection which is always nonempty.

Proposition 3.11. *Either all the quadrics of pencil $\mathcal{F}$ are degenerate or there are at most $n+1$ degenerate quadrics in $\mathcal{F}$. If $\mathbf{K} = \mathbf{C}$, there is always a degenerate quadric in $\mathcal{F}$.*

Proof. Let Q and Q' be the equations of two quadrics that determine the pencil $\mathcal{F}$, let A and A' be the matrices of these quadratic forms in a basis of E. The quadric of equation $\lambda Q + \lambda' Q'$ is degenerate if and only if

$$\det(\lambda A + \lambda' A') = 0.$$

The left-hand side is a homogeneous polynomial of degree $n+1$ in λ and λ'. If this polynomial is identically zero, all the quadrics of $\mathcal{F}$ are degenerate.

Otherwise, we factor out the highest possible power of λ' to put it in the form

$$\det(\lambda A + \lambda' A') = (\lambda')^k P(\lambda, \lambda')$$

where P is a homogeneous polynomial of degree $n+1-k$ and $P(\lambda, 0)$ is not identically zero. Then the ordered pairs (λ, λ') that are solutions are:

– the $(\lambda, 0)$ if $k \geqslant 1$,

– the $\left(\dfrac{\lambda}{\lambda'}, 1\right)$ such that $P\left(\dfrac{\lambda}{\lambda'}, 1\right) = 0$, thus there are at most $n+1-k$ solutions.

The result follows. □

For instance, one can investigate and classify the pencils of conics according to the number and the nature of the degenerate conics it contains. There are at most three degenerate conics, and, in the complex case, the degenerate conics are double lines or pairs of lines. See Exercise VI.43 or [**Ber77**].

Intersection of two conics. Two complex projective conics intersect at four points. Here $4 = 2 \times 2$ is the product of the degrees of the two curves and this property is a special case of Bézout's theorem[(12)]: two complex projective curves of degrees m and n intersect at mn points. The multiplicity of the intersections must of course be taken into account: a point at which the two curves are tangent to each other "counts more".

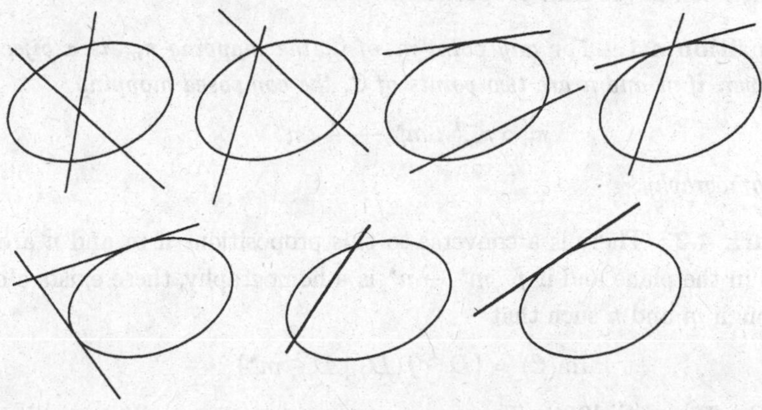

Fig. 15

Let us look at the case of conics. Assume that $\mathcal{C}$ and $\mathcal{C}'$ are two complex projective conics, defined by the two quadratic forms Q and Q'. Assume that

(12)See, *e.g.*, [**Per95**].

$\mathcal{C}$ is proper. Clearly, $\mathcal{C} \cap \mathcal{C}'$ is also the intersection of $\mathcal{C}$ with any other conic of the pencil generated by $\mathcal{C}$ and $\mathcal{C}'$:

$$Q(x, y, z) = Q'(x, y, z) = 0$$

if and only if

$$Q(x, y, z) = \lambda Q(x, y, z) + \lambda' Q'(x, y, z) = 0 \text{ for } \lambda' \neq 0.$$

Hence $\mathcal{C} \cap \mathcal{C}'$ is the intersection of $\mathcal{C}$ with a degenerate conic $\mathcal{C}''$ in the pencil. But $\mathcal{C}''$ consists of two secant lines or of a double line. Figure 15, in which the "heavy" lines represent double lines, shows the "four" expected intersection points (notice, however, that the pairs of conics shown on this figure do not define distinct pencils (see Exercise VI.43).

4. The cross-ratio of four points on a conic and Pascal's theorem

A conic $\mathcal{C}$ and a point m are given in a plane P. We consider the pencil of lines through m. Recall that this is the line $m^\star$ of $P^\star$.

Any point m of $\mathcal{C}$ defines a mapping

$$\pi_m : \mathcal{C} \longrightarrow m^\star$$

which, with any point n of $\mathcal{C}$, associates the line mn, with the convention that $\pi_m(m)$ is the tangent to $\mathcal{C}$ at m. Let us prove that π_m, or more exactly π_m^{-1} is a *parametrization* of the conic $\mathcal{C}$ by the projective line $m^\star$ (this is the first assertion in the next proposition).

Proposition 4.1. *For any point m of $\mathcal{C}$, the mapping π_m is a bijection. Moreover, if m and n are two points of $\mathcal{C}$, the composed mapping*

$$\pi_n \circ \pi_m^{-1} : m^\star \longrightarrow n^\star$$

is a homography.

Remark 4.2. There is a converse to this proposition: if m and n are two points in the plane and if $f : m^\star \to n^\star$ is a homography, there exists a conic $\mathcal{C}$ through m and n such that

$$\operatorname{Im}(\mathcal{C}) = \{D \cap f(D) \mid D \in m^\star\}$$

(see [**Ber77**] or [**Sid93**]). The set consisting of the proposition and its converse is often called the Chasles–Steiner theorem. The transformation $m^\star \to n^\star$ is shown in Figure 16.

Corollary 4.3. *Let $\mathcal{C}$ be a proper conic with nonempty image and let $m_1, \dots, m_4$ be four points of $\mathcal{C}$. The cross-ratio $[mm_1, mm_2, mm_3, mm_4]$ does not depend on the choice of the point m on $\mathcal{C}$.*

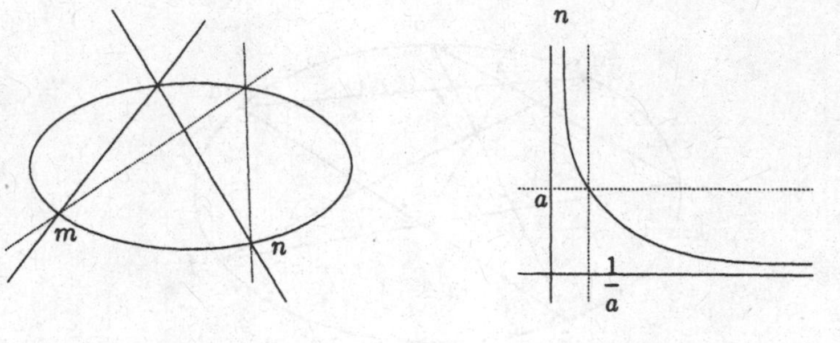

Fig. 16 Fig. 17

This cross-ratio is called the *cross-ratio* of the four points $m_1, \ldots, m_4$ on the conic $\mathcal{C}$.

Proof of the proposition. To avoid using technical results on homographies, let us use coordinates. All we have to do is to choose these cleverly. We choose the line mn as the line at infinity (assuming $m \neq n$, otherwise there is nothing to prove). In the remaining affine plane, the conic $\mathcal{C}$ is a hyperbola. We choose the origin at its center and the two basis vectors as directing vectors of the asymptotes, scaling them so that the equation of the affine conic is $xy = 1$. The points m and n are the points at infinity of the x- and y-axes respectively, so that the pencil $m^\star$ consists of the lines $y = a$ parallel to the x-axis and the pencil $n^\star$ of the lines $x = b$.

Thus the mapping $\pi_n \circ \pi_m^{-1}$ is (see Figure 17)

$$\begin{aligned} \mathbf{K} \cup \{\infty\} = m^\star &\longrightarrow n^\star = \mathbf{K} \cup \{\infty\} \\ a &\longmapsto \frac{1}{a} \end{aligned}$$

a projective transformation indeed... □

We are now going to prove Pascal's theorem, special cases of which we have already encountered (see Exercise III.51). Pascal has proved his theorem in a circle (the general case of conics being a consequence of the special case of circles) probably using Menelaüs' theorem (Exercise I.37).

Theorem 4.4 (Pascal's theorem). *Let $\mathcal{C}$ be (the image of) a proper conic and let a, b, c, d, e and f be six points of $\mathcal{C}$. Then the intersection points of ab and de, bc and ef, cd and fa are collinear.*

Proof. Let us give names to some of the intersection points:

$$\begin{cases} x = bc \cap ed \\ y = cd \cap ef \\ z = ab \cap de \\ t = af \cap cd \end{cases}$$

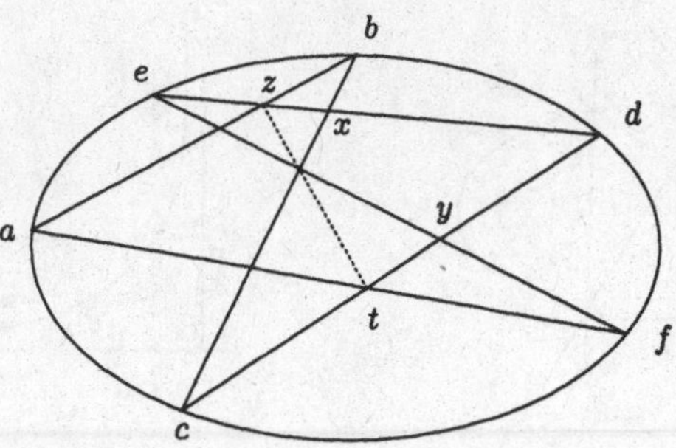

Fig. 18. The "mystic hexagram" of Pascal

(see Figure 18). We want to prove that the intersection point of the lines bc and ef lies on the line zt. To do this, let us calculate the cross-ratio $[z, x, d, e]$ (these four points are collinear on the line ed):

$$\begin{aligned}
[z, x, d, e] &= [bz, bx, bd, be] \\
&= [ba, bc, bd, be] \quad \text{(these are other names for the same lines)} \\
&= [fa, fc, fd, fe] \quad \text{(because } f \text{ is another point of } \mathcal{C}) \\
&= [t, c, d, y] \quad \text{(using the secant } cd).
\end{aligned}$$

We thus have the equality

$$[z, x, d, e] = [t, c, d, y].$$

Let m be the intersection point of zt and bc. The perspectivity[13] of center m (from the line ed to the line cd) maps z to t, x to c, d to d and thus, as it preserves the cross-ratio, it must map e to y, hence e, m and y are collinear. Therefore m, which belongs to bc and zt, also belongs to ef... and this is what we wanted to prove. □

5. Affine quadrics, *via* projective geometry

Let us consider now the real projective plane as the completion of some affine plane. We have said that any proper projective conic defines an affine conic. We have also seen that there is only one type of (nonempty) real proper conic over **R** and over **C**. The question of the affine classification is thus reduced to the question of the relative positions of the conic and the line at infinity. We have already noticed that the intersection of the conic and the line at infinity is the set of isotropic vectors of the dehomogenized quadratic form q.

Figure 19 shows the affine conics, from top to bottom:

---

(13) See if necessary Exercise V.12.

– in the projective completion of the affine plane $\mathcal{P}$, together with the line at infinity,
– in the affine plane $\mathcal{P}$ itself,
– in the 3-dimensional vector space $\mathcal{P} \times \mathbf{R}$, where the affine plane is the $z = 1$-plane.

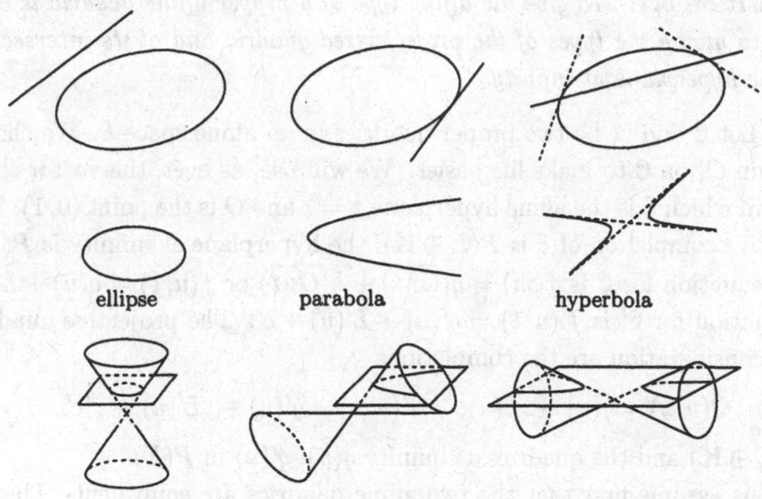

Fig. 19

For those of our readers who prefer equations to very clear pictures as those shown here, here are some equations. Let us choose the line of equation $z = 0$ as line at infinity and describe the conic by a homogeneous equation $Q(x, y, z) = 0$ for some quadratic form Q of signature $(2, 1)$.
There are three possibilities:

– The conic does not intersect the line at infinity, it is thus contained in the affine plane, this is an ellipse in this plane. In equations: the conic has an equation $x^2 + y^2 - z^2 = 0$, its affine part is the ellipse of equation $x^2 + y^2 = 1$.
– The conic is tangent to the line at infinity, its affine part is a parabola. In equations: take $x^2 - yz = 0$ for the conic; an equation of the affine part is $y = x^2$.
– The conic intersects the line at infinity at two points; its affine part is a hyperbola, the asymptotes are the tangents at the two points at infinity. In equations: $x^2 - y^2 - z^2 = 0$ for the projective conic, $x^2 - y^2 = 1$ for its affine part.

What we have just described can be rephrased, in a more pedantic way, as "two proper affine conics are (affinely) equivalent if and only if their points at infinity are two projectively equivalent 0-dimensional quadrics". We have used both the fact that there is only one type of projective conics and the affine classification: we knew, *e.g.*, that the affine conics with no point at infinity were all affinely equivalent... but the affine classification can also be deduced from the projective classification, as I show now, in any dimension.

Proposition 5.1. *To give the affine type of a proper affine quadric is equivalent to giving the types of the projectivized quadric and of its intersection with the hyperplane at infinity.*

Proof. Let $\mathcal{C}$ and $\mathcal{C}'$ be two proper quadrics in an affine space $\mathcal{E}$. We choose an origin O, on $\mathcal{C}$ to make life easier. We will use, as ever, the vector space $E \oplus \mathbf{K}$ in which $\mathcal{E}$ is the affine hyperplane $z = 1$ and O is the point $(0, 1)$. The projective completion of $\mathcal{E}$ is $P(E \oplus \mathbf{K})$, the hyperplane at infinity is $P(E)$.

An equation for $\mathcal{C}$ is $f(m) = q(\overrightarrow{OM}) + L(\overrightarrow{OM})$ or $f(u, 1) = q(u) + L(u)$. An equation for $\mathcal{C}'$ is $f'(u, 1) = q'(u) + L'(u) + c'$. The projective quadrics under consideration are the completions

$$Q(u, z) = q(u) + zL(u), \quad Q'(u, z) = q'(u) + zL'(u) + z^2c'$$

in $P(E \oplus \mathbf{K})$ and the quadrics at infinity $q(u)$, $q'(u)$ in $P(E)$.

Let us assume first that the two affine quadrics are equivalent. That is, there is an affine transformation

$$\psi : \mathcal{E} \longrightarrow \mathcal{E} \text{ such that } f' \circ \psi(M) = \lambda f(M)$$

(for some nonzero scalar λ). In our notation,

$$\psi(u, 1) = (v + \overrightarrow{\psi}(u), 1) \text{ where } v = \overrightarrow{O\psi(O)}$$

so that the condition is equivalent to

$$q' \circ \overrightarrow{\psi}(u) = \lambda q(u), \quad L' \circ \overrightarrow{\psi}(u) + 2\varphi'(v, \overrightarrow{\psi}(u)) = \lambda L(u), \quad q'(v) + L'(v) + c' = 0.$$

The first equation gives the equivalence of the quadrics at infinity, the last one just says that $\psi(O) \in \mathcal{C}'$.

Recall (from Proposition V-5.7) that ψ extends to a linear isomorphism

$$\begin{aligned} \Psi : E \oplus \mathbf{K} &\longrightarrow E \oplus \mathbf{K} \\ (u, z) &\longmapsto (zv + \overrightarrow{\psi}(u), z). \end{aligned}$$

Hence

$$\begin{aligned} Q'(\Psi(u, z)) &= q'(zv + \overrightarrow{\psi}(u)) + zL'(zv + \overrightarrow{\psi}(u)) + z^2c' \\ &= q'(\overrightarrow{\psi}(u)) + z(L'(\overrightarrow{\psi}(u)) + 2\varphi'(v, \overrightarrow{\psi}(u))) \\ &\qquad + z^2(q'(v) + L'(v) + c'). \end{aligned}$$

Eventually, $Q' \circ \Psi(u, z) = \lambda Q(u, z)$, so that the two projective quadrics are indeed equivalent.

Conversely, assume that the projective quadrics are equivalent and the quadrics at infinity too. That is, there are two linear isomorphisms

$$\Phi : E \oplus \mathbf{K} \longrightarrow E \oplus \mathbf{K} \text{ and } \psi : E \longrightarrow E$$

such that

$$Q' \circ \Phi = \lambda Q \text{ and } q' \circ \psi = \mu q$$

for some nonzero scalars λ, μ that we can (and will) take equal to 1. Our task is to deduce that there exists a linear isomorphism

$$\Theta : E \oplus \mathbf{K} \to E \oplus \mathbf{K}$$

such that $\Theta\mid_E = \psi$ and $Q' \circ \Theta = Q$. Then Θ will define a projective transformation of $P(E \oplus \mathbf{K})$ preserving the hyperplane at infinity and hence an affine transformation of $\mathcal{E}$. As it extends ψ, it will be of the form

$$\Theta(u, 1) = (v + \psi(u), 1)$$

so that the same computation as above gives

$$f' \circ \Theta(u, 1) = f(u, 1)$$

and the affine quadrics will be equivalent.

Now the existence of Θ is a consequence of Witt's theorem (Theorem 8.10 below): $Q = Q' \circ \Phi$ is a quadratic form on $E \oplus \mathbf{K}$ and $\psi : E \to E$ satisfies $q' \circ \psi = q$. Hence

$$F = \Phi^{-1} \circ \psi : E \longrightarrow E \longrightarrow \Phi^{-1}(E) = E'$$

satisfies $Q \circ F = Q\mid_E$, in other words, F is an isometry from $(E, Q\mid_E)$ to $(E', Q\mid_{E'})$. Witt's theorem asserts[14] that it can be extended to a linear isomorphism

$$\widetilde{F} : E \oplus \mathbf{K} \longrightarrow E \oplus \mathbf{K}$$

that is an isometry for Q. Now we have

$$Q' \circ (\Phi \circ \widetilde{F}) = Q \circ \widetilde{F} = Q.$$

Therefore $\Theta = \Phi \circ \widetilde{F}$ is the expected isomorphism. □

The classification of proper projective real quadrics with nonempty images in $\mathbf{P}_3(\mathbf{R})$ is very simple, as we have seen (Proposition 3.6): there are only two types,

$$\text{I}: \quad x^2 + y^2 + z^2 - t^2 = 0 \quad \text{and} \quad \text{II}: \quad x^2 + y^2 - z^2 - t^2 = 0.$$

Notice that, for the homogenized form $q + t(L + ct)$ to have rank 4, the initial form q must have rank at least 2, hence the conic at infinity can be:

---

[14]This is the place where we need the quadrics to be proper (namely Q to be nondegenerate).

- empty (a),
- proper and nonempty (b),
- or degenerate with image a point (c) or two intersecting lines (d).

We get the following five affine types:

- ellipsoid (Ia), $x^2 + y^2 + z^2 = 1$,
- hyperboloids of one (IIb) or two (Ib) sheets,
- hyperbolic (IId) or elliptic (Ic) paraboloids

that are depicted in Figure 20.

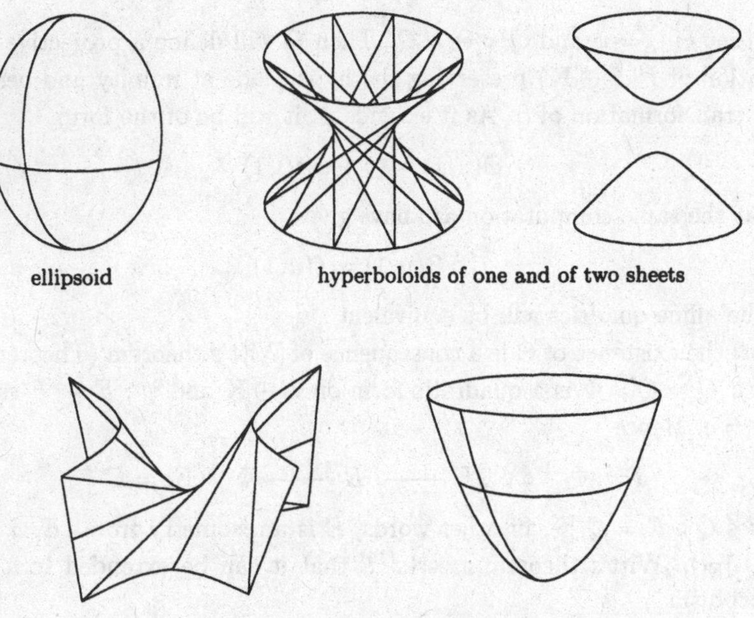

Fig. 20. Affine quadrics in $\mathbf{R}^3$

From the affine viewpoint, the only thing which is left to explain is the question of the center. We have:

Proposition 5.2. *The center of an affine quadric is the pole of the hyperplane at infinity.*

Remark 5.3. And this is why we were not able to find the center of a parabola: it was at infinity!

Proof. Let O be the center of the affine central quadric $\mathcal{C}$. This is a center of symmetry for $\mathcal{C}$. Let D be a line contained in the hyperplane at infinity. The point O and the line D span a projective plane that intersects $\mathcal{C}$ along a conic

$\mathcal{C}_D$. Let d be a line through O in this plane. It intersects $\mathcal{C}_D$ at two points m and m'. By symmetry, the tangent lines to $\mathcal{C}_D$ at m and m' are parallel, hence (see Remarks 3.5), the pole of d is on the line at infinity D. Thus the polar line of O with respect to $\mathcal{C}_D$ is D. Therefore the polar hyperplane of O with respect to $\mathcal{C}$ is indeed the hyperplane at infinity. □

6. Euclidean conics, *via* projective geometry

We are now going to find again the metric properties of affine Euclidean conics using projective geometry[15]. Therefore, our plane is now an affine Euclidean plane $\mathcal{P}$. As usual (see § V-3), we consider $\mathcal{P}$ as the affine plane of equation $z = 1$ in a vector space of dimension 3 (namely $\mathcal{P} \times \mathbf{R}$). Now, the Euclidean structure of $\mathcal{P}$ is important, we thus also endow E with a Euclidean structure that restricts to the given structure on $\mathcal{P}$:

– Concretely, an orthonormal affine frame of $\mathcal{P}$ being chosen, it defines a basis of the vector space E that we decide to be an orthonormal basis.
– Even more concretely,
  - the vector space E (with its basis) is $\mathbf{R}^3$ endowed, *e.g.*, with the standard Euclidean form $x^2 + y^2 + z^2$;
  - the affine plane $\mathcal{P}$, of equation $z = 1$, is directed by the (vector) plane P of equation $z = 0$, endowed with the standard Euclidean form $x^2 + y^2$;
  - we are now in the projective plane $P(E)$, the completion of $\mathcal{P}$.

In $P(E)$, the Euclidean form defines a conic, that of equation

$$x^2 + y^2 + z^2 = 0.$$

I have already mentioned that its image is empty... but that it will nevertheless play an important role. To begin with, let us give points to it: in order to do this, it suffices to complexify the space E. As we have already identified E with $\mathbf{R}^3$ by the choice of a basis, complexification is a benign process, we simply consider the coordinates (x, y, z) as being complex numbers.

The image of the projective projective conic of equation $x^2 + y^2 + z^2 = 0$ has become nonempty. Since we are interested in *affine* conics in $\mathcal{P}$, it is natural to wonder where our Euclidean conic intersects the line at infinity (that of equation $z = 0$). This is at the two points of homogeneous coordinates $(1, \pm i, 0)$.

(15) This exposition of the metric properties of conics *via* projective geometry is due to Plücker.

Definition 6.1. The two points I and J of $P(E)$ where the projective conic defined by the Euclidean structure intersects the line at infinity are called the *circular points.*

Remarks 6.2

- There is no use in making precise which is I and which is J.
- The coordinates of the points I and J are conjugated. In particular, if I satisfies an equation with real coefficients (for instance if I lies on a real conic), then J satisfies the same equation (J lies on the same conic).
- These points are imaginary *and* at infinity, two good reasons why some of our (young) readers have never met them before.

Before coming to conics, let us notice that the orthogonality (in the usual Euclidean sense) of the affine lines of $\mathcal{P}$, a relation between two lines $\mathcal{D}$ and $\mathcal{D}'$, can be expressed in terms of the circular points.

Proposition 6.3. *The affine lines $\mathcal{D}$ and $\mathcal{D}'$ are orthogonal if and only if, on the line at infinity, one has*

$$[I, J, \infty_{\mathcal{D}}, \infty_{\mathcal{D}'}] = -1.$$

Proof. We have already proved above (Remark 2 about polarity and duality) that, if a and b (here I and J) are the intersection points of a conic (here that defined by the Euclidean form) with a line (here the line at infinity), two points m and n (here $\infty_{\mathcal{D}}$ and $\infty_{\mathcal{D}'}$) represent orthogonal lines in the 3-dimensional vector space (here, thus, for the Euclidean scalar product) if and only if the equality

$$[a, b, m, n] = -1$$

holds. We just need now to remember that the lines in the vector plane of equation $z = 0$ that represent the points at infinity of the affine lines $\mathcal{D}$ and $\mathcal{D}'$ in the affine plane of equation $z = 1$ are the directions of $\mathcal{D}$ and $\mathcal{D}'$. They are thus orthogonal if and only if $\mathcal{D}$ and $\mathcal{D}'$ are. □

Let us come now to conics. To begin with, we characterize circles among the conics.

Proposition 6.4. *Circles are the proper conics that pass through the circular points.*

Proof. Circles are the conics that have an equation of the form[16]

$$x^2 + y^2 + L(x, y) + c = 0$$

in an orthonormal frame, that is, conics such that the quadratic part of one of their equations is the Euclidean form itself. The points at infinity of a circle are given by the equations

$$z = 0 \text{ and } x^2 + y^2 = 0.$$

These are indeed the circular points.

Conversely, if a conic has an affine equation

$$q(x, y) + L(x, y) + c = 0,$$

to say that it passes through the circular points is to say that the quadratic form q vanishes at I and J and thus that it is proportional to the Euclidean form $x^2 + y^2$: the form

$$ax^2 + bxy + cy^2$$

vanishes at $(1, \pm i)$ if and only if

$$a \pm bi - c = 0,$$

that is, if and only if $b = 0$ and $a = c$, that is, if and only if $q(x, y) = a(x^2 + y^2)$. □

Remarks 6.5

- Compare Corollary 3.9 "five points of the plane always lie on a conic" with the fact that three points of the plane always lie on a circle.
- If a (real) conic contains the point I, it must contain the conjugate point J as well. This is thus a circle. Circles are actually the conics containing one of the circular points.
- A pencil of circles (in the sense of § III-4) is a pencil of conics containing the circular points. See also Exercise VI.52.
- While two conics can intersect at four points, two circles intersect at at most two points. This can be understood by a calculation: since the two equations have the same highest degree terms $(x^2 + y^2)$, to looking for the intersection of two circles amounts to look for the intersection of a circle and a line... the radical axis of the two circles. This can also be understood geometrically: the two circles intersect at the two circular points[17] and at their intersection points at finite distance.

---

[16] There are circles consisting of a single point, or even empty circles, in this family, but this does not matter.

[17] The circular points lie on all circles; they can thus be considered as "universally cyclic".

Tangents through the circular points. Consider now a real affine conic $\mathcal{C}$ and its tangents through the circular points. They will give us all the expected metric properties.

Let us assume first that I (and thus J as well) lies on $\mathcal{C}$.

Proposition 6.6. *The tangents to a circle at the circular points intersect at the center of this circle.*

Proof. The tangents to $\mathcal{C}$ at I and J intersect at a point F which is a real point: their equations are conjugated to each other, thus their intersection point is real. As it lies at the intersection of the tangents at I and J, this point is the pole of the line IJ, alias the line at infinity (Figure 21). Hence, thanks to Proposition 5.2, F is the center of the circle! □

Let us assume now that neither I, nor J lies on $\mathcal{C}$, but that the line IJ, namely the line at infinity, is tangent to $\mathcal{C}$. We know that $\mathcal{C}$ is then a parabola.

Proposition 6.7. *Let $\mathcal{C}$ be a parabola. The line IJ is tangent to $\mathcal{C}$. The second tangent to $\mathcal{C}$ through I and the second tangent to $\mathcal{C}$ through J intersect at the focus F of $\mathcal{C}$.*

Let us finally assume that I and J do not lie on $\mathcal{C}$ and that the line IJ is not tangent to $\mathcal{C}$. Then $\mathcal{C}$ is an ellipse or a hyperbola.

Proposition 6.8. *Let $\mathcal{C}$ be a central conic that is not a circle. Let D_1 and D_2 be the tangents to $\mathcal{C}$ through I, D_1' and D_2' the respective conjugated complex lines. The lines D_1' and D_2' are the tangents to $\mathcal{C}$ through J. The intersection points F_1 and F_2 of D_1 and D_1', D_2 and D_2' (resp.) are the foci of $\mathcal{C}$. Their polars with respect to $\mathcal{C}$ are the corresponding directrices.*

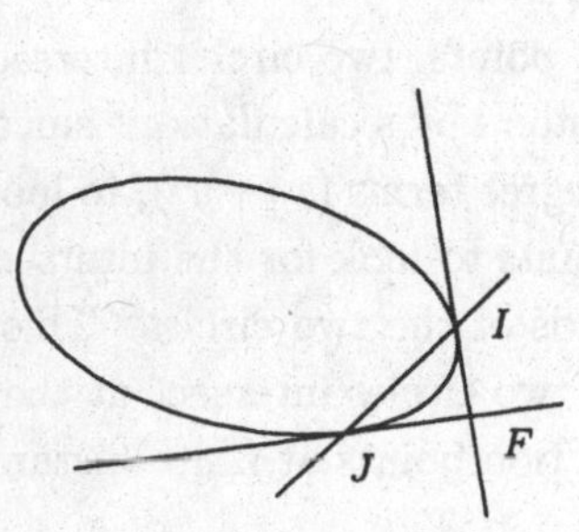

Fig. 21

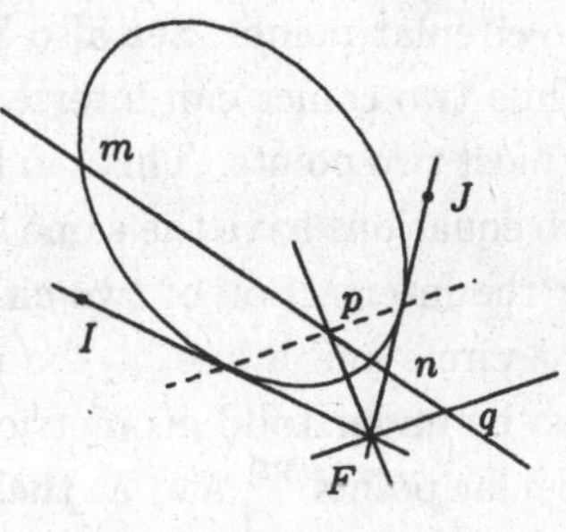

Fig. 22

Proof of Propositions 6.7 and 6.8. In the two cases, consider one of the tangents D through I and the conjugate line D', which is a tangent through J. Let F be the intersection of D and D' and let Δ be the polar of F with respect to $\mathcal{C}$. Choose any two points m and n on $\mathcal{C}$; let p be the intersection point of mn with Δ and q the point of mn defined by the equality

$$[m,n,p,q] = -1.$$

Thus p and q are orthogonal with respect to $\mathcal{C}$. Hence both F and q are on the polar of p with respect to $\mathcal{C}$, so that this polar is the line Fq (Figure 22). In particular, if $q' = Fq \cap \Delta$ and if a and b are the points of contact of D and D' with $\mathcal{C}$, we have

$$[a,b,p,q'] = -1, \text{ that is, } [FI,FJ,Fp,Fq] = -1.$$

Viewed on the line at infinity, this equality simply expresses the fact that the lines Fp and Fq are orthogonal. As the pencil (Fm, Fn, Fp, Fq) is harmonic, this means that Fp and Fq are the bisectors of (Fm, Fn) (see if necessary Exercise V.21). We then have:

$$\begin{aligned} \frac{Fm}{Fn} &= \frac{pm}{pn} \text{ (see Exercise III.26)} \\ &= \frac{d(m,\Delta)}{d(n,\Delta)} \text{ by similarity.} \end{aligned}$$

We fix the point n on $\mathcal{C}$ and we deduce $Fm = ed(m,\Delta)$, and this is what we wanted to prove. □

Remark 6.9. This is also why we have not been able to find a directrix for the circle: this was the line at infinity!

7. Circles, inversions, pencils of circles

In § III-4, we considered equations of circles to describe the pencils of circles in the affine Euclidean plane. In this chapter, we have considered more general equations of conics. I explain what was happening in § III-4 using (at last!) the "good" space of circles, the one that contains circles, lines... together with the points of the plane, a space in which a pencil of circles is, simply, a line.

More or less all what is explained here may be generalized to the space of spheres of an affine Euclidean space of dimension n (see [**Ber77**, Chapter 20]).

The space of circles. We are now in an affine Euclidean plane $\mathcal{P}$. Circles are the conics that have an equation whose quadratic part is a nonzero multiple of the Euclidean norm:

$$f(M) = \lambda\|\overrightarrow{OM}\|^2 + L(\overrightarrow{OM}) + c$$

or

$$f(M) = \lambda \|\overrightarrow{OM}\|^2 + 2\overrightarrow{OM} \cdot u + c$$

for some scalars λ (nonzero) and c and some vector u of the vector plane P directing $\mathcal{P}$.

Let us denote by $\mathcal{C}(\mathcal{P})$ and call *space of circles* of $\mathcal{P}$ the projective space deduced from the vector space $P \times \mathbf{R}^2$ of these equations. The $\mathbf{R}^2$ summand is the one where the ordered pair (λ, c) lives, the P summand the one where the vector u leaves. The space of circles is a real projective space of dimension 3, a projective subspace of the space of all affine conics in $\mathcal{P}$.

If such an equation indeed describes a circle, it must have the form

$$f(M) = \lambda \left(\|\overrightarrow{AM}\|^2 - R^2 \right)$$

for some point A, with λ nonzero and $R^2 > 0$. As

$$R^2 = \|u\|^2 - \lambda c,$$

we must have

$$\lambda \neq 0 \text{ and } \|u\|^2 - \lambda c > 0.$$

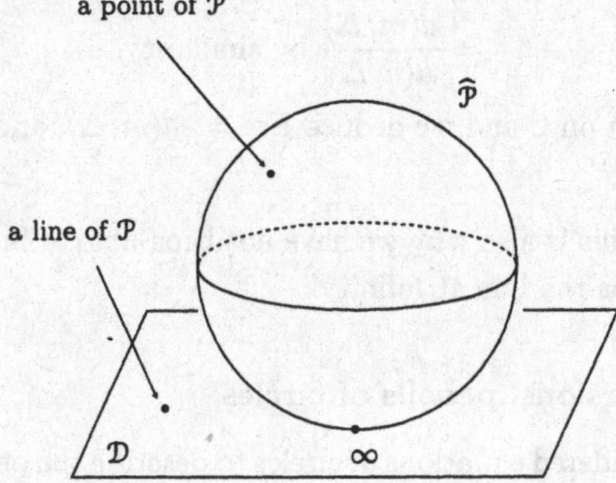

Fig. 23. The fundamental quadric in the space of circles

The space $\mathcal{C}(\mathcal{P})$ thus consists of equations of:

- genuine circles (corresponding to $\lambda \neq 0$ and $\|u\|^2 - \lambda c > 0$),
- "point-circles" (corresponding to $\lambda \neq 0$ and $\|u\|^2 - \lambda c = 0$),
- circles of imaginary radius[18], with empty image (corresponding to $\lambda \neq 0$ and $\|u\|^2 - \lambda c < 0$),
- affine lines of $\mathcal{P}$ (corresponding to $\lambda = 0$ and $u \neq 0$)

[18] Sometimes called "pseudo-real circles" (see [**DC51**]). These are real circles, since they have a real equation, without any real point, since their image is empty.

– and of the equation corresponding to $\lambda = 0$ and $u = 0$ (the image of the corresponding conic is empty).

The fundamental quadric. An important role is played by the projective quadric of equation

$$r(u, \lambda, C) = \|u\|^2 - \lambda c.$$

The image of this quadric consists of point-circles, to which the point $(0,0,1)$ has to be added. In other words, it is identified with the union of the set of points of $\mathcal{P}$ and an additional point, that we shall not fail to call a point at infinity.

This quadric is indeed our old $\widehat{\mathcal{P}} = \mathcal{P} \cup \{\infty\}$, even as a topological space: the signature of the quadratic form r is $(3,1)$ since

$$r(u, \lambda, c) = \|u\|^2 + \frac{1}{4}\left[(\lambda - c)^2 - (\lambda + c)^2\right],$$

its image is contained in the affine subspace[19] $\lambda + c \neq 0$ of $\mathcal{P}$... where it (the image quadric) is clearly homeomorphic to a sphere of dimension 2.

Remark 7.1. This way, we have *embedded* the sphere $\widehat{\mathcal{P}} = \mathcal{P} \cup \{\infty\}$, which has already proved to be useful for the study of circles in § V-7, in the space of circles.

Let $\mathcal{D}$ be the projective hyperplane of equation $\lambda = 0$. We have seen that it consists of the affine lines of $\mathcal{P}$ ($\lambda = 0$ and $u \neq 0$), a set to which the point ∞ should be added.

One should notice that $\widehat{\mathcal{P}}$ and $\mathcal{D}$ are tangent at ∞ (Figure 23).

A summary. The space $\mathcal{C}(\mathcal{P})$ contains both the set $\mathcal{D}$ of the affine lines[20] of $\mathcal{P}$ (a projective plane) and the set of points of $\mathcal{P}$ (a quadric $\widehat{\mathcal{P}}$) tangent at a point (denoted ∞). This is another way to explain why the plane $\mathcal{P}$ has been completed, in § V-7, by the addition of a point rather than a line.

Circles and their centers. The quadric $\widehat{\mathcal{P}}$ divides the projective space $\mathcal{C}(\mathcal{P})$ into two components. I shall let the *interior* of $\widehat{\mathcal{P}}$ denote the set of points satisfying $r(u, \lambda, c) < 0$, namely, the set of circles of imaginary radii, and the *exterior* of $\widehat{\mathcal{P}}$ the set of genuine circles.

Any circle has a center... in other words, it is possible to associate, with any point C of $\mathcal{C}(\mathcal{P})$, the other point A where the line $C\infty$ intersects $\widehat{\mathcal{P}}$ (we are going to use systematically and intensively the intersection of the lines

[19] This is the affine space of dimension 3 in which all the picture of this section are drawn.

[20] The set of the affine lines $\mathcal{P}$ is here $\mathcal{D} - \{\infty\}$. This is the complement of a point in a projective plane (see Exercise V.41).

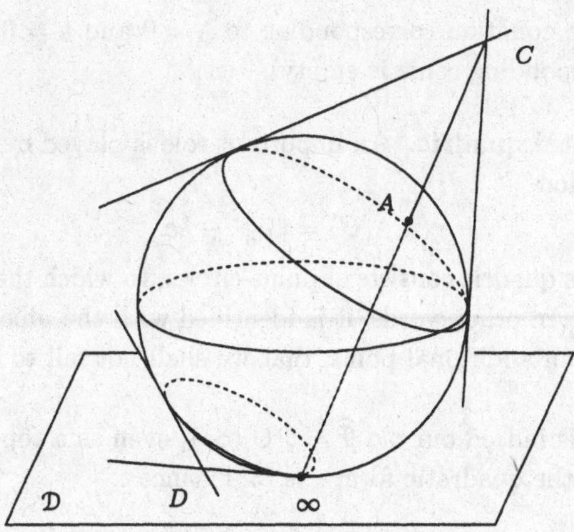

Fig. 24. Circles, their centers, lines

with the quadric). It should be noticed that the circles of imaginary radius have a center and that the lines also have a center... at ∞ (Figure 24).

However, the point C of $\mathcal{C}(\mathcal{P})$ is a circle of $\mathcal{P}$, hence a subset of $\widehat{\mathcal{P}}$ (something which is also shown by Figure 24): the points of C are indeed the points M of $\widehat{\mathcal{P}}$ such that the line CM is tangent to $\widehat{\mathcal{P}}$. The circle C of $\widehat{\mathcal{P}}$ is thus the intersection of $\widehat{\mathcal{P}}$ with the polar plane of the point C with respect to the quadric $\widehat{\mathcal{P}}$, a plane that I will denote by $C^{\perp}$.

One should also see in this description a confirmation of the fact that the imaginary circles have no points and that all the lines pass through ∞.

Inversions. Any circle C which is not a point defines an inversion, namely a transformation of $\widehat{\mathcal{P}}$. This is the involution that associates, with the point M of $\widehat{\mathcal{P}}$, the second intersection point M' of CM with $\widehat{\mathcal{P}}$ (Figure 25). It should be noticed that:

- his transformation indeed exchanges the center A of C, pole of the inversion, and the point ∞;
- the lines also define inversions, with pole at infinity (these are, of course, the reflections).

An inversion transforms any circle into a circle; we should also see it act on $\mathcal{C}(\mathcal{P})$. With Γ, the inversion of circle C associates the point Γ' such that

$$[C, N, \Gamma, \Gamma'] = -1,$$

where N denotes the intersection point of $C\Gamma$ with the plane $C^{\perp}$ defined by C (Figure 25). The inversion defined this way is a *projective transformation* of $\mathcal{C}(\mathcal{P})$.

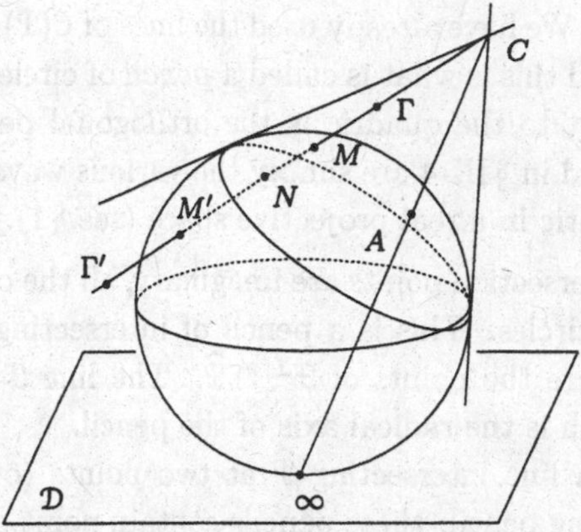

Fig. 25. Images of circles and points by an inversion

Orthogonal circles. The polarity with respect to $\widehat{\mathcal{P}}$ describes the orthogonality of circles: the circles C and C' are orthogonal if C' is in the polar plane $C^\perp$ of C (and conversely). For example, the points of C (if it has points) are point-circles orthogonal to C'.

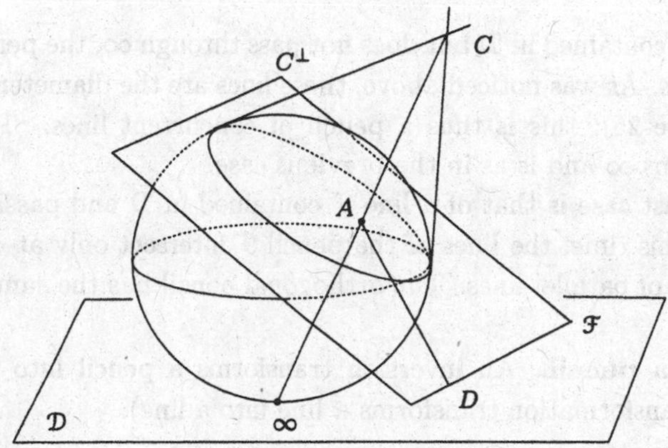

Fig. 26. Diameters of a circle

The plane $C^\perp$ intersects the space $\mathcal{D}$ of lines along a projective line $\mathcal{F}$ that is the set of diameters of C (Figure 26).

It should also be noticed that:

- the circles orthogonal to C are the fixed points of the inversion of circle C acting on the space of circles;
- the two notions of orthogonality we have for lines coincide: on $\mathcal{D}$, we have $\lambda = 0$ and the two orthogonalities are defined by the Euclidean scalar product in the plane.

Pencils of circles. We have already used the lines of $\mathcal{C}(\mathcal{P})$ quite a lot. A line is a set of circles, and this is what is called a *pencil* of circles. The orthogonal line $\mathcal{F}^\perp$ with respect to the quadric is the orthogonal pencil. The various types of pencils listed in § III-4 are simply the various ways in which a line $\mathcal{F}$ can intersect a quadric in a real projective space (see § 1).

- If the two intersection points are imaginary, all the circles of the pencil are genuine circles. This is a pencil of intersecting circles. The two base points are the points of $\mathcal{F}^\perp \cap \widehat{\mathcal{P}}$. The line $\mathcal{F}$ intersects $\mathcal{D}$ at a point D which is the radical axis of the pencil.
- Conversely, a line intersecting $\widehat{\mathcal{P}}$ at two points (other that ∞) is a nonintersecting pencil: these pencils contain point-circles and imaginary radius circles. One would have noticed that $\mathcal{F}$ intersects $\widehat{\mathcal{P}}$ at two points if and only if $\mathcal{F}^\perp$ does not intersect $\widehat{\mathcal{P}}$.
- A line $\mathcal{F}$ tangent to $\widehat{\mathcal{P}}$ at a point not equal to ∞ is a pencil of tangent circles.
- If $\mathcal{F}$ intersects $\widehat{\mathcal{P}}$ at ∞ and at another point A, all the circles of $\mathcal{F}$ have their center at A; the pencil $\mathcal{F}$ is thus a pencil of concentric circles. There is no radical axis, it has gone to infinity (and more precisely to ∞).
- If $\mathcal{F}$ is contained in $\mathcal{D}$ but does not pass through ∞, the pencil consists of lines. As was noticed above, these lines are the diameters of a circle (Figure 26); this is thus a pencil of concurrent lines. The line $\mathcal{F}^\perp$ contains ∞ and is as in the previous case.
- The last case is that of a line $\mathcal{F}$ contained in $\mathcal{D}$ and passing through ∞. This time, the lines of the pencil $\mathcal{F}$ intersect only at ∞; this is a pencil of parallel lines. The orthogonal pencil has the same nature.

Inverse of a pencil. An inversion transforms a pencil into a pencil (a projective transformation transforms a line into a line).

The circular group. Consider now the group $\mathcal{G}$ of projective transformations of $\mathcal{C}(\mathcal{P})$ that preserve the quadric $\mathcal{C}(\mathcal{P})$. They act on $\widehat{\mathcal{P}}$ and the restriction defines a homomorphism Φ from $\mathcal{G}$ to the group of all transformations of $\widehat{\mathcal{P}}$.

Theorem 7.2. *The restriction to $\widehat{\mathcal{P}}$ defines an isomorphism of $\mathcal{G}$ onto the circular group of $\mathcal{P}$.*

Proof. Notice first that $\widehat{\mathcal{P}}$ contains projective frames of $\mathcal{C}(\mathcal{P})$ and thus that any element of $\mathcal{G}$, which is a projective transformation, is well determined by its restriction to $\widehat{\mathcal{P}}$. Hence Φ is injective.

On the other hand, the circular group is generated by the inversions, of which we know that they come from $\mathcal{G}$. The image of Φ thus contains all the circular group.

We still need to check that this image is contained in the circular group. By definition, the elements of $\Phi(\mathcal{G})$ preserve the set of circles of $\widehat{\mathcal{P}}$. Theorem V-7.12 thus allows us to conclude. □

8. Appendix: a summary of quadratic forms

We have already used examples of quadratic forms, the scalar products, in the Euclidean chapters of this book. In this appendix, **K** denotes one of the fields **R** or **C**, E is a vector space of finite dimension over **K**, $\varphi : E \times E \to \mathbf{K}$ is a symmetric bilinear form and eventually $q : E \to \mathbf{K}$ is the quadratic form associated with φ. It is said that φ is the *polar form* of q.

Recall that q is defined from φ by

$$q(x) = \varphi(x, x)$$

or φ from q by

$$\varphi(x, y) = \frac{1}{2}[q(x + y) - q(x) - q(y)]$$

or even by

$$\varphi(x, y) = \frac{1}{4}[q(x + y) - q(x - y)].$$

A nonzero vector x is *isotropic* if $q(x) = 0$. The set of isotropic vectors is a *cone* (if x is isotropic, λx is isotropic for any scalar λ), the *isotropy cone.*

Calculus. The polar form is essentially the differential of the quadratic form: fix a norm $\|\cdot\|$ on E and write

$$q(x + h) = q(x) + 2\varphi(x, h) + q(h).$$

The term $q(h)$ is quadratic (!) in h and hence is an $o(\|h\|)$. Thus q is differentiable at x (for all x) and its differential is the linear form

$$dq_x : h \longmapsto 2\varphi(x, h).$$

Calculation in a basis. Let $(e_1, \ldots, e_n)$ be a basis of E. The form φ is determined by the values of the $\varphi(e_i, e_j)$, a square table of numbers that can be put in a matrix $A = (\varphi(e_i, e_j)_{1 \leqslant i,j \leqslant n})$. The matrix A is symmetric, which means that it satisfies the equality ${}^tA = A$. It is possible to write $\varphi(x, y)$ as a product of matrices (writing x for the column matrix that gives the coordinates of the vector x in the basis $(e_1, \ldots, e_n)$):

$$\varphi(x, y) = {}^t x A y.$$

Notice that A is *not* a matrix of endomorphisms. The best proof of this is the way it transforms under a change of basis: if $(e'_1, \ldots, e'_n)$ is another basis and if S is the matrix of $(e'_1, \ldots, e'_n)$ in the basis $(e_1, \ldots, e_n)$, with obvious notation, $x = Sx'$, $y = Sy'$ and

$$^t xAy = {}^t(Sx')A(Sy') = {}^t x'({}^t SAS)y'$$

which is indeed of the form ${}^t x'A'y'\ldots$ but for a transformed matrix $A' = {}^t SAS$.

Nondegeneracy. The bilinear form φ defines a linear mapping $\widetilde{\varphi}$:

$$\begin{aligned} \widetilde{\varphi} : E &\longrightarrow E^\star \\ x &\longmapsto (y \mapsto \varphi(x, y)). \end{aligned}$$

Recall that φ (or q) is *nondegenerate* when $\widetilde{\varphi}$ is injective. The vector space E having finite dimension, it is equivalent to say that $\widetilde{\varphi}$ is an isomorphism.

In mathematical symbols, to say that φ is nondegenerate is to say that

$$(\forall y \in E, \quad \varphi(x, y) = 0) \quad \Longrightarrow \quad x = 0.$$

Orthogonality. Two vectors x and y are said to be *orthogonal* if $\varphi(x, y) = 0$. A basis consisting of mutually orthogonal vectors is said to be *orthogonal.*

If F is a subspace of E, its orthogonal $F^\perp$ is the subspace

$$F^\perp = \{x \in E \mid \varphi(x, y) = 0 \quad \forall y \in F\}$$

consisting of all the vectors of E that are orthogonal to all the vectors in F. If φ is nondegenerate, $F^\perp$ is mapped, by $\widetilde{\varphi}$, to the annihilator F° of F

$$F^\circ = \{f \in E^\star \mid f\,|_F = 0\}.$$

Under this nondegeneracy assumption, we have

$$\dim F + \dim F^\perp = \dim E$$

(using a basis of F, completed into a basis of E, and the dual basis of $E^\star$). Notice that F and $F^\perp$ may intersect. For instance, to say that x is isotropic is to say that $\langle x \rangle \subset \langle x \rangle^\perp$.

Classification

Proposition 8.1. *For any quadratic form q, there exists an orthogonal basis.*

Proof. If $q = 0$, any basis works. Otherwise, let x be a nonzero vector such that $q(x) \neq 0$. The linear form $\widetilde{\varphi}(x) \in E^\star$ is nonzero, since

$$\widetilde{\varphi}(x)(x) = q(x) \neq 0.$$

Its kernel is thus a hyperplane of E, denoted by $x^\perp$ (this is the orthogonal of x). Hence E is the orthogonal direct sum (for q) of $x^\perp$ and the line generated by x.

It is thus possible to conclude by induction on the dimension of E, the result obviously holding in dimension 1. □

Corollary 8.2. *There exists an integer R, scalars λ_i (all nonzero), and a basis of E in which one has*

$$q(x) = \sum_{i=1}^{R} \lambda_i x_i^2$$

(equality in which $(x_1, \dots, x_n)$ denote the coördinates of the vector x). Moreover, if $\mathbf{K} = \mathbf{C}$, there exists a basis in which one has

$$q(x) = \sum_{i=1}^{R} x_i^2.$$

If $\mathbf{K} = \mathbf{R}$, there exists a basis in which

$$q(x) = \sum_{i=1}^{s} x_i^2 - \sum_{i=s+1}^{R} x_i^2.$$

Proof. One simply writes q in an orthogonal basis. In the complex case, one then replaces x_i by $\mu_i x_i$ (where μ_i is a square root of λ_i) to obtain the expected formula. In the real case, one uses in the same way square roots of those of the λ_i that are nonnegative and square roots of the opposites of those that are nonpositive. □

The integer R that appears in this statement is called the *rank* of the quadratic form q. This is also the rank of the linear mapping $\widetilde{\varphi} : E \to E^\star$ (see Exercise VI.7). Let us say that two quadratic forms q on E and q' on E' are *equivalent* if there exists a linear isomorphism $f : E \to E'$ such that $q' \circ f = q$.

Remark 8.3. Let q and q' be two equivalent forms. Then q is nondegenerate if and only if q' is. Indeed, if they are equivalent by an isomorphism f, their polar forms are related by the isomorphism f. Then, we notice the equality of the mappings

$${}^t f \circ \widetilde{\varphi}' \circ f = \widetilde{\varphi}$$

from E to $E^\star$: one has

$${}^t f \circ \widetilde{\varphi} \circ f(x) = {}^t f \circ \widetilde{\varphi}'(f(x)),$$

and $\widetilde{\varphi}'(f(x))$ is the linear form E' that maps y' to $\varphi'(f(x), y)$, so that ${}^t f \circ \widetilde{\varphi}' \circ f(x)$ is the linear form on E which maps y, to $\varphi'(f(x), f(y))$ and this is, by definition, equal to $\varphi(x, y)$. We have thus the expected equality. As f is an isomorphism, $\widetilde{\varphi}$ is an isomorphism if and only if $\widetilde{\varphi}'$ is.

If we want more precise results, we must specify which field we consider. We concentrate here on **R** and **C**. The case of $\mathbf{F}_q$ is easy too (see Exercise VI.59).

Corollary 8.4. *Two complex quadratic forms are equivalent if and only if they are defined on spaces of the same dimension and if they have the same rank.* □

Corollary 8.5 (The inertia law of Sylvester). *Two real quadratic forms are equivalent if and only if they are defined on spaces of the same dimension and can be written*

$$q(x) = \sum_{i=1}^{s} x_i^2 - \sum_{i=s+1}^{s+r} x_i^2, \qquad q'(y) = \sum_{i=1}^{s'} y_i^2 - \sum_{i=s'+1}^{s'+r'} y_i^2$$

with $r = r'$ and $s = s'$.

Remark 8.6. In particular, the ordered pair (s, r) depends only on the quadratic form q and not on the basis in which it is written. It is called the *signature* of the quadratic form. The rank of the quadratic form is $R = s + r$.

Proof. If $r = r'$ and $s = s'$, it is clear that the two forms are equivalent. Conversely, to prove that these numerical data coincide when the forms are equivalent, it suffices to characterize them geometrically. Let us define

$$\begin{cases} \sigma = \sup\{\dim F \mid F \text{ subspace of } E \text{ and } q|_F \text{ is positive definite}\} \\ \rho = \sup\{\dim F \mid F \text{ subspace of } E \text{ and } q|_F \text{ is negative definite}\} \end{cases}$$

and let $(e_1, \ldots, e_n)$ be a basis in which the quadratic form q is written

$$q(x) = \sum_{i=1}^{s} x_i^2 - \sum_{i=s+1}^{s+r} x_i^2.$$

Let A be the subspace spanned by $(e_1, \ldots, e_s)$. The form q is positive definite on A thus one has $\sigma \geqslant s$ and similarly $\rho \geqslant r$.

Let B be the subspace spanned by $(e_{s+1}, \ldots, e_r, \ldots, e_n)$ and let F be any subspace such that the restriction $q|_F$ is positive definite. Then the intersection $F \cap B$ is 0. We thus have

$$\dim F + \dim B \leqslant n.$$

In other words, $\sigma + n - s \leqslant n$, that is $\sigma \leqslant s$. We thus have $\sigma = s$. The equality $\rho = r$ is proved in the same way. □

The method of Gauss. This is a constructive (computational) proof of the existence of orthogonal bases.

We prove the result by induction on the dimension n of the space E. Let us assume it is proved for all the quadratic forms in (at most) $n-1$ variables and consider a quadratic form q in n variables. Write it in a basis:

$$q(x) = \sum_{i,j} a_{i,j} x_i x_j.$$

Then, either there is a nonzero diagonal term, or there is none.

- If there is a nonzero diagonal term, up to a renumbering of the basis vectors, it can be assumed that $a_{1,1} = \lambda_1 \neq 0$. One then writes

$$\begin{aligned} q(x) &= \lambda_1 x_1^2 + 2A(x_2, \dots, x_n)x_1 + B(x_2, \dots, x_n) \\ &= \lambda_1 \left(x_1 + \frac{A}{\lambda_1} \right)^2 + \left(B - \frac{A^2}{\lambda_1} \right), \end{aligned}$$

  formulas in which A is a linear form and B a quadratic form. That is to say, one has written

$$q(x) = \lambda_1 x_1'^2 + C(x_2, \dots, x_n)$$

  where x_1' is a linear form in $x_1, \dots, x_n$ independent of $x_2, \dots, x_n$ and C a quadratic form in the $n-1$ variables $x_2, \dots, x_n$.
- If $a_{i,i} = 0$ for all i, as $q \neq 0$, there is a nonzero coefficient $a_{i,j}$. One can assume that this is $\lambda = a_{1,2}$. As above, one writes

$$q(x) = \lambda x_1 x_2 + A x_1 + B x_2 + C$$

  where A and B are linear forms and C is a quadratic form in the $n-2$ variables $x_3, \dots, x_n$. One factors out (by force):

$$q(x) = \lambda \left(x_1 + \frac{B}{\lambda} \right) \left(x_2 + \frac{A}{\lambda} \right) + C - \frac{AB}{\lambda},$$

  that is,

$$\frac{\lambda}{4} \left(\left(x_1 + x_2 + \frac{A+B}{\lambda} \right)^2 - \left(x_1 - x_2 - \frac{A-B}{\lambda} \right)^2 \right) + C - \frac{AB}{\lambda}.$$

  What is obtained is the difference of the squares of two independent linear forms (why are they independent?) in $x_1, \dots, x_n$ to which a quadratic form $(C - AB/\lambda)$ in which the variables x_1 and x_2 do not appear is added.

This study allows us to conclude by induction on n. □

Remark 8.7. This is indeed a practical method of reduction: this is the way you proceed when you need to write a given quadratic form as a linear combination of squares of independent linear forms.

Simultaneous orthogonalization[21]. We assume now that $\mathbf{K} = \mathbf{R}$ and that E is a *Euclidean* vector space. This means that it is already endowed with a positive definite quadratic form q. We investigate the possibility to "diagonalize" another form q' in an orthonormal (for q) basis. In other words, we are looking for a basis in which both the forms q and q' are written as sums of squares. The next result is very useful, for instance this is the one that says that the axes of a central conic are orthogonal (Proposition 2.1). It will also be used in Chapter VIII in relation to the curvature of surfaces.

Theorem 8.8. *If q is a positive definite quadratic form and q' any quadratic form, there exists an orthonormal basis for q which is orthogonal for q'.*

Proof. The proof I present here works by induction on the dimension n of the vector space E. There is no doubt that the result holds in dimension 1. Assume we have proved it in dimension $n-1$.

We are now in a space of dimension n. The form q is positive definite, therefore it defines a scalar product. Denote thus $q = \|\cdot\|^2$ and $\varphi(x,y) = x\cdot y$.

To begin with, we shall look for a unit vector e_1 in E such that, if x is orthogonal to e_1 for q (that is to say, if $e_1 \cdot x = 0$), then they are also orthogonal for q' (that is to say $\varphi'(e_1, x) = 0$).

Then it will be possible to decompose E as an orthogonal (for q) direct sum:

$$E = \langle e_1 \rangle \oplus \langle e_1 \rangle^\perp.$$

Using the property satisfied by e_1, the orthogonal of e_1 for q' contains the hyperplane $\langle e_1 \rangle^\perp$. It is thus equal, either to the whole space E, or to the orthogonal of e_1. In any case, if u is a vector in $\langle e_1 \rangle^\perp$, we have

$$q'(xe_1 + u) = \lambda x^2 + q'(u)$$

for some λ (which may be zero) and we can apply the induction hypothesis to $e_1^\perp$.

Let us thus prove that such a vector e_1 exists. As e_1 should have norm 1, it is natural to consider the unit sphere S of our Euclidean space. Define a function $f : E - \{0\} \to \mathbf{R}$ by

$$f(x) = \frac{q'(x)}{\|x\|^2}.$$

The sphere S being compact (the vector space E being finite dimensional) and the mapping $f|_S$ being continuous (why?), it reaches its maximum (for

[21] It is traditional to call the orthogonalization of quadratic forms "diagonalization", although this operation is different from the diagonalization of matrices and this terminology may be confusing. See Exercise VI.5.

instance) at a point e_1 of S. As we have

$$f(\lambda x) = f(x) \text{ for all } x,$$

the vector e_1 also gives the maximum of f on $E - \{0\}$.

Now, f is a differentiable mapping (why?) defined on the open subset $E - \{0\}$ of the normed vector space E. If it reaches an extremum at a point e_1, its differential must vanish at this point. Hence we have

$$\begin{aligned} 0 &= df_{e_1}(x) = \frac{1}{\|e_1\|^4}\left(2\varphi'(e_1,x)\|e_1\|^2 - 2q'(e_1)(e_1\cdot x)\right) \\ &= 2\varphi'(e_1,x) - 2q'(e_1)(e_1\cdot x) \quad \text{for all } x\in E. \end{aligned}$$

Thus, for any x in E, the orthogonality of x and e_1 (namely, the equality $e_1 \cdot x = 0$) implies the equality $\varphi'(e_1, x) = 0$. □

Remark 8.9. The function f reaches its maximum at e_1 and one has $f(e_1) = q'(e_1)$. The previous proof constructs by induction an orthonormal basis $(e_1, \ldots, e_n)$ which is orthogonal for q' and the vectors of which are critical points of f. In this basis, the form is written

$$q'(x) = \sum_{i=1}^{n} q'(e_i)x_i^2.$$

Witt's theorem. Another useful result on quadratic forms is the theorem of Witt, that asserts that, if q is nondegenerate and if its restrictions $q|_F$ and $q|_{F'}$ to two subspaces are equivalent by an isomorphism $F \to F'$, this isomorphism can be extended to the whole space.

Theorem 8.10 (Witt). *Let q be a nondegenerate quadratic form on a vector space E. Let F, F', be two subspaces of E and $f : E \to E'$ be a linear mapping such that $q \circ f(x) = q(x)$ for all $x \in F$ (f is an isometry from $q|_F$ to $q|_{F'}$). Then there exists a linear isomorphism $\widetilde{f} : E \to E$ that extends f and such that $q \circ \widetilde{f} = q$ ($\widetilde{f}$ is an isometry of q).*

This theorem has a lot of applications (see, *e.g.*, [**Ber77, Per96, Ser73**]). There is an application to quadrics in this chapter (Proposition 5.1). See also Exercise VI.64.

Proof. We first prove the theorem in the case where $q|_F$ is nondegenerate. The proof is by induction on the dimension r of F.

If $r = 1$, F is spanned by a vector x such that $q(x) \neq 0$. We put $y = f(x)$; we have $q(y) = q(f(x)) = q(x)$, so that

$$q(x + y) + q(x - y) = 4q(x) \neq 0.$$

Therefore, at least one of the two vectors $x+\varepsilon y$ (for $\varepsilon=\pm1$) is not isotropic and $H=(x+\varepsilon y)^{\perp_q}$ is a hyperplane containing $x-\varepsilon y$. If s_H is the reflection about H,

$$s_H(x+\varepsilon y)=-(x+\varepsilon y),\qquad s_H(x-\varepsilon y)=x-\varepsilon y$$

thus $-\varepsilon s_H$ extends f.

We assume now the theorem to be true for all subspaces of dimension $\leqslant r$ and consider a subspace F of dimension $r+1$ (such that $q\mid_F$ is nondegenerate). Let x_1 be a nonzero vector in F and let F_1 be its orthogonal in F. This is an r-dimensional subspace of E, $q\mid_{F_1}$ is nondegenerate, so that we can apply the induction hypothesis and find an extension f_1 of $f\mid_{F_1}$ to E. Notice that $f_1^{-1}\circ f\mid_{F_1}=\operatorname{Id}_{F_1}$. To extend f, it suffices to extend $f_1^{-1}\circ f$, so that we can (and will) assume that $f\mid_{F_1}=\operatorname{Id}_{F_1}$. We look at $x_1\in F_1^{\perp}$. Using the $r=1$ case or the induction hypothesis for $\langle x_1\rangle\subset F_1^{\perp}$, we deduce an isometry $f_2\circ f$ of $q\mid_{F_1^{\perp}}$ and the mapping $\operatorname{Id}_{F_1}\oplus f_2$ is the expected extension of Id_{F_1}.

This ends the proof of the case where $q\mid_F$ is nondegenerate. Let us assume now that this is not the case and let K_0 be the kernel of $q\mid_F$, and F_0 a complementary subspace

$$F=K_0\oplus F_0.$$

Let $x_1,\dots,x_s$ be a basis of K_0. We use the following lemma:

Lemma 8.11. *There exist vectors $y_1,\dots,y_s\in E$ such that on each plane $P_i=\langle x_i,y_i\rangle$, the restriction of q has a matrix $\begin{pmatrix}0&1\\1&0\end{pmatrix}$. The subspaces $F_0,P_1,\dots,P_s$ are pairwise orthogonal and in direct sum in E.*

Delaying the proof of the lemma, let us end the proof of Witt's theorem: we put $x_i'=f(x_i)\in F'$. Since f is an isometry of $q\mid_F$, the subspace $\langle x_1',\dots,x_s'\rangle$ is the kernel of $q\mid_{F'}$. Applying the lemma to F', we get vectors $y_1',\dots,y_s'$ and we extend f to $P_1\oplus\cdots\oplus P_s\oplus F_0$ simply by declaring that $\overline{f}(y_i)=y_i'$. Now we have an isometry of the restrictions of q

$$\overline{f}:\overline{F}=P_1\oplus\cdots\oplus P_s\oplus F_0\longrightarrow\overline{F'}=P_1'\oplus\cdots\oplus P_s'\oplus F_0'$$

and now, $q\mid_{\overline{F}}$ is nondegenerate by construction, so that we can apply the first case. □

Proof of the lemma. We construct the vectors $y_1,\dots,y_s$ by induction. We start with y_1. The vector x_1 is a nonzero vector of K_0. Notice that the quadratic form q restricted to $F_0^{\perp}$ is nondegenerate, so that there exists a vector y in $F_0^{\perp}$ such that $\varphi(x_1,y)=a\neq0$, thus there exists a vector y' such that $\varphi(x_1,y')=1$ and $y_1=y'-\dfrac{q(y)}{2}x_1$ has the expected properties. Let

$P_1 = \langle x_1, y_1 \rangle$, we consider then

$$F_1 = P_1 \oplus \langle x_2, \dots, x_s \rangle \oplus F_0.$$

This is an orthogonal direct sum, and now the kernel of $q|_{F_1}$ is $\langle x_2, \dots, x_s \rangle$ and the result follows by induction. □

Exercises and problems

Easy exercises on quadratic forms[22]

Exercise VI.1. Prove that

$$\varphi(x, y) = \frac{1}{2} \sum_{i=1}^{n} y_i \frac{\partial q}{\partial x_i}(x_1, \dots, x_n)$$

and that

$$\varphi(x, y) = \frac{1}{2}(d^2 q)_0(x, y).$$

Exercise VI.2. Let $Q(E)$ be the set of all quadratic forms on the n-dimensional vector space E. Prove that $Q(E)$ is a vector space of dimension $n(n+1)/2$.

Exercise VI.3. Let f and g be two linear forms on a vector space E of dimension n. Prove that, for $n \geqslant 3$, the quadratic form

$$q(x) = f(x)g(x)$$

is degenerate.

Exercise VI.4. Let $(e_1, \dots, e_n)$ be a basis of E and let φ be a symmetric bilinear form whose matrix in this basis is A. What is the matrix of $\widetilde{\varphi} : E \to E^\star$ if $E^\star$ is endowed with the basis dual to $(e_1, \dots, e_n)$?

Prove that φ is nondegenerate if and only if the matrix A is invertible.

Exercise VI.5 (Diagonalization of real symmetric matrices). Let A be an $n \times n$ real symmetric matrix. Prove that there exists an orthonormal basis in which it is diagonal. To avoid any ambiguity: we are indeed speaking of diagonalizing A as the matrix of an endomorphism, namely of finding an invertible matrix P such that $P^{-1}AP$ is diagonal.

---

(22) Other exercises on quadratic forms can be found among the "more theoretical" exercises at the end of this chapter.

Exercise VI.6 (Another case of simultaneous orthogonalization). Let q and q' be two quadratic forms on an n-dimensional vector space E. Assume that the composed endomorphism $\widetilde{\varphi}^{-1} \circ \widetilde{\varphi}'$ of E has n distinct eigenvalues. Prove that there exists a basis of E that is orthogonal both for q and q'.

Exercise VI.7. Prove that the rank of the quadratic form q is the rank of the linear mapping $\widetilde{\varphi}$. Determine the ranks of the quadratic forms

- in an n-dimensional vector space, with $n \geqslant 2$, x_1^2, $x_1^2 - x_2^2$, $x_1^2 + x_2^2$, and $2x_1x_2$,
- in an n-dimensional vector space, with $n \geqslant r$, $\sum_{i=1}^{r} \lambda_i x_i^2$.

Exercise VI.8. Write the polar forms, then reduce over $\mathbf{R}$, the following quadratic forms:

- in $\mathbf{R}^4$, $x^2 + y^2 + 2(z^2 + t^2) + xz + zt + tx$,
- in $\mathbf{R}^3$, $xy + yz + zx$.

Exercises on affine and Euclidean conics and quadrics[23]

Exercise VI.9 (Hyperbolic paraboloid). Does the quadric in $\mathbf{R}^3$ of equation $z = xy$ have a center? Draw it and prove that it is the union of a family of lines[24].

Exercise VI.10. Construct in $\mathbf{R}^3$ a central affine quadric with infinitely many centers. Same question with no center at all.

Exercise VI.11 (Affine quadrics in 3-dimensional space). Assume that the quadratic part q of an affine quadric is a nondegenerate quadratic form. Let Ω be the center of this quadric. Complete the following table, making precise what the dots represent[25].

$q = 1$ and $\Omega \notin \mathcal{C}$	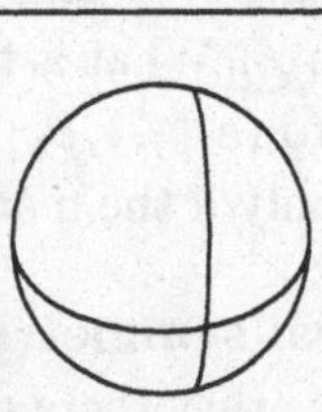	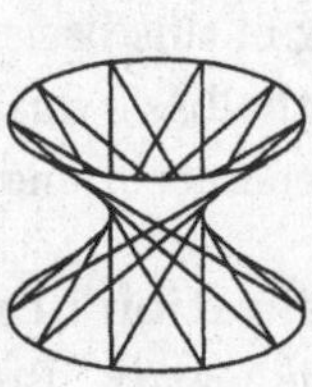	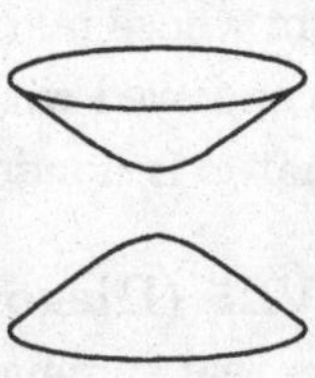	$\varnothing$
signature of q	$(3,0)$	$(2,1)$	$(1,2)$	$(0,3)$
$\Omega \in \mathcal{C}$	.	.	.	.
signature of q	$(3,0)$	$(2,1)$	$(1,2)$	$(0,3)$

---

(23) Other exercises on conics can be found in Chapter VII and many others in [**LH61**].

(24) This is a *ruled* surface; see Chapter VIII.

(25) The proper quadrics are shown in Figure 20.

Filling is the blank spaces in the next table, complete the affine classification of quadrics in 3-dimensional space.

rank of q	Ω, center	position of Ω
2	a line D of Ω	$D \not\subset \mathcal{C}$
		$D \subset \mathcal{C}$
	no Ω	
1	a plane P of Ω	$P \not\subset \mathcal{C}$
		$P \subset \mathcal{C}$
	no Ω	

Exercise VI.12. What would be the classification of conics in a *complex* affine plane?

Exercise VI.13. We are in a Euclidean plane. Describe the sets defined in an orthonormal frame by the equations

$$x^2 - 2xy + y^2 + \lambda(x+y) = 0, \quad x^2 + xy + y^2 = 1, \quad xy + \lambda(x+y) + 1 = 0,$$

$$y^2 = \lambda x^2 - 2x, \quad x^2 + xy - 2y + \lambda x + 1 = 0.$$

Exercise VI.14. A hyperbola is given. Write its equation in an orthonormal frame with origin at the center and axes the asymptotes.

Exercise VI.15. In an affine Euclidean plane endowed with an orthonormal frame with origin at O, is it possible to write the equation of a parabola, knowing that:

- its vertex is O, its axis is the x-axis and its parameter is 2?
- its focus is $F = (4, 3)$, its directrix $D : y = -1$? Determine its vertex and its parameter.

Exercise VI.16. In a real affine plane, a proper conic is given. Determine the points of the plane from which two (*resp.* one, zero) tangents to $\mathcal{C}$ can be drawn.

Exercise VI.17. What can be said of the image of an ellipse under an affine mapping? of that of a circle?

Exercise VI.18. A circle is given in a plane of the Euclidean space of dimension 3. The space is projected onto a plane. What is the image of the circle?

Exercise VI.19. Prove that the line from the center of an ellipse to the midpoint of a chord MM' passes through the common point of the tangents to the ellipse at M and M'.

Exercise VI.20 (Conjugated diameters of an ellipse). In the Euclidean plane endowed with an orthonormal frame of origin O, consider the quadratic form

$$q(x,y) = \frac{x^2}{a^2} + \frac{y^2}{b^2}$$

(with $0 < b < a$). If (u, v) is an orthonormal basis for q, it is said that u and v are *conjugated diameters* of the ellipse $\mathcal{C}$ of equation $q = 1$. Assume that $\overrightarrow{OM}$ and $\overrightarrow{OM'}$ are conjugated diameters of $\mathcal{C}$. Let P be the intersection point of the tangents to $\mathcal{C}$ at M and M'. Prove that:

- The quadrilateral $OMPM'$ is a parallelogram of (constant) area equal to ab (first theorem of Apollonius).
- The quantity $OM^2 + OM'^2$ is constant, equal to $a^2 + b^2$ (second theorem of Apollonius).

Exercise VI.21. Parametrize a branch of the hyperbola of equation

$$\frac{x^2}{a^2} - \frac{y^2}{b^2} = 1$$

with the help of hyperbolic functions.

Exercise VI.22. Prove that a parabola is contained in one of the half-planes determined by any of its tangents.

Exercise VI.23. Two parabolas are given in an affine Euclidean plane. Prove that they are (directly) similar.

Two proper conics $\mathcal{C}$ and $\mathcal{C}'$ are given in an affine Euclidean plane. To which condition(s) are they similar?

Exercise VI.24. A proper conic $\mathcal{C}$ is given in an affine Euclidean plane. What is the group of isometries that preserve $\mathcal{C}$?

Exercise VI.25. Let M be a point on a parabola $\mathcal{P}$ of vertex S. The normal[(26)] to $\mathcal{P}$ at M intersects the axis at N, the tangent intersects it at T. Let m be the orthogonal projection of M on the axis.

Prove that mN does not depend on M. What is its value?

Prove that S is the midpoint of mT. What is the midpoint of NT?

(26) The line through M that is perpendicular to the tangent.

Exercise VI.26. Prove that any proper conic can be represented in an orthonormal frame by an equation

$$y^2 = 2px + qx^2$$

with q real and $p > 0$.

Exercise VI.27. Write the equation of a conic of focus F in polar coordinates with origin at F.

Exercise VI.28. In an affine Euclidean plane, a line D, a point F not on D and a nonnegative number e are given. Describe the set

$$\{M \mid MF \leqslant ed(M, D)\}.$$

Exercise VI.29. In an affine Euclidean plane, two points F and F' are given. Let $a = \frac{1}{2}FF'$. What is the set of points M such that $MF + MF' = 2a$? such that $MF - MF' = 2a$?

Exercise VI.30. In an affine Euclidean plane, two points F and F' and a positive real number a are given. Describe the sets

$$\{M \mid MF + MF' \leqslant 2a\}, \qquad \{M \mid MF' - MF \leqslant 2a\}.$$

Exercise VI.31. A circle $\mathcal{C}$ of center F and a point F' inside this circle are given. What is the set of centers of the circles that are tangent to $\mathcal{C}$ and passing through F'? Same question with F' outside $\mathcal{C}$.

Exercise VI.32. Let $\mathcal{C}$ be a hyperbola of eccentricity e, F one of its foci and D the corresponding directrix. Let Δ be a line parallel to an asymptote of $\mathcal{C}$. Let N be a point of Δ and n be its projection on D. Prove that, when N tends to infinity on Δ, the ratio NF/Nn tends to e.

Let M be a point of $\mathcal{C}$. The perpendicular to MF at F intersects D at P. Prove that MP is not parallel to an asymptote[27] of $\mathcal{C}$.

Exercise VI.33 (The motion of planets). According to Kepler's laws, the planets move on plane trajectories that are described in polar coordinates (ρ, θ) by an equation $\rho = f(\theta)$, where the function f satisfies the differential equation

$$\frac{1}{f} + \frac{d^2}{d\theta^2}\left(\frac{1}{f}\right) = \text{constant}$$

(the constant depends on the masses and of universal constants; it is nonzero). Determine the nature of these trajectories.

---

(27) This completes the "other" proof of Proposition 2.13.

Exercise VI.34. Three lines in general position (not concurrent and any two of them not parallel) are given in an affine Euclidean plane. Assume that a parabola $\mathcal{P}$ is tangent to these three lines. Prove that the three projections of the focus F of $\mathcal{P}$ on the three lines are collinear on the tangent at the vertex. Deduce that F is on the circumcircle of the triangle determined by the three lines. What can be said of the tangent at the vertex? of the directrix?

What is the locus of the foci of the parabolas tangent to the three lines (be careful with the vertices of the triangle)?

Exercise VI.35. Four lines in general position are given in an affine Euclidean plane. Prove that there exists a unique parabola which is tangent to these four lines.

Exercise VI.36 (Conic sections). Investigate the intersection of an affine plane with the cone(28) of revolution of equation $x^2 + y^2 = z^2$.

How many spheres are inscribed in the cone and tangent to the plane? How can we interpret the contact points of these spheres with the plane (another Belgian story, this time a good one)?

Exercises on projective conics and quadrics

Exercise VI.37. Let D be a line in an affine space that intersects a quadric at a (unique and simple) point. Prove that, in the projective completion, it intersects the quadric at infinity as well. In an affine plane, check that these secants are:

- the parallels to the asymptotes if the conic is a hyperbola,
- the parallels to the axis if this is a parabola.

Exercise VI.38. Prove that the intersection of an affine quadric $\mathcal{C}$ with the hyperplane at infinity is a quadric of equation the quadratic part of an equation of $\mathcal{C}$. Reinterpret the equation of the asymptotes of a hyperbola.

Exercise VI.39 (Nullstellensatz). Assume that $\mathbf{K} = \mathbf{C}$. Prove that the mapping which, to a projective quadric associates its image, is injective(29) (from $PQ(E)$ into the set of subsets of $P(E)$). One can begin with the case where $P(E)$ is a line then use the intersection of the quadric with the lines in the general case.

What happens when $\mathbf{K} = \mathbf{R}$?

---

(28) This is the way which Apollonius used to define conics, and this explains their name. This is the definition which gives rise to the magnificent woodcuts of [**Dür95**].

(29) This is a special case of a theorem of Hilbert, the *theorem of zeroes*, in German *Nullstellensatz*, that asserts that the analogous property holds for more general equations. See, *e.g.*, [**Per95**].

Exercise VI.40. Prove that a proper nonempty quadric of a real projective space of dimension 3 is homeomorphic to the 2-dimensional sphere S^2 or to the Cartesian product $\mathbf{U} \times \mathbf{U}$ of two circles.

Exercise VI.41 (Lines contained in a quadric). Let $\mathcal{C}$ be a proper quadric in a complex projective space of dimension 3. Prove that the quadric $\mathcal{C}$ is the union of a family of lines[30]. What can be said of the real quadrics?

Prove that $\mathcal{C}$ is homeomorphic to $\mathbf{P}_1(\mathbf{C}) \times \mathbf{P}_1(\mathbf{C})$.

Prove that any (affine or projective) complex quadric contains (affine or projective!) lines.

Exercise VI.42. Three lines are given in the space of dimension 3. Prove that they are contained in a quadric.

Exercise VI.43 (Pencils of conics). Consider the pencils of conics defined by the seven pairs of conics of Figure 15. Among these pictures, two pairs define (respectively) the same pencil. Which ones? For any of the five different remaining pencils, draw (at least) one proper conic and *all* the degenerate conics.

Draw (at least) one pencil of *real* conics that does not appear in this list... By the way, find again the classification of pencils of circles of § III-4 (see also Exercise VI.52).

Exercise VI.44 (Equations of degree 4). We want to prove and to explain that, to solve a degree-4 equation

$$x^4 + ax^3 + bx^2 + cx + d = 0$$

amounts to solving a degree-3 and a few degree-2 equations (this allowing you to solve the equations of degree-4 by radicals once you know how to do it for those of degree 2 and 3). Put $y = x^2$ and notice that the solutions of the equation are the abscissas of the intersection points of two plane conics. Think of the pencil generated by these two conics and conclude.

Exercise VI.45. Let $\mathcal{C}$ be a degenerate conic consisting of two secant lines in the plane. Let m be a point in the plane. What can be said of the orthogonal of m with respect to $\mathcal{C}$?

Exercise VI.46 (Construction of the polar). Let $\mathcal{C}$ be a conic and m be a point of the plane. Two lines through m, that intersect $\mathcal{C}$ at x, y for the first one, z, t for the second, are drawn. The lines xz and yt intersect at a point n. Prove that n belongs to the polar of m with respect to $\mathcal{C}$ (Figure 27).

---

(30) Of two, actually.

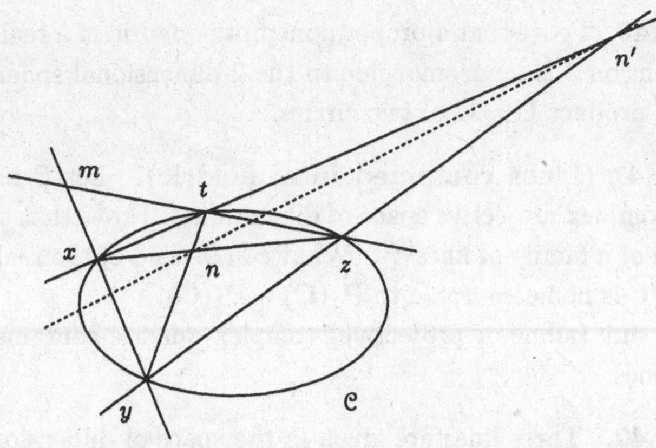

Fig. 27

Let n' be the intersection point of xt and yz. Prove that the line nn' is the polar of m.

Exercise VI.47 (Lamé's theorem). Let $\mathcal{F}$ be a pencil of conics and let A be a point of the plane. Prove that the polars of A with respect to the conics of $\mathcal{F}$ are concurrent.

Exercise VI.48. Let $\mathcal{C}$ be the conic that is the union of the two lines $\mathcal{D}$ and $\mathcal{D}'$. What does Pascal's theorem say if a, c and e are chosen on $\mathcal{D}$, and b, d and f on $\mathcal{D}'$?

Exercise VI.49. Let $\mathcal{C}$ be a conic[31] of an affine plane and let D_1, D_2, D_3 be three directions of lines. Let M_0 be a point of $\mathcal{C}$, M_1 the other intersection point of the parallel to D_1 passing through M_0 and $\mathcal{C}$, M_2 the other intersection point of the parallel to D_2 passing through M_1 and $\mathcal{C}$, *etc.* This way, points M_i are defined for $i \geqslant 0$. Prove that $M_6 = M_0$.

Exercise VI.50. Five points a, b, c, d and e are given in a plane. Assume there is a unique proper conic through these five points. Construct (with a ruler) an arbitrary point f of $\mathcal{C}$ and the tangent to $\mathcal{C}$ at f.

Exercise VI.51. Four points A, B, C and D of a plane, any three of them not collinear, and a scalar k are given. Describe the set

$$\{M \in P \mid [MA, MB, MC, MD] = k\}.$$

Exercise VI.52. In a pencil of circles, there is a line, the radical axis. In a pencil of conics, there are only conics. Is this a contradiction?

[31] The case where $\mathcal{C}$ is a circle in a Euclidean plane was proposed as an exercise in Chapter III (Exercise III.31).

Exercise VI.53. Let $\mathcal{C}$ be a conic of an affine Euclidean plane. What can be said of $\mathcal{C}$ knowing that the circular points I and J are conjugated with respect to $\mathcal{C}$ (that is, knowing that the line IJ intersects $\mathcal{C}$ at A and B with $[I, J, A, B] = -1$)?

Exercise VI.54. Prove all the affirmations on circles, pencils and inversions made in § 7 of which you think that the proofs given there are not complete.

Exercise VI.55. Investigate, in the terms of § 7, the various possibilities for the image of a pencil by an inversion listed in § III-4.

Exercise VI.56 ("Anallagmatic invariant", continuation). Let C and C' be two circles in the plane $\mathcal{P}$, of respective radii R and R'. The pencil $\mathcal{F}$ they define contains two point-circles Γ and Γ', the two (possibly imaginary, possibly equal) intersection points of $\mathcal{F}$ with the quadric $\widehat{\mathcal{P}}$. Calculate the cross-ratio of the four points $[C, C', \Gamma, \Gamma']$ of the line $\mathcal{F}$ in terms of R, R' and the distance d of the centers of C and C'. Deduce that the ratio

$$\frac{R^2 + R'^2 - d^2}{2RR'}$$

is invariant by inversion (see also Exercise III.54).

More theoretical exercises

Exercise VI.57 (Over Q). Consider the field $\mathbf{Q}$ of rational numbers. Prove that, on a vector space of dimension 1 over $\mathbf{Q}$, there are infinitely many nonequivalent quadratic forms.

Exercise VI.58 (Discriminant). Let q be a nondegenerate quadratic form on a finite-dimensional vector space over a field $\mathbf{K}$. Show that the determinant of the matrix of q in a basis defines an element $\delta(q) \in \mathbf{K}^\star/(\mathbf{K}^\star)^2$ which is an invariant of the isomorphism type of q.

Exercise VI.59 (Over $\mathbf{F}_q$). Let $\mathbf{F}_q$ be a finite field of characteristic different from 2. How many squares are there in $\mathbf{F}_q$? Deduce that, for all $a, b \in \mathbf{F}_q^\star$, there exist two elements x, y of $\mathbf{F}_q$ such that

$$ax^2 + by^2 = 1.$$

Now let E be an n-dimensional vector space over $\mathbf{F}_q$. Let Q be a nondegenerate quadratic form on E. Let $a \in \mathbf{F}_q^\star$ be any element which is *not* the square of an element of $\mathbf{F}_q^\star$. Prove that Q is equivalent, either to

$$x_1^2 + \cdots + x_{n-1}^2 + x_n^2$$

or to

$$x_1^2 + \cdots + x_{n-1}^2 + ax_n^2$$

and that the two possibilities are mutually exclusive. Given a nondegenerate quadratic form, how do you decide which type it has?

Exercise VI.60 (Square root of a positive definite symmetric real matrix). Let A be a real symmetric matrix. Assume that the associated bilinear form, namely the form φ defined by

$$\varphi(x, y) = {}^t x A y$$

is positive definite.

(1) Prove that the eigenvalues $\lambda_1, \ldots, \lambda_n$ of A are positive real numbers.

(2) Let P be a polynomial such that

$$P(\lambda_i) = \sqrt{\lambda_i} \qquad 1 \leqslant i \leqslant n.$$

Prove that $S = P(A)$ is a positive definite symmetric matrix such that $S^2 = A$.

Exercise VI.61 (Polar decomposition of $GL(n; \mathbf{R})$). Prove that any invertible real matrix M can be written as a product

$$M = \Omega S$$

where $\Omega \in O(n)$ is an orthogonal matrix and S is a positive definite symmetric matrix. Prove that this decomposition is unique.

Exercise VI.62 (Decomposition "of Cartan"). Prove that any invertible real matrix can be written as a product

$$M = \Omega_1 D \Omega_2$$

where D is a diagonal matrix with positive eigenvalues and Ω_1, Ω_2 are orthogonal matrices.

Exercise VI.63. Let Q be a quadratic form on a vector space E. Let F be a vector subspace of E. Assume that F does not consist of vectors isotropic for Q. Define the intersection of the projective quadric defined by Q with $P(F)$. What happens when all the vectors of F are isotropic?

Let D be a projective line. What can be said of the intersection of D with the quadric defined by Q?

Exercise VI.64. In $\mathbf{R}^3$, consider the quadratic form $q(x, y, z) = x^2 + y^2 - z^2$. Let $O(q)$ be its isometry group, that is, $O(q) = \{f \in GL(3; \mathbf{R}) \mid q \circ f = q\}$. What are the orbits of its action on $\mathbf{P}_2(\mathbf{R})$ (as a subgroup of $GL(3; \mathbf{R})$)?

Exercise VI.65 (Quadrics and projective duality). If $\mathcal{C}$ is a curve in a projective plane $P(E)$ one defines $\widetilde{\mathcal{C}} \subset E$ by

$$v \in \widetilde{\mathcal{C}} \text{ if and only if } p(v) \in \mathcal{C}.$$

(1) Check that $\widetilde{\mathcal{C}}$ is a *cone*[32]. It is said that a projective line d of $P(E)$ is tangent to $\mathcal{C}$ at m if $d = P(F)$ for some plane F of E tangent[33] to $\widetilde{\mathcal{C}}$ along the line m of E. Check that, if $\mathcal{C}$ is a conic, the tangent at m to $\mathcal{C}$ thus defined coincides with that defined in § 3.

(2) The set of projective lines tangent to $\mathcal{C}$ constitute a curve $\mathcal{C}^\star$ of $P(E^\star)$. Prove that if $\mathcal{C}$ is a proper conic, $\mathcal{C}^\star$ is also a conic[34]. More generally, if $\mathcal{C}$ is a quadric of a projective space $P(E)$ of dimension n, prove that the family of all the projective hyperplanes tangent to $\mathcal{C}$ is a quadric of the projective space $P(E^\star)$ (see if necessary Exercise V.2).

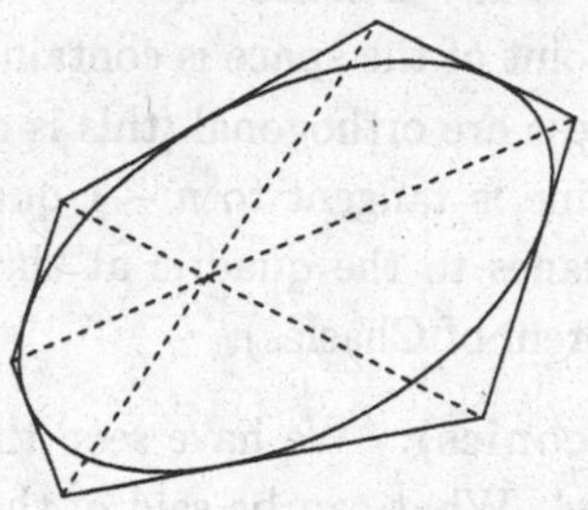

Fig. 28. Brianchon's theorem

(3) Coming back to the case of the plane, how do the intersection properties of $\mathcal{C}^\star$ with the lines of $P(E^\star)$ translate in $P(E)$? Prove Brianchon's theorem: if all the sides of a hexagon are tangent to a proper conic, then its diagonals are concurrent (Figure 28).

(4) Five general lines are given in a projective plane. How many conics tangent to these lines are there? Four general lines are given in an affine plane. How many parabolas tangent to these four lines are there (see also Exercise VI.35)?

Exercise VI.66 (Confocal families). In a Euclidean plane endowed with an orthonormal frame, two real numbers α and β are given (assuming that $0 < \alpha < \beta$). Consider the conic $\mathcal{C}_\lambda$ of equation

$$\frac{x^2}{\alpha - \lambda} + \frac{y^2}{\beta - \lambda} = 1.$$

Draw on the same figure the curves $\mathcal{C}_\lambda$ for $\lambda < \alpha$, $\alpha < \lambda < \beta$, $\beta < \lambda$. Prove that all the $\mathcal{C}_\lambda$ have the same foci. The conics $\mathcal{C}_\lambda$ are said to be *confocal*.

---

[32] That is to say, if $\widetilde{\mathcal{C}}$ contains v, it contains the whole line generated by v.

[33] See if necessary Chapter VIII, at least in the real case.

[34] Figure 1 of Chapter VII represents the curve $\mathcal{C}^\star$ when $\mathcal{C}$ is an ellipse.

Complete the plane in a projective plane and consider the family of conics dual to that of $\mathcal{C}_\lambda$ (as in Exercise VI.65). Prove that this is a (linear) pencil of conics.

More generally, consider, in a Euclidean affine space of dimension n, the quadrics $\mathcal{C}_\lambda$ of equations

$$\frac{x_1^2}{\alpha_1-\lambda}+\cdots+\frac{x_n^2}{\alpha_n-\lambda}=1,$$

where the α_i's are fixed real numbers such that $0<\alpha_1<\cdots<\alpha_n$. Check that the dual family is a pencil of quadrics. By analogy with the case of conics, it is said that this family is *confocal*.

Prove that a general point of the space is contained in exactly n quadrics of the family and that these are orthogonal (this is a theorem of Jacobi).

Prove that a general line is tangent to $n-1$ quadrics of the family and that the tangent hyperplanes to the quadric at the points of tangency are orthogonal (this is a theorem of Chasles).

Exercise VI.67 (Real conics). We have seen that the complement of a line in $\mathbf{P}_2(\mathbf{R})$ is connected. What can be said of the complement of a conic in $\mathbf{P}_2(\mathbf{R})$?

Exercise VI.68 (Real cubics). Consider, in $\mathbf{R}^2$, the curve $\mathcal{C}$ of equation $y^2=P(x)$, where P is a polynomial of degree 3 which we assume has no multiple root. Draw $\mathcal{C}$. How many connected components does it have? Prove that one of the components is unbounded and homeomorphic to a line and that, if there exists another component, it is homeomorphic to a circle.

Consider now the curve $\widehat{\mathcal{C}}$ of homogeneous equation $y^2z=\widehat{P}(x,z)$ (where $\widehat{P}$ is the homogeneous polynomial of degree 3 in two variables obtained by homogenization of P). Prove that $\widehat{\mathcal{C}}$ is obtained by adding a point to $\mathcal{C}$. How many connected components does $\widehat{\mathcal{C}}$ have? To what are they homeomorphic?

Prove that the complement in $\mathbf{P}_2(\mathbf{R})$ of the component of $\widehat{\mathcal{C}}$ that contains the point at infinity is connected. If $\mathcal{C}$ has a second connected component, what can be said of the complement of this component in $\mathbf{P}_2(\mathbf{R})$?

Exercise VI.69. A bilinear form φ is *alternated* if it satisfies

$$\varphi(x,x)=0 \text{ for all } x.$$

Check that this property is equivalent to the skew-symmetry of φ, that is,

$$\varphi(x,y)=0 \text{ for all } x \text{ and } y$$

(assuming that the characteristic of the ground field is not 2).

Let φ be a *nondegenerate* alternated bilinear form. Prove that there exists a basis

$$(e_1,\ldots,e_n,f_1,\ldots,f_n)$$

of E such that, for all i and j,

$$\varphi(e_i, e_j) = \delta_{i,j} \qquad \varphi(e_i, e_j) = \varphi(f_i, f_j) = 0.$$

What can be said of the dimension of E? Let F be an isotropic vector subspace[35] of E. What can be said of the dimension of F?

Let A be a skew-symmetric matrix. Prove that the rank of A is an even number.

(35) That is, $\varphi(x, y) = 0$ for all x and y of F.

Chapter VII

Curves, Envelopes, Evolutes

Curves appear in a wide variety of mathematical problems and in very many ways. Here is a list of examples (in a real affine space):

- in kinematics, *e.g.*, as integral curves of vector fields, in other words as solutions of differential equations;
- a variant: as envelopes of families of lines, or of families of other curves;
- they also appear, in space, at the intersection of two surfaces (*e.g.*, of a surface and a plane), therefore, they are useful for the study of surfaces (see Chapter VIII);
- as the solutions of an algebraic equation $P(x, y) = 0$ (as conics in the plane, for instance), or of several algebraic equations ($P_1(x, y, z) = P_2(x, y, z) = 0$ in space[(1)] of dimension 3, *etc.*).

They are described in very different ways in these examples, either as parametrized curves, namely as images of mappings $t \mapsto f(t)$ defined on an interval of $\mathbf{R}$, or by Cartesian equations.

It is important be able to pass from one presentation to another. For instance, given a curve of equation $P(x, y) = 0$ in the plane, is it possible to parametrize it? How? Is there a parametrization of the form $(t, g(t))$? The answers to these questions can be found in the implicit function theorem. I am not going to explain all that in this chapter, as there will be an analogous study, for the case of surfaces, in Chapter VIII.

There are nevertheless special cases where it is needed to be able to pass from a Cartesian equation to a parametrization without hesitation. For example, a straight line intersects a conic at two points, and this gives a parametrization of the conic by the family of lines through any of its points. The

(1) This is also an intersection of surfaces in this case.

same geometrical "trick" can be used in slightly different situations (see Exercises VII.1 and VII.2).

Local/global. There are local properties: singular points, position of the curve with respect to its tangent line, *etc.*, but there are also global properties, as for example, Jordan's theorem, that asserts that a simple closed curve divides the plane in two components, only one of them being bounded. In this book, we will consider only local properties, global properties being harder to prove(2). For the purpose of local study, it is enough to consider parametrized curves.

Affine/metric. Even staying at the local level, there are affine properties, as is, *e.g.*, the position of the curve with respect to its tangents, and there are metric properties, those in which the length and the curvature of a curve show up. We will look at these two types of properties.

The most beautiful theorems are both global and metric(3); these are the four vertex theorem, the isoperimetric inequality, which states that, among all the plane curves of a given length, the one that encircles the largest area is the circle, *etc.* The interested reader may look at [**BG88**].

In this chapter, I look first at envelope problems. Then, I remind the reader of a few results on the curvature of plane curves, in order to come back to envelopes in the framework of evolutes. The chapter is concluded by an appendix, giving a brief summary of a few useful definitions related to curves.

1. The envelope of a family of lines in the plane

We have already met families of lines in this text. For instance the Simson lines of a triangle (Exercise III.33): there is one Simson line for each point of the circumcircle. Another example is that of the family of all the tangents to a curve: there is a line associated with each point of the curve.

The problem of envelopes is precisely, given a family of lines in the plane, to decide whether there exists a curve, *the envelope*, the family of tangents of which is the given family.

This is not at all an artificial problem: the reader will have already met envelopes—possibly without knowing it. The bright curve the sun draws in

(2) Jordan's theorem is an example of a theorem which is easy to state, very intuitive, but not that easy to prove.

(3) It should be noticed that all the examples of global results given here are theorems on *plane* curves. The global properties of the simple closed curves in space are very interesting and are the object of a hard and lively theory, the theory of knots.

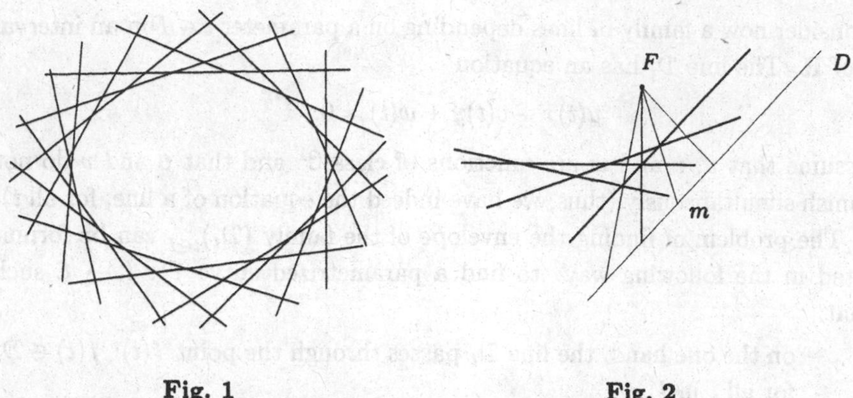

Fig. 1 Fig. 2

their cup of coffee(4) is one of them; the rainbow is another manifestation of the existence of envelopes(5).

Imagine that the lines of the family are drawn as in Figure 1. Although the envelope has not been drawn in this figure, it is visible: this is where the ink concentrates. Another example is given by Figure 38 of Dürer's book [**Dür95**]. The artist explains to his reader how to construct a "useful" curve(6)... by drawing a family of lines (this will be found in Exercise VII.6) and even explains how to make a wooden tool allowing one to draw the lines of this family.

The optical phenomena known as *caustics* have exactly the same nature: the light (and therefore the heat(7)) concentrates along the envelope.

As an introduction, let us begin with a very simple examples, given by the tangential properties of conics studied in § VI-2.

The parabola as an envelope. Let D be a line in an affine Euclidean plane and let F be a point out of D. What is the envelope of the family of perpendicular bisectors of the segments Fm when m belongs to D (Figure 2)? Using Corollary VI-2.16, it is clear that the parabola of focus F and directrix D is the solution. Several examples of this situation will be found in the exercises.

Envelope of a parametrized family. The affine plane $\mathcal{E}$ is endowed with a frame, so that an affine line $\mathcal{D}$ is described by an equation of the form

$$ux + vy + w = 0.$$

(4)The nature of liquid does not matter, at least for mathematical purposes.

(5)See Exercises VII.13, VII.14, VII.15.

(6)This is a parabola.

(7)Caustic: that burns.

Consider now a family of lines depending on a parameter $t \in I$ in an interval I of $\mathbf{R}$. The line $\mathcal{D}_t$ has an equation

$$u(t)x + v(t)y + w(t) = 0.$$

Assume that u, v and w are functions of class $\mathcal{C}^1$ and that u and v do not vanish simultaneously (thus, we have indeed the equation of a line, for all t).

The problem of finding the envelope of the family $(\mathcal{D}_t)_{t\in I}$ can be formulated in the following way: to find a parametrized curve $f : I \to \mathcal{E}$ such that:

- on the one hand, the line $\mathcal{D}_t$ passes through the point $f(t)$: $f(t) \in \mathcal{D}_t$ for all t in I
- and on the other hand, it is tangent to the curve at $f(t)$: the vector $f'(t)$ directs the line $\mathcal{D}_t$ for all t in I.

As we have described the lines by equations written in a frame, let us look for $f(t)$ *via* its coordinates ($f(t) = (x(t), y(t))$). The two conditions above can be translated as the system of equations

$$\begin{cases} u(t)x(t) + v(t)y(t) + w(t) &= 0 \\ u(t)x'(t) + v(t)y'(t) &= 0. \end{cases}$$

Let us differentiate the first equation with respect to t. Using the second one, we see that the system is equivalent to the following one:

$$\begin{cases} u(t)x(t) + v(t)y(t) + w(t) &= 0 \\ u'(t)x(t) + v'(t)y(t) + w'(t) &= 0. \end{cases}$$

This is nothing other than a *linear* system[8] in $x(t)$ and $y(t)$. There is a unique solution $(x(t_0), y(t_0))$ exactly when the determinant does not vanish:

$$\begin{vmatrix} u(t_0) & u'(t_0) \\ v(t_0) & v'(t_0) \end{vmatrix} \neq 0.$$

The system has solutions $(x(t), y(t))$ for all t in a neighborhood $I_0 \subset I$ of such a point t_0 (why?), this defining a parametrized curve that is a solution of our problem, the envelope of the family $(\mathcal{D}_t)_{t\in I_0}$.

Parallel lines, concurrent lines. It is *a priori* clear that a family of parallel lines cannot have an envelope, and similarly that the envelope of a family of concurrent lines is the "constant" curve that is their common point. Let us check that this is indeed what our equations say.

To say that all the lines $\mathcal{D}_t$ are parallel is to say that $u(t)$, $v(t)$ are proportional to some constants u, v:

$$u(t) = \lambda(t)u, \qquad v(t) = \lambda(t)v.$$

---

(8) Another interpretation of this system can be found in Exercise VIII.14.

Then

$$u'(t) = \frac{\lambda'(t)}{\lambda(t)} u(t), \qquad v'(t) = \frac{\lambda'(t)}{\lambda(t)} v(t)$$

and the determinant of the system is identically zero. The system can have a solution only if $w(t) = \lambda(t)w$, that is, if all the lines in the family coincide: $\mathcal{D}_t = \mathcal{D}_{t_0}$ and then the line $\mathcal{D}_{t_0}$ is itself the envelope.

To say that they are concurrent is to say that one can write for $\mathcal{D}_t$ an equation of the form

$$u(t)(x - x_0) + v(t)(y - y_0) = 0,$$

in other words that one has $w(t) = -x_0 u(t) - y_0 v(t)$. The linear system can be written

$$\begin{cases} u(t)(x(t) - x_0) + v(t)(y(t) - y_0) &= 0 \\ u'(t)(x(t) - x_0) + v'(t)(y(t) - y_0) &= 0, \end{cases}$$

so that it has the constant solution $(x(t), y(t)) = (x_0, y_0)$ and this is the only solution when the determinant is nonzero.

Examples 1.1

(1) Let $\mathcal{D}_t$ be the line of equation

$$3tx - 2y - t^3 = 0.$$

To find the envelope of the family of lines $\mathcal{D}_t$, one solves the system

$$\begin{cases} 3tx - 2y - t^3 = 0 \\ 3x - 3t^2 = 0. \end{cases}$$

This gives the parametric equations

$$\begin{cases} x = t^2 \\ y = t^3 \end{cases}$$

of the cuspidal cubic (see if necessary § 4)[9]. It should be noticed that there are solutions for all t, but this does not prevent the envelope from having a singular point.

(2) Look now at the family of lines $\mathcal{D}_t$ of equation

$$3t^2x - y - 2t^3 = 0,$$

giving the system

$$\begin{cases} 3t^2x - y - 2t^3 = 0 \\ 6tx - 6t^2 = 0. \end{cases}$$

[9] This example is very artificial: the line $\mathcal{D}_t$ has been constructed precisely as the tangent to the cuspidal cubic.

This system has a unique solution only if $t \neq 0$, in which case the solution is

$$\begin{cases} x = t \\ y = t^3. \end{cases}$$

This is the subset $x \neq 0$ of the graph of the function $f(x) = x^3$. Notice, however, that the tangent to this graph at the point corresponding to $t = 0$ is indeed the line $\mathcal{D}_0$ (here the x-axis).

Cycloidal curves. This is a very beautiful family of examples, some of them arising very naturally in numerous problems of geometry. The plane is now Euclidean and oriented (and identified with **C**). A circle $\mathcal{C}$ of center O and radius R is given. Two points M_1 and M_2 describe the circle with constant angular velocities ω_1 and ω_2 (Figure 3). One looks for the envelope of the family of lines M_1M_2.

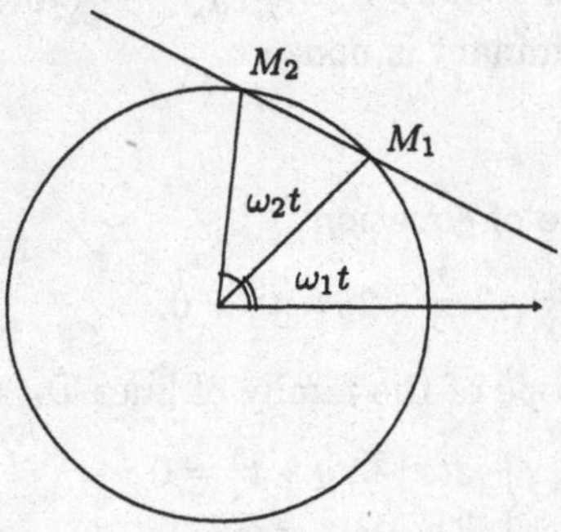

Fig. 3

Assume that $\omega_1 + \omega_2 \neq 0$. Let G be the barycenter of $((M_1, \omega_2), (M_2, \omega_1))$, so that

$$(\omega_1 + \omega_2)\overrightarrow{OG} = \omega_2\overrightarrow{OM_1} + \omega_1\overrightarrow{OM_2} = \omega_2 Re^{i\omega_1 t} + \omega_1 Re^{i\omega_2 t}.$$

Differentiating this relation with respect to t, one finds

$$(\omega_1 + \omega_2)\overrightarrow{V_G} = \omega_2\overrightarrow{V_{M_1}} + \omega_1\overrightarrow{V_{M_2}}$$

where $\overrightarrow{V_M}$ is the velocity of the point M. For instance, here

$$\overrightarrow{OM_1} = \begin{pmatrix} R\cos(\omega_1 t) \\ R\sin(\omega_1 t) \end{pmatrix} \text{ and } \overrightarrow{V_{M_1}} = \omega_1 \begin{pmatrix} -R\sin(\omega_1 t) \\ R\cos(\omega_1 t) \end{pmatrix},$$

so that the vectors $\omega_2\overrightarrow{V_{M_1}}$ and $\omega_1\overrightarrow{V_{M_2}}$ correspond to each other in the rotation of center O that maps M_1 to M_2 and that $\overrightarrow{V_G}$ and $\overrightarrow{M_1M_2}$ are collinear.

When t varies, the point G describes a curve whose tangent is parallel to the line M_1M_2 and, of course, it belongs to this line, being a barycenter of M_1 and M_2. This curve is thus the envelope we were looking for.

Putting $\theta = \omega_1 t$ and $m = \omega_2/\omega_1$, we get the equality

$$\overrightarrow{OG} = \frac{R}{m+1}\left(me^{i\theta} + e^{im\theta}\right),$$

and this is a representation, in polar coordinates, of the envelope.

Remarks 1.2

- In the case where $\omega_1 + \omega_2 = 0$ (or $m = -1$), this is a family of parallel lines, hence they have no envelope.
- The parameter θ is proportional to t; this is a measure of the angle from the x-axis to the vector $\overrightarrow{OM_1}$. The number m is positive if the two points go in the same direction, negative otherwise. The case where $m = 1$ corresponds to $M_1 = M_2$, in which case the line M_1M_2 is the tangent to the circle. One finds indeed $\overrightarrow{OG} = Re^{i\theta}$ in this case (the envelope is the circle $\mathcal{C}$). Therefore, we assume now that $m \neq \pm 1$.
- The singular points of the envelope are those at which the derivative
$$\frac{d}{d\theta}\overrightarrow{OG} = \frac{iRm}{m+1}\left(e^{i\theta} + e^{im\theta}\right)$$
vanishes. One finds $\theta = \dfrac{2k+1}{m-1}\pi$ for some integer k. For instance, if $m \in \mathbf{Z}$, there are $|m-1|$ points (the condition to have a finite number of points is that m is rational).
- The change of variables $\theta = \varphi + \dfrac{\pi}{m-1}$ gives
$$me^{i\theta} + e^{im\theta} = \exp\left(i\frac{\pi}{m-1}\right)\left(me^{i\varphi} - e^{mi\varphi}\right).$$
Cycloidal curves are often described by an equation of this form (here, there is a cusp point for $\varphi = 0$).
- The envelope is located in the interior of the circle $\mathcal{C}$ when $m > 0$—it is then said that this is an *epicycloid*—and in the exterior otherwise—this is then a *hypocycloid*.

Here is a list of examples corresponding to the small integral values of m.

- The three-cusped hypocycloid is obtained for $m = -2$. The line M_1M_2 is tangent three times to the circle $\mathcal{C}$ at points that are the vertices of an equilateral triangle; its envelope is tangent to the circle at the very same points!

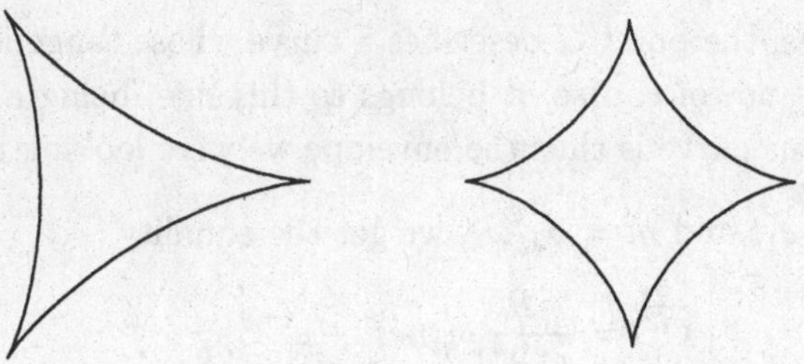

Fig. 4. A three-cusped hypocycloid and an astroid

- The four-cusped hypocycloid is obtained for $m = -3$; it is tangent four times to the circle $\mathcal{C}$. A glance at Figure 4 will be enough to understand why this curve is also called an *astroid*.

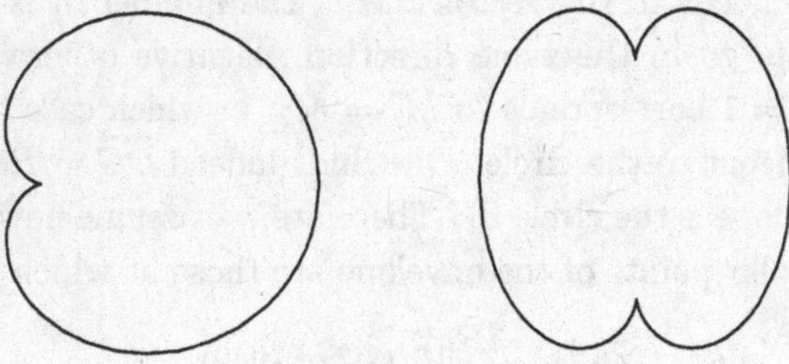

Fig. 5. A cardioid and a nephroid

- There is no one- or two-cusped hypocycloid (why?) but there are one-cusped epicycloids, obtained for $m = 2$. These are *cardioids* (Figure 5), curves having the shape of a heart.
- The two-cusped epicycloids are *nephroids* (Figure 5), curves having the shape of a kidney.

2. The curvature of a plane curve

In this section, we consider curves of an affine Euclidean space and we look at their metric properties.

Curvature of plane curves. We assume now that $\mathcal{E}$ is an *oriented* affine Euclidean *plane*.

Let $s \mapsto f(s)$ be a curve parametrized by arc length (see if necessary § 4). The tangent vector at the point defined by s is a unit vector, that we denote $\tau(s)$. Let $n(s)$ be the unique unit vector such that $(\tau(s), n(s))$ is a direct orthonormal basis of E. It is called the *normal* vector.

Differentiating the relation $\|\tau(s)\|^2 = 1$ with respect to s, it is seen that $\tau'(s)$ is perpendicular to $\tau(s)$, therefore this vector is collinear to $n(s)$. There exists thus a scalar $K(s)$ such that

$$\tau'(s) = K(s)n(s).$$

The number $K(s)$ is the *algebraic curvature* of the curve at the point of parameter s. The word "algebraic" simply means "signed". The sign depends on the orientation of the plane and of that given to the curve by the parameter s.

Since $\tau(s) \cdot n(s) = 0$, we get by differentiation the equality

$$\tau'(s) \cdot n(s) + \tau(s) \cdot n'(s) = 0.$$

Since $\|n(s)\|^2 = 1$, we see similarly that $n'(s)$ is collinear with $\tau(s)$, and thus also that

$$n'(s) = -K(s)\tau(s).$$

Remark 2.1. This result can also be formulated in the following way: the matrix that gives $d\tau/ds$ and dn/ds in the orthonormal basis (τ, n) is skew symmetric (see also Exercise VII.20).

The curvature $K(s)$ is positive when the curve lies (locally) in the half-plane defined by the tangent that contains the normal vector $n(s)$, that is, the curve "turns to the left". The curvature $K(s)$ vanishes when $\tau'(s) = 0$, in particular when the curve has a flex for the value s of the parameter (see § 4). For instance, the curvature of a line is constant and equal to zero.

Example 2.2. Let us consider a circle of radius R. The origin is chosen at its center; the affine Euclidean plane $\mathcal{E}$ is identified with $\mathbf{C}$. A parametrization by arc length is

$$s \longmapsto Re^{is/R},$$

for which $\tau(s) = ie^{is/R}$. Using the orientation of the plane given by $(1, i)$, we have

$$n(s) = -e^{is/R} \text{ and } \frac{d\tau}{ds} = -\frac{1}{R}e^{is/R} = \frac{1}{R}n(s).$$

The curvature of a circle is a constant, equal to the inverse of the radius (up to a sign depending on the orientation).

Curvature radius, curvature center. In analogy with the radius of a circle, we define, when $K(s) \neq 0$, the *curvature radius* $\rho(s)$ by the equality

$$\rho(s) = \frac{1}{K(s)},$$

and the *curvature center* $C(s)$, by

$$C(s) = g(s) + \rho(s)n(s).$$

We have used the notation relating points and vectors introduced in Remark I-1.3, especially suitable here. It is equivalent to define $C(s)$ by the vector equality

$$\overrightarrow{g(s)C(s)} = \rho(s)n(s).$$

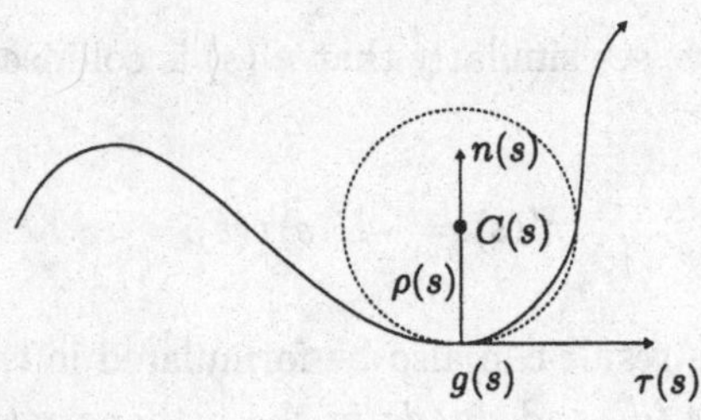

Fig. 6

The circle of center $C(s)$ and radius $|\rho(s)|$ is the *osculating circle*; this is the circle that best approximates[10] the curve at $g(s)$ (see Figure 6)—in a very precise sense; see Exercise VII.21.

Remarks 2.3

(1) While $K(s)$ and $\rho(s)$ indeed depend on the orientation chosen for the plane and that chosen on the curve, the same is not true of $C(s)$: to change the orientation changes both the sign of $n(s)$ and that of $\rho(s)$.

(2) When the curvature $K(s)$ is not zero, the curve and its curvature center are, locally in a neighborhood of $g(s)$, on the same side of the tangent (see also § 4).

3. Evolutes

As above, we are in an affine Euclidean plane. We are now interested in the envelope of the family of normals to a curve.

[10] This is why, in order to draw the curve in the neighborhood of $g(s)$, you should put your hand at the point $C(s)$.

Proposition 3.1. *Let $f : I \to \mathcal{E}$ be a parametrization of class $\mathcal{C}^3$ of a plane curve. Assume that the curvature has no zero on I. Let $\mathcal{D}_t$ be the normal to the curve at the point of parameter t. The envelope of the family of lines $(\mathcal{D}_t)_{t\in I}$ is the set of curvature centers.*

Proof. It can be assumed that f is a parametrization by the arc length s. The plane is oriented and we use the notation τ and n as above. A point P of the plane lies on the line $\mathcal{D}_s$ if and only if it satisfies the equation

$$\overrightarrow{f(s)P}\cdot\tau(s) = 0$$

(this is an equation of $\mathcal{D}_s$). Differentiating with respect to s, we get

$$-\tau(s)\cdot\tau(s) + \overrightarrow{f(s)P}\cdot\tau'(s) = 0.$$

We replace $\tau'(s)$ by its value, so that

$$K(s)(\overrightarrow{f(s)P}\cdot n(s)) = \|\tau(s)\|^2 = 1.$$

As $K(s)$ is not zero, the two linear equations in the unknown P we have got are

$$\begin{cases} \overrightarrow{f(s)P}\cdot\tau(s) = 0 \\ \overrightarrow{f(s)P}\cdot n(s) = \dfrac{1}{K(s)} = \rho(s). \end{cases}$$

The unique solution is the point P on the normal such that $\overrightarrow{f(s)P} = \rho(s)n(s)$, that is, the point $P = C(s)$. □

The envelope of normals is called the *evolute*; this is also the locus of curvature centers.

Parallel curves. If $s \mapsto f(s)$ is a parametrization of a curve γ by arc length, we shall consider the *parallel* curves γ_a, the ones that are defined by the parametrization f_a:

$$f_a(s) = f(s) + an(s)$$

for some fixed real number a (Figure 7).

Proposition 3.2 (Huygens' principle). *The curve f_a has a singular point for any value of s such that $\rho(s) = a$. The set of these singular points (when a varies) is the evolute of the curve γ.*

Proof. For this calculation, we choose an origin for the affine plane, so that we consider that $f(s)$ and $f_a(s)$ are vectors. We are looking for the singular points of γ_a, that is, for the values of s for which df_a/ds is zero:

$$0 = \frac{d}{ds}(f(s) + an(s)) = \tau(s) + an'(s) = (1 - aK(s))\tau(s),$$

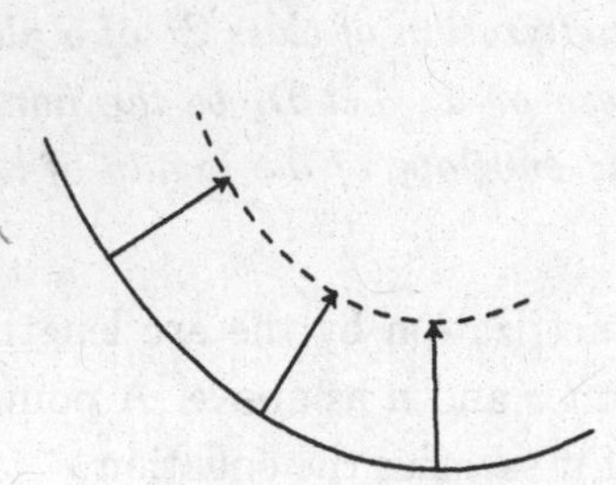
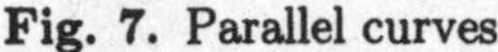

Fig. 7. Parallel curves

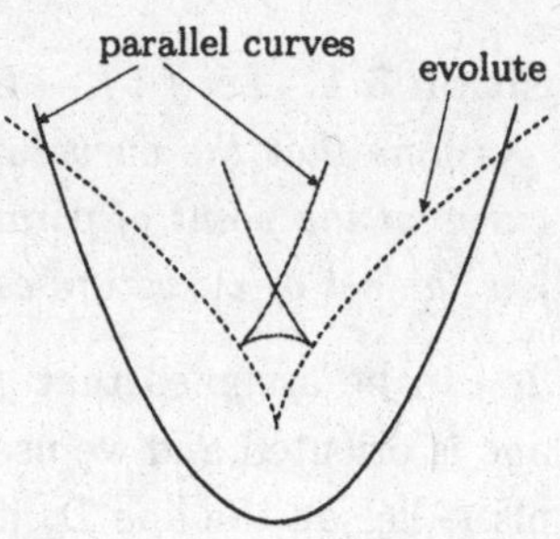

Fig. 8. Huygens' principle

this giving $\rho(s) = a$. The set of singular points is obtained by replacing a by $\rho(s)$ in $f_a(s)$; it is thus parametrized by

$$s \mapsto f(s) + \rho(s)n(s).$$

This is indeed the evolute of γ. □

Remark 3.3. This proposition, illustrated by Figure 8, is known in geometrical optics under the name of "Huygens' principle". The curve is considered as (the boundary of) a light source: it emits straight light rays that are perpendicular to the source. The parallel curve γ_a is the place reached, at the time a, by the waves emitted at the time 0; it is called a *wave-front*. The evolute, envelope of the normals, is the *caustic* of the light curve, on which the intensity of the light concentrates. The principle of Huygens asserts that the caustic consists of all the singular points of the wave-fronts.

4. Appendix: a few words on parametrized curves

This is a very brief summary. Let us begin with the affine properties.

Singularities. A *parametrized curve* of class $\mathcal{C}^k$ is a mapping of class $\mathcal{C}^k$ from an open interval I of $\mathbf{R}$ to $\mathcal{E}$. A point $t_0 \in I$ is *regular* if the vector $f'(t_0)$ is not zero, *singular* otherwise. A parametrized curve is *regular* if all the values of the parameter are regular points. If t_0 is regular, the vector $f'(t_0)$ is *tangent* to the curve, the affine line through $f(t_0)$ directed by $f'(t_0)$ is called the *tangent* to the parametrized curve at $f(t_0)$.

Example 4.1. In the plane $\mathcal{E} = \mathbf{R}^2$, consider the parametrized curve given by

$$t \longmapsto (t^2, t^3)$$

(Figure 9). It is easily checked that all the points of $\mathbf{R}$ are regular, except 0. The tangent at (t_0^2, t_0^3) is directed by the vector $(2t_0, 3t_0^2)$. The point 0 is a *cusp point*.

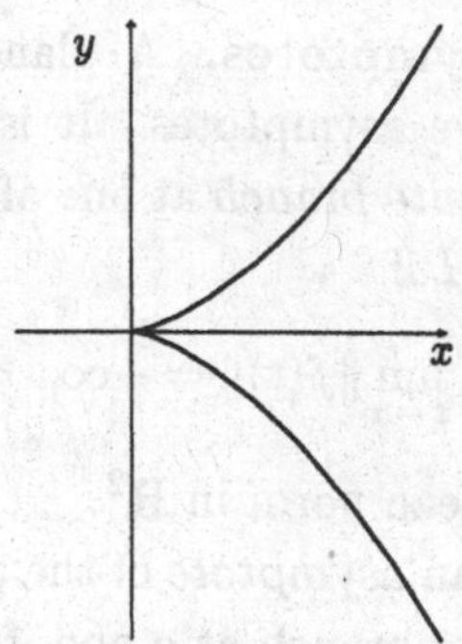

Fig. 9. A cuspidal cubic

Position of a plane curve with respect to its tangent. Assume that $k \geqslant 2$. We look at a regular point (so that $f'(t_0) \neq 0$). Expand f in a neighborhood of t_0 by the Taylor formula

$$f(t) = f(t_0) + (t - t_0) f'(t_0) + \frac{1}{2}(t - t_0)^2 f''(t_0) + o((t - t_0)^2).$$

Hence, if $f'(t_0)$ and $f''(t_0)$ are two independent vectors, the curve is located, in a neighborhood of $f(t_0)$, on one side of the tangent, it is contained in the half-plane defined by the tangent and containing $f''(t_0)$ (see Figure 10, which shows the curve together with the basis $(f'(t_0), f''(t_0))$).

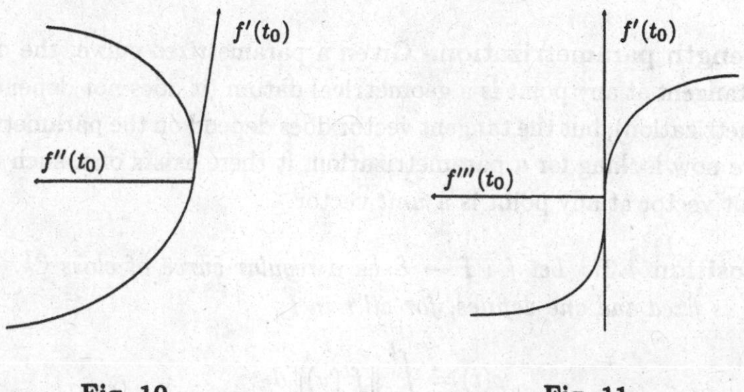

Fig. 10 **Fig. 11**

If the two vectors are collinear, if $k \geqslant 3$ and if, for instance, the vector $f'''(t_0)$ is independent of $f'(t_0)$, the expansion gives

$$f(t) = f(t_0) + (t - t_0)\left(1 + \lambda(t - t_0)\right) f'(t_0) + \frac{1}{6}(t - t_0)^3 f'''(t_0) + o((t - t_0)^3)$$

so that the curve crosses its tangent line (Figure 11); this is a *flex* or an *inflexion point*.

Infinite branches and asymptotes. A plane curve may have infinite branches and these may have asymptotes. It is said that a parametrized curve $f : I \to \mathbf{R}^2$ has an *infinite branch* at one of the ends a (which may be infinite) of the open interval I if

$$\lim_{t \to a} \|f(t)\| = +\infty$$

where $\|\cdot\|$ denotes the Euclidean norm in $\mathbf{R}^2$.

It is said that a line D is an *asymptote* of the parametrized curve $f : I \to \mathbf{R}^2$ if the curve has an infinite branch at a and if, moreover,

$$\lim_{t \to a} d(f(t), D) = 0$$

where d is the usual distance on $\mathbf{R}^2$.

Remark 4.2 (Important). These are *affine* notions even if we have used the Euclidean norm of $\mathbf{R}^2$ to define them. Indeed, it is clear that any equivalent norm would give the same infinite branches and the same asymptotes[11]. But, on $\mathbf{R}^2$, all the norms are equivalent (see, *e.g.*, [**Cho66**]).

On the other hand, it is easy to check that these notions do not depend on the chosen parametrization.

Let us come now to the metric properties. The affine space $\mathcal{E}$ is now Euclidean.

Arc length parametrization. Given a parametrized curve, the direction of its tangent at any point is a geometrical datum (it does not depend on the parametrization), but the tangent vector does depend on the parametrization. We are now looking for a parametrization, it there exists one, such that the tangent vector at any point is a *unit* vector.

Proposition 4.3. *Let $f : I \to \mathcal{E}$ be a regular curve of class $\mathcal{C}^1$. A point $t_0 \in I$ is fixed and one defines, for all t in I,*

$$\varphi(t) = \int_{t_0}^{t} \|f'(u)\| \, du.$$

Then φ is a diffeomorphism of class $\mathcal{C}^1$ from I to the interval $J = \varphi(I)$. The mapping $g = f \circ \varphi^{-1}$ is a parametrization of the same curve and satisfies

$$\|g'(s)\| = 1 \quad \forall s \in J.$$

---

[11] For the "projective" reader: an asymptote is a tangent at infinity, therefore the notion of asymptote should be as affine as the notion of tangent is.

Proof. The vector $f'(t)$ is never zero and the function $x \mapsto \|x\|$ is of class $\mathcal{C}^\infty$ on $E - \{0\}$, so that the function φ is of class $\mathcal{C}^1$. Its derivative

$$\varphi'(t) = \|f'(t)\|$$

is positive for all t. Therefore, φ is strictly increasing and it admits an inverse function, which is also of class $\mathcal{C}^1$.

Finally, we compute the derivative of g. This is

$$g'(s) = f' \circ \varphi^{-1}(s)\,(\varphi^{-1})'(s) = \frac{f' \circ \varphi^{-1}(s)}{\varphi' \circ \varphi^{-1}(s)} = \frac{f' \circ \varphi^{-1}(s)}{\|f' \circ \varphi^{-1}(s)\|},$$

which has, indeed, length 1. □

A parameter such as the s just obtained is an *arc length parameter* because

$$s_2 - s_1 = \varphi(t_2) - \varphi(t_1) = \int_{t_1}^{t_2} \|f'(t)\|\, dt$$

is the *length* of the arc from $g(s_1)$ to $g(s_2)$.

Remark 4.4. This is a *definition* of the length of an arc. It is justified, for instance, by approximating the curve by inscribed polygonal lines (see, *e.g.*, [**Dix76**]).

Proposition 4.3 gives the unique parametrization by arc length such that $g(0) = f(t_0)$ and $g'(s) = \lambda f'(t)$ with $\lambda > 0$. Any other arc length parameter is of the form

$$s' = \pm s - a.$$

Exercises and problems

Exercise VII.1. Let $\mathcal{C}$ be a conic in the plane $\mathbf{R}^2$ and let A be a point of $\mathcal{C}$. The line $\mathcal{D}_t$ of slope t through A intersects in general $\mathcal{C}$ at another point, denoted by M_t, having coordinates $(x(t), y(t))$. Prove that the mapping

$$\begin{array}{ccc} \mathbf{R} & \longrightarrow & \mathbf{R}^2 \\ t & \longmapsto & (x(t), y(t)) \end{array}$$

parametrizes $\mathcal{C} - \{A\}$ by rational functions.

Find a parametrization of the circle $x^2 + y^2 = 1$ (minus a point) by rational functions.

Exercise VII.2. Draw the curve $\mathcal{C}$ of the Cartesian equation $y^2 = x^2(x+1)$. At how many points does a line through 0 intersect it? Find a parametrization of $\mathcal{C}$ by rational functions. Same questions for the cuspidal cubic $y^2 = x^3$ (Figure 9).

Exercise VII.3 (Bernoulli's lemniscate). Let $\mathcal{C}$ be the curve parametrized by

$$\begin{cases} x = \dfrac{t}{1+t^4} \\ y = \dfrac{t^3}{1+t^4}. \end{cases}$$

Find a Cartesian equation of $\mathcal{C}$ and draw it (the curve).

Exercise VII.4. What is the curve image of a rectangular hyperbola under an inversion of pole the center of the hyperbola? Describe the form of the curve, then find both a Cartesian equation and an equation in polar coordinates.

Exercise VII.5. Draw the plane curve parametrized by

$$\begin{cases} x = t + \dfrac{1}{2t^2} \\ y = t^2 + \dfrac{2}{t} \end{cases}$$

and determine its singular points and its behavior at infinity (the curve is asymptote to two parabolas).

Exercise VII.6. Let $\mathcal{D}$ and $\mathcal{D}'$ be two (distinct) lines in the plane. Two points M and M' describe $\mathcal{D}$ and $\mathcal{D}'$ with proportional velocities. Find the envelope of the line MM'. One might consider the center F of the direct similarity mapping each point M to the corresponding point M' and show that the envelope is a parabola of focus F.

Exercise VII.7. In a Euclidean plane, two lines D and D' perpendicular at a point O and endowed with unit vectors are given. Two points A and A' of abscissas x and x' vary on D and D' in such a way that

$$x + x' = a$$

is a fixed positive number.

- Find the envelope of the perpendicular bisector of AA'.
- What is the midpoint of AA'?
- What is the envelope of the lines AA'?

Exercise VII.8. With the notation of Exercise I.41, the point P is fixed, but M varies on AP. What is the envelope of the perpendicular to $B'C'$ at M? One might prove that the intersection point F of BC with $B'C'$ is fixed.

Exercise VII.9. In an affine Euclidean plane, a circle $\mathcal{C}$ and a point F out of $\mathcal{C}$ are given. A point M describes $\mathcal{C}$. What is the envelope of the perpendicular bisector of MF?

Exercise VII.10. Prove that the curve of equation

$$\rho = 2(1 - \cos\varphi)$$

in polar coordinates (ρ, φ) is a cardioid.

Exercise VII.11. What is the group of isometries that preserve a three-cusped hypocycloid?

Exercise VII.12 (Discriminant). Let $\mathcal{E}$ be the (real affine 2-dimensional) space of polynomials $P \in \mathbf{R}[X]$ of the form

$$P(X) = X^3 - 3aX + 2b.$$

Describe by parametric equations the curve of $\mathcal{E}$ consisting of polynomials having a multiple root and draw this curve. What does its singular point correspond to? Characterize the two components of the complement in terms of the roots of the polynomials.

Exercise VII.13 (Caustics, parabolic aerials). A family of lines parallel to the axis of a parabola is reflected by this parabola (so that the angle between the reflected ray and the normal is equal to the angle between this normal and the incident ray). What is the envelope of the reflected rays?

Exercise VII.14 (Caustics, a heart in a bowl). A family of parallel lines is reflected by a circle (see Figure 12). Prove that the envelope of the

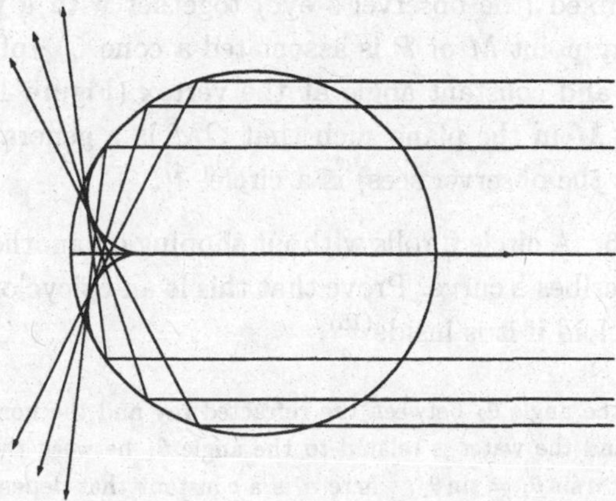

Fig. 12

family of reflected lines is a nephroid[12].

[12] In the real life, where light rays are reflected on the inner surface of a bowl, only half a circle counts, this is why what you see every morning in your bowl is half a nephroid, hence the heart shape.

Exercise VII.15 (Caustics, rainbow). "There is a rainbow in my heart" [**Her36**]. A line $\mathcal{D}$ is first reflected by a circle and then refracted according to Snell's law (in French Descartes' law)[(13)] getting out of this circle (Figure 13). The output line is called $\mathcal{D}'$. We are interested in the envelope of the lines $\mathcal{D}'$ when $\mathcal{D}$ describes a family of parallel lines. Write the equations and prove that the curve has an asymptote.

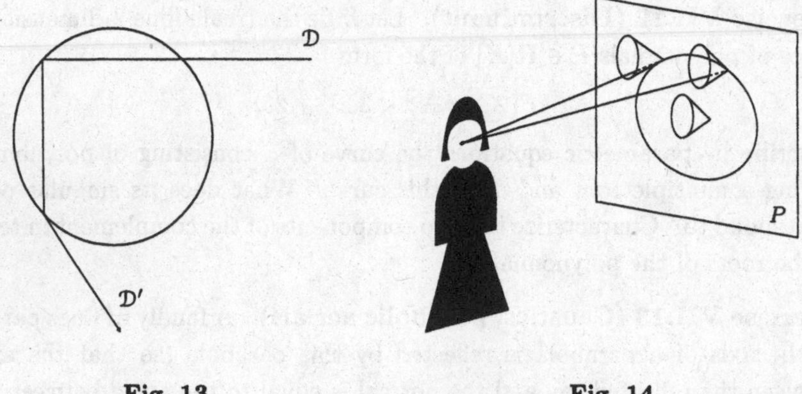

Fig. 13 **Fig. 14**

Sun rays meet a small raindrop, reflect and are refracted according to the above rule. Prove that the transformed rays envelop a surface that has an asymptote cone whose axis is parallel to the direction of the incident rays.

A point O is fixed (the observer's eye) together with a plane P (a rain screen). With any point M of P is associated a cone C_M of vertex M, axis orthogonal to P and constant angle at the vertex (Figure 14). Check that the set of points M in the plane such that OM is a generatrix of the cone C_M (this is what the observer sees) is a circle[(14)].

Exercise VII.16. A circle $\mathcal{C}$ rolls without slipping on another circle $\mathcal{C}'$. The point M of $\mathcal{C}$ describes a curve. Prove that this is an epicycloid if $\mathcal{C}$ is outside $\mathcal{C}'$ and a hypocycloid if it is inside[(15)].

---

(13) That is to say, the angle θ_2 between the refracted ray and the normal to the surface separating the air and the water is related to the angle θ_1 between the incident ray and the same normal by $n \sin\theta_1 = \sin\theta_2$, where n is a constant that depends on the medium (water, here) and on the wavelength of the light.

(14) What we have described is a monochromatic rainbow—this is less beautiful than a real one! The refraction index, and thus the cone we get, depends on the color, each color producing its own cone so that the observer sees several concentric circles. On the other hand, some rays reflect twice (or even several times) before finding their way out of the raindrop. This is what produces the second rainbow that can sometimes be seen. This description is due to Newton. See [**Boy87**] for other viewpoints on the rainbow and, for instance, the comic strip [**Ste82**] for general explanations on caustics.

(15) This is what justifies the prefixes *epi* and *hypo*.

Exercise VII.17. Consider a cycloid $\mathcal{C}$ (Figure 15), namely the curve described by a point on a circle that rolls without slipping on a line. Prove that it can be parametrized by

$$\begin{cases} x = R(t - \sin t) \\ y = R(1 - \cos t) \end{cases}$$

and prove that the evolute of $\mathcal{C}$ is also a cycloid, image of $\mathcal{C}$ by a translation.

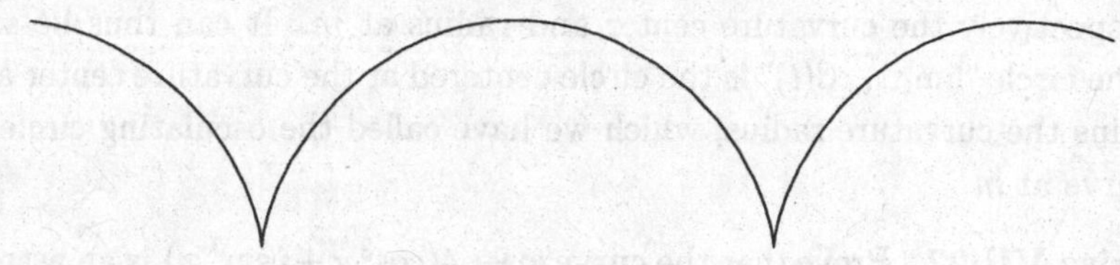

Fig. 15. A cycloid

Exercise VII.18. A circle, a point A on this circle and a direction of a line D are given. With any point M of the circle, is associated the line $\mathcal{D}_M$ through M such that D is parallel to a bisector of the angle $(AM, \mathcal{D}_M)$. Prove that the envelope of the lines $\mathcal{D}_M$ when M describes the circle is a three-cusped hypocycloid.

Exercise VII.19 (Curvature and torsion in dimension 3). Let $s \mapsto f(s)$ be an arc length parametrized curve in a 3-dimensional oriented Euclidean space. Assume that the vector $\tau'(s)$ is never zero. Define the vectors n and b by

$$\tau(s) = f'(s), \qquad n(s) = \tau'(s) / \|\tau'(s)\|, \qquad b(s) = \tau(s) \wedge n(s),$$

so that (τ, n, b) is a direct orthonormal basis. Prove that there exist numbers $K(s)$ and $T(s)$ (curvature and torsion) such that

$$\begin{cases} \tau'(s) &= K(s)n(s) \\ n'(s) &= -K(s)\tau(s) - T(s)b(s) \\ b'(s) &= T(s)n(s) \end{cases}$$

(Frenet's formulas). Find the curvature and the torsion when the curve is contained in a plane; when this is a circular helix ($t \mapsto (a\cos t, a\sin t, kt)$).

Exercise VII.20. Let $s \mapsto A(s)$ be a differentiable one parameter family of orthogonal n times n matrices (a curve in $O(n)$, if one prefers). Prove that there exists a skew symmetric matrix $B(s)$ such that, for all s,

$$\frac{dA}{ds} = A(s)B(s).$$

What is the relation of this formula to Remark 2.1? to Exercise VII.19?

Exercise VII.21 (Osculating circle). Let $t \mapsto f(t)$ be a parametrized curve of class $\mathcal{C}^2$ and let $m = f(t_0)$ be a point at which the curvature is not zero. For $t \neq t_0$ in a neighborhood of t_0, prove that there exists a unique circle $\mathcal{C}(t)$ through $f(t)$ and tangent to the curve at m. Let $O(t)$ be its center and $R(t)$ its radius. Prove that

$$\lim_{t \to t_0} O(t) \text{ and } \lim_{t \to t_0} R(t)$$

are respectively the curvature center and radius at m. It can thus be said that the circle "$\lim_{t \to t_0} \mathcal{C}(t)$" is the circle centered at the curvature center and of radius the curvature radius, which we have called the osculating circle to the curve at m.

Exercise VII.22. Prove that the curve $\varphi \mapsto 4(\cos^3 \varphi + i \sin^3 \varphi)$ is an astroid (four-cusped hypocycloid). Prove that the evolute of an ellipse is the curve transformed of an astroid by an affinity.

Exercise VII.23. Prove that the singular points of the evolute correspond to the extrema of the curvature[16].

Exercise VII.24. What is the evolute of the parabola $y^2 = 2px$?

Exercise VII.25 (Envelope of the Simson lines). Let ABC be a

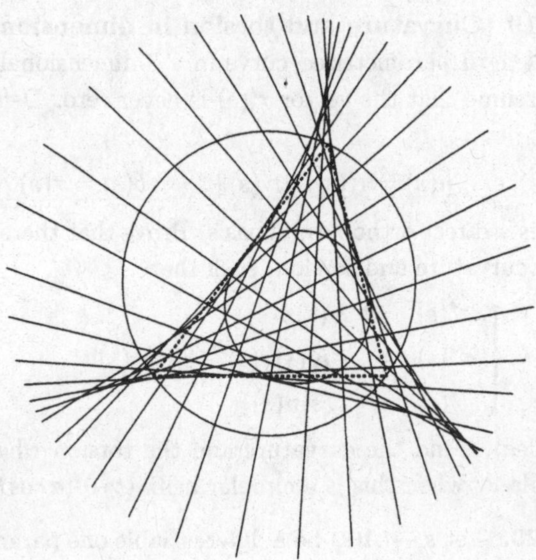

Fig. 16

[16] For instance, in Exercise VII.22, the four singular points of the astroid correspond to the four vertices of the ellipse.

triangle of orthocenter H and of circumcircle $\mathcal{C}$. Let M be a point of $\mathcal{C}$. Its projections on the three sides of the triangle are called M_A (on BC), M_B (on CA) and M_C (on AB). They are collinear on the Simson line of M, denoted S_M (Exercise III.33). Let m_A, m_B and m_C be the images of M by the reflections about BC, CA and AB. They are collinear on a line D_M, the Steiner line of M (Exercise III.34).

What is the envelope of the Steiner lines D_M when M describes $\mathcal{C}$?

Let μ be the midpoint of HM. Prove that μ lies on the Euler circle of ABC (Exercise II.20) and on the Simson line S_M. Prove that the envelope of the Simson lines is a three-cusped hypocycloid(17) that is tritangent to the Euler circle.

---

(17)Naturally associated with *any* triangle, a figure with the same symmetries as an equilateral triangle! There is also the triangle of Morley; see **[Ber94]** or **[Cox69]**.

Chapter VIII

Surfaces in 3-dimensional Space

This chapter is an introduction to the local properties of the surfaces in 3-dimensional space. Before coming to (necessarily) heavy definitions, I give a few simple examples of objects which I am sure the reader will agree should be called surfaces: surfaces of revolution, ruled surfaces, *etc.* I then come to the definitions and to the affine properties, tangent plane and position with respect to the tangent plane, in particular. The last section is devoted to the metric properties of surfaces in a Euclidean space, in particular to the Gauss curvature.

Although there is no global result in this chapter, I hope it contains enough examples and applications not to be boring. Two good references for the subject are the article "Classical differential geometry" in the *Encyclopædia Universalis* [**Lib74**] and volumes 2 and 3 of [**Spi75**].

1. Examples of surfaces in 3-dimensional space

This is a section of examples; it is deliberately vague: I do not insist on the regularity assumptions, in order to lighten the presentation for the reader.

Quadrics. We have already met a few surfaces in this book: spheres and more generally quadrics in the 3-dimensional space, namely ellipsoids, hyperboloids and paraboloids (see § VI-5 and Figure 20 in Chapter VI).

Here are other examples of surfaces. All of them are constructed from curves: we make a plane curve turn about an axis, and we look at a family of lines given along a curve.

Surfaces of revolution. Consider a plane curve C and make it turn about a line D of its plane. The line D is called the *axis of revolution* (Figure 1).

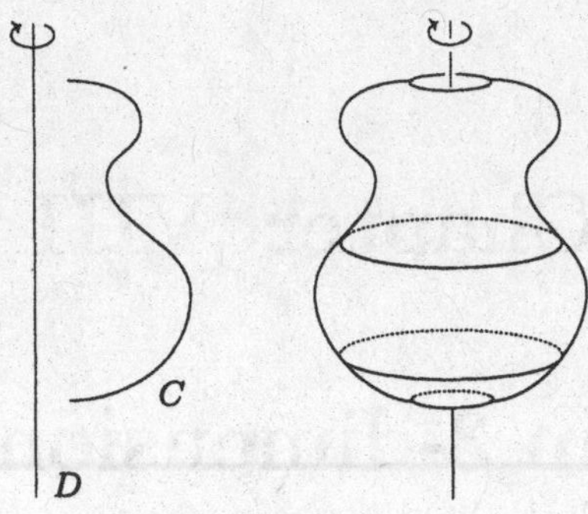

Fig. 1

Examples 1.1

- If the curve C is a straight line, the surface obtained is a *cylinder* if the lines C and D are parallel, a plane if they are perpendicular and a *cone* otherwise.
- If the curve C is a circle, the surface obtained is a sphere if the line D is a diameter of C, or a *torus of revolution* (a tire without valve) if D and C do not intersect.

By definition, the surface of revolution S is invariant under the group of rotations about the axis D. If M is a point of S, the whole circle of axis D through M is contained in S. It is called a *parallel* in analogy with the case of the Earth. The same analogy justifies the term *meridians* for the intersections of S with the half-planes limited by D. Any meridian is the image of C by a rotation of axis D.

It is very easy to describe the surface S by parametric, *resp.* Cartesian equations, starting from equations of the same type for C (Exercise VIII.1).

Ruled surfaces. A curve Γ is given together with a nonzero vector $w(p)$ for any point p of Γ. Through any point p of Γ passes a line, denoted by D_p,

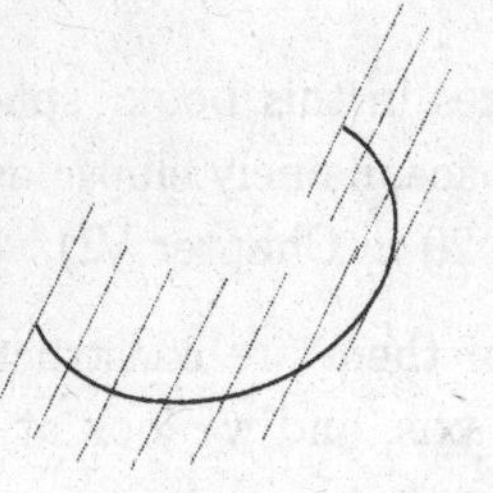

Fig. 2

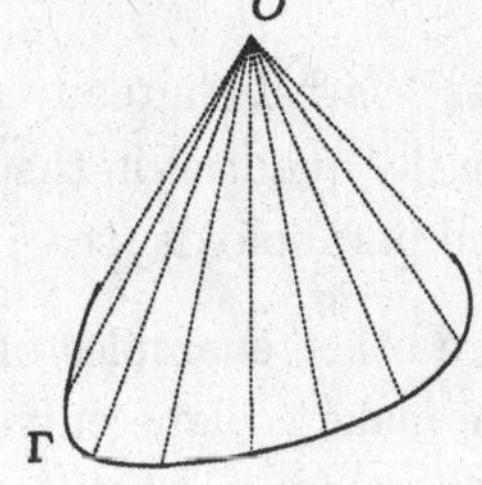

Fig. 3

that is directed by $w(p)$. The ruled surface is the union of the lines D_p.

The simplest examples are those of:

- *cylinders*, corresponding to the case where the vector $w(p)$ is *constant*: $w(p) \equiv w$ (Figure 2);
- *cones*, where $w(p) = \overrightarrow{Op}$ for some fixed point O, called the *vertex* (or *apex*) of the cone (Figure 3).

There are other examples of ruled surfaces among quadrics[(1)]: the hyperboloids of one sheet are ruled surfaces (see Exercise VIII.5). One of them is

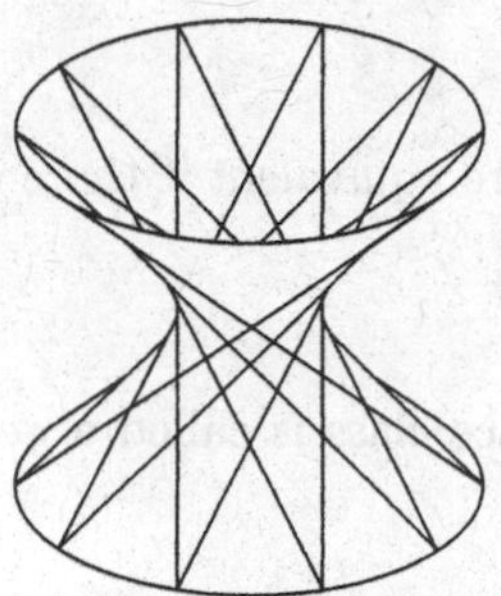

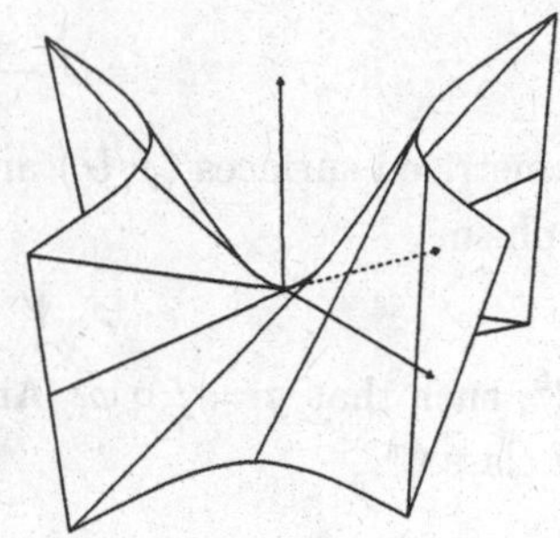

Fig. 4

shown in Figure 4 together with a hyperbolic paraboloid, an example belonging to the same family. The curve Γ is the x-axis and the vector $w(x)$ has coordinates $(0, 1, x)$, so that

$$(x, t) \longmapsto (x, 1 + t, x + tx)$$

is a parametrization of the ruled surface Γ and w define. A Cartesian equation of this surface is easily found, this is simply $z = xy$. This is a hyperbolic paraboloid (§ VI-5), also called a "saddle" or a "mountain pass" (Figure 4).

Examples of ruled surfaces and surfaces of revolution will be found in the exercises at the end of this chapter. All these examples are classical. They are very conveniently grouped in the third volume of [**Spi75**].

2. Differential geometry of surfaces in space

Definitions. After these examples, or families of examples, it is time to give the definition of what a surface could (or should) be. In the same way as a

---

(1) All the quadrics contain straight lines (Exercise VI.41), this is part of their nature, even if these lines are often imaginary, as this is the case for ellipsoids or hyperboloids of two sheets. Therefore quadrics show up among the ruled surfaces.

curve is defined by a mapping from an open interval of $\mathbf{R}$ to an affine space, we shall see that a surface is defined by a mapping

$$f : U \longrightarrow \mathcal{E}$$

from an open subset U of $\mathbf{R}^2$ to an affine space $\mathcal{E}$ of dimension 3. The reader will guess that some additional differentiability assumptions will be needed.

Definition 2.1. A *parametrized surface* of class $\mathcal{C}^k$, for $k \geqslant 1$, is a mapping f of class $\mathcal{C}^k$ defined on an open subset U of $\mathbf{R}^2$:

$$f : U \longrightarrow \mathcal{E}.$$

Two parametrized surfaces (f, U) and (g, V) are *equivalent* if there exists a diffeomorphism

$$\varphi : V \longrightarrow U,$$

of class $\mathcal{C}^k$, such that $g = f \circ \varphi$. An equivalence class is called a *geometric surface* of class $\mathcal{C}^k$.

The reader should of course check that this equivalence is indeed an equivalence relation.

What we really have in mind, when we think of a surface, is rather the common image of the various parametrizations of a geometric surface, the *support* of the geometric surface. We will often abuse the terminology, as the reader can imagine.

We will generally let *surface* denote a subspace of our space that is, at least in the neighborhood of each of its points, a geometric surface. For instance a sphere, and, more generally, a quadric, are surfaces (why?).

Regular points. To be quite honest, there is no reason yet why the image of a parametrization should be something we really want to call a surface. For instance, the image of the mapping

$$\begin{aligned} f : \quad \mathbf{R}^2 &\longrightarrow \mathbf{R}^3 \\ (u, v) &\longmapsto (u, 0, 0) \end{aligned}$$

is something we would rather call a curve! We shall require that f be "a little bit" injective. If there are quotation marks, this is because injectivity is not really needed. We could allow self-intersections, as in Figure 5, but we need to have enough points $m_0 \in \mathbf{R}^2$ such that the image of a small neighborhood of m_0 looks very much like a piece of a plane (Figure 6). Now, near a point m_0, f looks like its differential df_{m_0}, a linear mapping $\mathbf{R}^2 \to E$. A good way to ensure that the image of a neighborhood of m_0 looks like a piece of a plane is to require that df_{m_0} be injective.

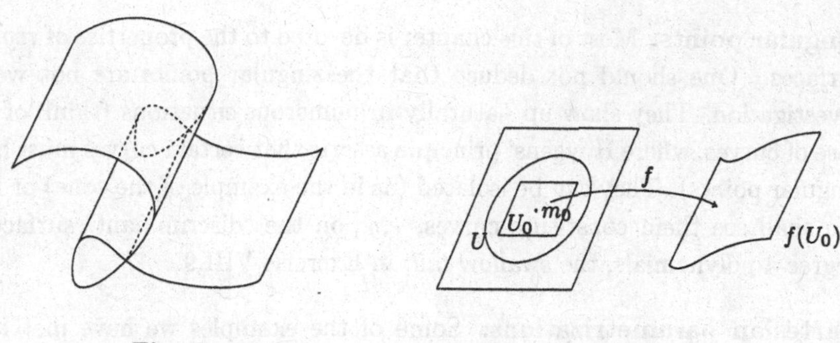

Fig. 5 Fig. 6

Definition 2.2. A geometric surface Σ defined by a parametrization $f : U \to \mathcal{E}$ is said to be *regular* if the differential of f at any point of U has rank 2.

Remark 2.3. It is easy, but important, to check that this is a property of the *geometric* surface Σ and not of the particular chosen parametrization f. Indeed, if $g : V \to \mathcal{E}$ is a parametrization that is equivalent to f by a diffeomorphism $\varphi : V \to U$, we have, at any point $n = \varphi^{-1}(m) \in V$

$$dg_n = df_{\varphi(n)} \circ d\varphi_n = df_m \circ d\varphi_n.$$

But φ is a diffeomorphism, therefore $d\varphi_n$ is a linear isomorphism $\mathbf{R}^2 \to \mathbf{R}^2$. Thus the rank of dg_n is equal to that of df_m. □

Notation. To make things easier to understand, I have used the notation m with coordinates (u, v) for the points of $\mathbf{R}^2$ and p with coordinates (x, y, z) for the points of $\mathbf{R}^3$.

Example 2.4. Consider the cone of revolution parametrized by the mapping

$$\begin{array}{rccc} f: & \mathbf{R}^2 & \longrightarrow & \mathbf{R}^3 \\ & (\theta, z) & \longmapsto & (kz\cos\theta, kz\sin\theta, z). \end{array}$$

We expect, if the definition is good, that this surface is regular, except at the vertex of the cone. And, indeed, the matrix of the differential is

$$df_{(\theta,z)} = \begin{pmatrix} -kz\sin\theta & k\cos\theta \\ kz\cos\theta & k\sin\theta \\ 0 & 1 \end{pmatrix}.$$

It has rank 2 as long as the first column vector is nonzero, namely as long as $z \neq 0$. The point $(0, 0, 0)$, the vertex of the cone, is a *singular* point.

Remark 2.5. As we have just seen in this example, f has rank 2 at the point (u, v) of U, if and only if the two vectors

$$\frac{\partial f}{\partial u}(u, v) \text{ and } \frac{\partial f}{\partial v}(u, v)$$

are independent.

Singular points. Most of this chapter is devoted to the properties of *regular* surfaces. One should not deduce that the singular points are not worth investigation. They show up naturally in numerous situations (think of the case of curves, where Huygens' principle asserts that certain curves must have singular points). They can be isolated (as in the example of the cone) or not. We shall see them constitute curves, *e.g.*, on the "discriminant" surface of degree-4 polynomials, the *swallow tail*, in Exercise VIII.9.

Cartesian parametrizations. Some of the examples we have met have equations of the form $z = h(x, y)$ (for instance the saddle with $h(x, y) = xy$). Such an equation can be considered as describing a surface parametrized by

$$f: \begin{array}{ccc} \mathbf{R}^2 & \longrightarrow & \mathbf{R}^3 \\ (x,y) & \longmapsto & (x,y,h(x,y)). \end{array}$$

One should notice that such a parametrization *always* has rank 2. The *Cartesian* parametrizations are obviously injective. They have another remarkable property: the subspace $df_m(\mathbf{R}^2)$, spanned by $\partial f/\partial x$ and $\partial f/\partial y$ never contains the z-axis. There is a converse to these remarks, expressed in the next theorem.

Theorem 2.6. *Let Σ be a surface of class $\mathcal{C}^k$ defined by a parametrization $f : U \to \mathbf{R}^3$ and let m_0 be a point of U such that df_{m_0} has rank* 2. *Then there exists a neighborhood U_0 of m_0 in U and a system of coordinates in $\mathbf{R}^3$ such that the surface parametrized by the restriction $f|_{U_0}$ has a Cartesian parametrization in these coordinates.*

Before proving this theorem, it might be useful to make a few remarks, illustrating the various aspects of the statement.

Remarks 2.7

(1) In particular, the differential of the restriction $f|_{U_0}$ has rank 2 at every point (this is an open condition and thus this is true on a whole neighborhood of m_0).
(2) In particular also, $f|_{U_0}$ is injective. Consider an example. Let C be a regular plane curve with a double point, parametrized by $g : t \mapsto (x(t), y(t))$, and let S be the cylinder parametrized by $f : (t, u) \mapsto (x(t), y(t), u)$. Since C is regular, the surface S is also regular (immediate verification), however, f is not injective. Let t_0 and t_1 be the values of t giving the double point of C. There is a neighborhood I_0 of t_0 such that $g|_{I_0}$ is injective and similarly $f|_{I_0 \times \mathbf{R}}$ (Figure 7). The theorem gives a property of local injectivity in this sense.
(3) Even if the rank of f is 2 everywhere and f is injective, it may very well happen that the geometric surface it defines has no global Cartesian

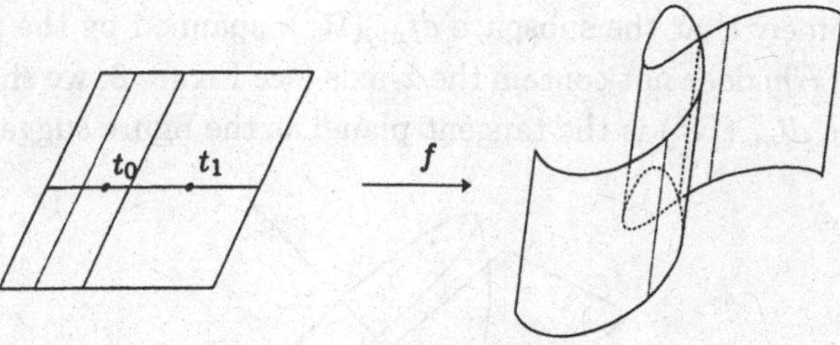

Fig. 7

parametrization. Indeed, in coordinates (x, y, z) where a surface is defined by a Cartesian parametrization, it is the graph of a function, and in particular, its intersection with a line parallel to the z-axis contains at most one point, the projection of the surface on the (x, y)-plane is injective. Here is an example. The mapping f defined on the open set $]-\pi, \pi[\times]-\frac{\pi}{2}, \frac{\pi}{2}[$ of $\mathbf{R}^2$ by

$$f(\theta, \varphi) = (\cos\theta\cos\varphi, \sin\theta\cos\varphi, \sin\varphi)$$

parametrizes a (subset of the) unit sphere. It has rank 2 everywhere and it is injective. However, there exists no plane of $\mathbf{R}^3$ such that the projection of S on this plane is injective. The sphere cannot be *globally* described by a Cartesian parametrization.

Proof of Theorem 2.6. Let $f : U \to \mathbf{R}^3$ be a parametrization of Σ. We write it in the form

$$(u, v) \longmapsto (\alpha(u, v), \beta(u, v), \gamma(u, v)).$$

The matrix of the differential df_{m_0} at the point $m_0 = (u_0, v_0)$ has rank 2 by assumption, and this is

$$\begin{pmatrix} \dfrac{\partial\alpha}{\partial u}(u_0, v_0) & \dfrac{\partial\alpha}{\partial v}(u_0, v_0) \\[2ex] \dfrac{\partial\beta}{\partial u}(u_0, v_0) & \dfrac{\partial\beta}{\partial v}(u_0, v_0) \\[2ex] \dfrac{\partial\gamma}{\partial u}(u_0, v_0) & \dfrac{\partial\gamma}{\partial v}(u_0, v_0) \end{pmatrix}.$$

Up to a permutation of the coordinates in $\mathbf{R}^3$, it can be assumed that the minor

$$\begin{vmatrix} \dfrac{\partial\alpha}{\partial u}(u_0, v_0) & \dfrac{\partial\alpha}{\partial v}(u_0, v_0) \\[2ex] \dfrac{\partial\beta}{\partial u}(u_0, v_0) & \dfrac{\partial\beta}{\partial v}(u_0, v_0) \end{vmatrix}$$

is nonzero, namely that the subspace $df_{m_0}(\mathbf{R}^2)$, spanned by the two vectors $\partial f/\partial u$ and $\partial f/\partial v$ does not contain the z-axis (see Figure 8; we shall see later that the plane $df_{m_0}(\mathbf{R}^2)$ is the tangent plane, as the figure suggests).

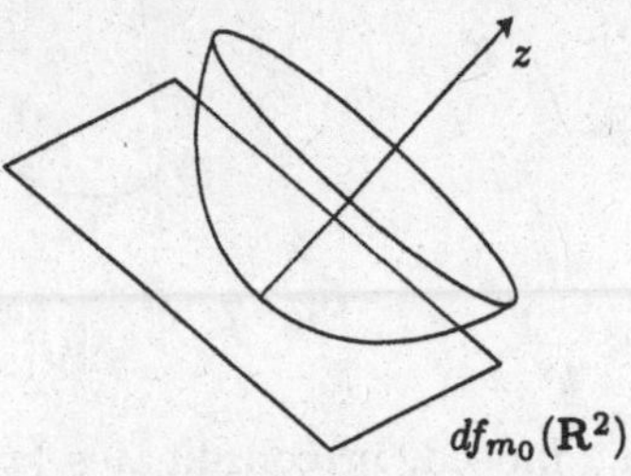

Fig. 8

Consider the mapping

$$\begin{aligned} g: \quad U &\longrightarrow \mathbf{R}^2 \\ (u,v) &\longmapsto (\alpha(u,v), \beta(u,v)). \end{aligned}$$

It has rank 2 at (u_0, v_0) and the inverse function theorem says that this is a local diffeomorphism[(2)], that is, there exists a neighborhood U_0 of (u_0, v_0) in U such that

$$h = g|_{U_0} : U_0 \longrightarrow \mathbf{R}^2$$

is a diffeomorphism onto its image W. Now, the composition $f \circ h^{-1}$ maps the point (x,y) to $(x, y, \gamma \circ h^{-1}(x,y))$:

$$(x,y) \xmapsto{h^{-1}} (u,v) \xmapsto{f} (\alpha(u,v), \beta(u,v), \gamma(u,v)) = (x, y, \gamma \circ h^{-1}(u,v)),$$

where $x = \alpha(u,v)$ and $y = \beta(u,v)$. This is thus, indeed, a Cartesian parametrization of the surface parametrized by $f|_{U_0}$. □

Surfaces defined by equations. It often happens that a surface is given by an equation, as is the case, *e.g.*, for the unit sphere, that consists of the points satisfying

$$x^2 + y^2 + z^2 - 1 = 0.$$

More generally, a function

$$F : \mathbf{R}^3 \longrightarrow \mathbf{R}$$

is given and the (geometric) "surface" Σ is

$$\Sigma = \left\{(x,y,z) \in \mathbf{R}^3 \mid F(x,y,z) = 0\right\}.$$

Under suitable assumptions, this is indeed a surface, as is asserted by the next proposition.

(2) See, for example, [**Car67**] for this variant of the implicit function theorem.

Proposition 2.8. *Let $F : \mathbf{R}^3 \to \mathbf{R}$ be a function of class $\mathcal{C}^1$. Let $p_0 = (x_0, y_0, z_0)$ be a point such that $F(p_0) = 0$ and such that the linear mapping*

$$(dF)_{p_0} : \mathbf{R}^3 \longrightarrow \mathbf{R}$$

is surjective. Then there exists a parametrization that makes a neighborhood of p_0 in

$$\Sigma = \{(x, y, z) \mid F(x, y, z) = 0\}$$

a regular surface.

Proof. This is merely a reformulation of the implicit function theorem. Under the assumption that the mapping $(dF)_{p_0}$ is surjective, at least one of the partial derivatives of F does not vanish at p_0. Up to a permutation of the coordinates, it can be assumed that

$$\frac{\partial F}{\partial z}(p_0) \neq 0.$$

Then the equation $F(x, y, z) = 0$ can be solved in the neighborhood of p_0: there exists an open subset U of $\mathbf{R}^2$ and a function φ of class $\mathcal{C}^1$ defined over U such that the equivalence

$$((x, y) \in U \text{ and } F(x, y, z) = 0) \iff z = \varphi(x, y)$$

holds. We have thus obtained a Cartesian parametrization, which is in particular a regular parametrization, of the "surface" Σ in a neighborhood of p_0. □

Tangent planes. The plane $df_m(\mathbf{R}^2)$, the image of the differential of a parametrization, has played a rather important role in the proof of Theorem 2.6. We have also seen that this plane does not depend on the regular parametrization used to define it. If f parametrizes a regular surface Σ and if $p = f(m)$, we would like to *define* the tangent plane to Σ at p by

$$T_p\Sigma = df_m(\mathbf{R}^2).$$

Surely, thanks to the regularity, $df_m(\mathbf{R}^2)$ is indeed a plane in $\mathbf{R}^3$. It might depend on m if f is not injective: if $p = f(m_1) = f(m_2)$ (as is the case, *e.g.*, in Figure 7), we do not know which plane to choose. When f is injective, the tangent plane to Σ at p is defined by this formula.

Remark 2.9. We have not defined the tangent plane for a noninjective parametrization. Look, however, at Figure 7: if it is clear that there is not *a* tangent plane at a double point, it is not hard to imagine that there are two tangent planes

At a *singular* point p_0, that is, at the image of a point m_0 such that the differential $(df)_{m_0}$ is *not* injective, the situation is much more dramatic. It can even happen that the tangent planes $T_p\Sigma$ at the regular close points have no limit at p_0. See Exercise VIII.10.

It is easily proved, in the case where the surface is defined by an equation $F(x,y,z)=0$:

Proposition 2.10. *Let $F : \mathbf{R}^3 \to \mathbf{R}$ be a function of class $\mathcal{C}^1$ and let p_0 be a point such that $F(p_0)=0$ and $(dF)_{p_0}$ is onto. Then the tangent plane at p_0 to the surface defined by the equation $F(p)=0$ in a neighborhood of p_0 is the kernel of the differential $(dF)_{p_0}$.*

Proof. Let us use the same notation as in the proof of Proposition 2.8. The tangent plane at p_0 is spanned by the partial derivatives at p_0 of the parametrization

$$(x,y) \longmapsto (x,y,\varphi(x,y)).$$

Of course, we have

$$(dF)_{p_0}\left(1,0,\frac{\partial \varphi}{\partial x}\right) = \frac{\partial F}{\partial x} + \frac{\partial F}{\partial z}\frac{\partial \varphi}{\partial x}$$

and

$$(dF)_{p_0}\left(0,1,\frac{\partial \varphi}{\partial y}\right) = \frac{\partial F}{\partial y} + \frac{\partial F}{\partial z}\frac{\partial \varphi}{\partial y}.$$

Thanks to the expression of the partial derivatives of φ given by the implicit function theorem, these two vectors are in the kernel of $(dF)_{p_0}$. This subspace has dimension 2, so we are done. □

Remarks 2.11

(1) As we have defined, $T_p\Sigma$ is a linear subspace of $\mathbf{R}^3$. It is often considered as an affine subspace, putting its origin at p.

(2) It does not depend on the chosen parametrization f: if $g = f \circ \varphi$ and $m = \varphi(n)$, $dg_n = df_m \circ d\varphi_n$ so that dg_n and df_m have the same image, as the reader certainly has already noticed.

(3) This is the subspace spanned by the vectors

$$\left(\frac{\partial f}{\partial u}\right)(m) \text{ and } \left(\frac{\partial f}{\partial v}\right)(m).$$

Examples 2.12

- The tangent plane to a surface of revolution at a point is spanned by the vectors that are tangent to the meridian and to the parallel through this point (Exercise VIII.7).
- The tangent plane to a ruled surface at a point contains the generatrix through this point (Exercise VIII.8).

Tangents to the curves drawn on a surface. Consider a surface Σ parametrized by an injective mapping

$$f : U \longrightarrow \mathbf{R}^3$$

and a curve Γ of U together with its image $f(\Gamma) \subset \Sigma$. If $t \mapsto \gamma(t)$ is a parametrization of Γ in U, the curve $f(\Gamma)$ is parametrized by $f \circ \gamma$ and

$$(f \circ \gamma)'(t) = df_{\gamma(t)}(\gamma'(t)).$$

The tangent vector to $f(\Gamma)$ at $f \circ \gamma(t)$, is the image by $df_{\gamma(t)}$ of a vector $\gamma'(t)$ of $\mathbf{R}^2$. In particular, this is a vector of the tangent plane to Σ at $f \circ \gamma(t)$.

The tangent line to a curve drawn on Σ is thus contained in the tangent plane. Conversely, if Δ is a line in $T_{p_0}\Sigma$, let X be one of its directing vectors and write it in the basis $\partial f/\partial u, \partial f/\partial v$ of this plane:

$$X = \lambda \frac{\partial f}{\partial u}(u_0, v_0) + \mu \frac{\partial f}{\partial v}(u_0, v_0).$$

Consider the intersection δ of the line spanned by (λ, μ) through (u_0, v_0) with U: this is a parametrized curve; its tangent line at (u_0, v_0) is spanned by (λ, μ). Then we have

$$df_{(u_0,v_0)}(\lambda, \mu) = X,$$

so that the line Δ is the tangent to the image of δ, an arc drawn on Σ.

The tangent plane at p_0 can thus be considered as the unique plane containing the tangents at p_0 to the curves drawn on Σ.

Example 2.13. A parametrized curve $t \mapsto \gamma(t)$ is drawn on the unit sphere if and only if the vector $\gamma(t)$ is a unit vector for all t:

$$\|\gamma(t)\|^2 = 1 \text{ for all } t.$$

Differentiating this relation with respect to t, it is seen that $\gamma'(t)$ is orthogonal to $\gamma(t)$ for all t. This way, it is checked that the tangent space to the unit sphere at p is the plane $p^\perp$. We shall consider, as we have said, the affine plane through p that is orthogonal to p as the tangent plane to the sphere at p.

Tangent mapping. Consider now a mapping h, defined on the support of the surface Σ of class $\mathcal{C}^k$, and taking its values in a space $\mathbf{R}^n$. It is said that h is *differentiable* if, for some parametrization $f : U \to \mathbf{R}^3$ of Σ, the mapping

$$h \circ f : U \longrightarrow \mathbf{R}^n,$$

a honest mapping from an open subset of $\mathbf{R}^2$ to $\mathbf{R}^n$, is differentiable. This property does not depend on the chosen parametrization (the verification is straightforward but must be done and is left to the reader). The mappings of class $\mathcal{C}^k$ are defined analogously.

In the same way as the differentiable mapping $h \circ f$ has a *differential* at any point of U, there is a linear mapping associated with h, its *tangent mapping*, at any point of Σ. This is a linear mapping

$$T_p h : T_p\Sigma \longrightarrow \mathbf{R}^n,$$

that can be denoted by dh_p as well and whose definition is given below.

If $f : U \to \mathbf{R}^3$ is a parametrization of a *regular* surface Σ with $f(u, v) = p$, any vector X of $T_p\Sigma$ is the image under $df_{(u,v)}$ of a unique vector ξ of $\mathbf{R}^2$. One defines

$$T_p h(X) = d(h \circ f)_{(u,v)}(\xi)$$

(the mapping $h \circ f$, always honest, has a differential $d(h \circ f)$ at any point of U).

For this definition to be meaningful, the result must not depend on the chosen parametrization. Consider therefore another parametrization $g : V \to \mathbf{R}^3$. Let $\varphi : V \to U$ be the change of parametrization, so that $g = f \circ \varphi$. Let $(u, v) = \varphi(s, t)$. Let p be a point of the surface Σ and let ξ, η be vectors of $\mathbf{R}^2$ such that:

- the point p satisfies $p = g(s, t) = f(u, v)$;
- the vectors ξ and η are related by the equality

$$\xi = d\varphi_{(s,t)}(\eta).$$

Let X be the vector of $\mathbf{R}^3$, tangent to Σ, defined by $X = dg_{(s,t)}(\eta) = df_{(u,v)}(\xi)$. Let us compute:

$$\begin{aligned} d(h \circ g)_{(s,t)}(\eta) &= d((h \circ f) \circ \varphi)_{(s,t)}(\eta) \\ &= d(h \circ f)_{\varphi(s,t)}\left(d\varphi_{(s,t)}(\eta)\right) \\ &= d(h \circ f)_{(u,v)}(\xi). \end{aligned}$$

This proves that $T_p h(X)$ does not depend on the choice of the parametrization of Σ.

Example 2.14. If Σ is a surface in $\mathbf{R}^3$ and if $h : \mathbf{R}^3 \to \mathbf{R}$ is a function, the tangent mapping to the restriction of h to Σ is the restriction of the differential of h, in symbols:

$$T_p\left(h|_\Sigma\right) = (dh)_p|_{T_p\Sigma}.$$

Intersection of two surfaces. If they are not tangent to each other(3), two surfaces intersect along a curve (Figure 9). This is what the next statement asserts, with precisions on the relation between the tangent planes to the surfaces and the tangent line to the curve.

---

(3) This is a "general position" assumption.

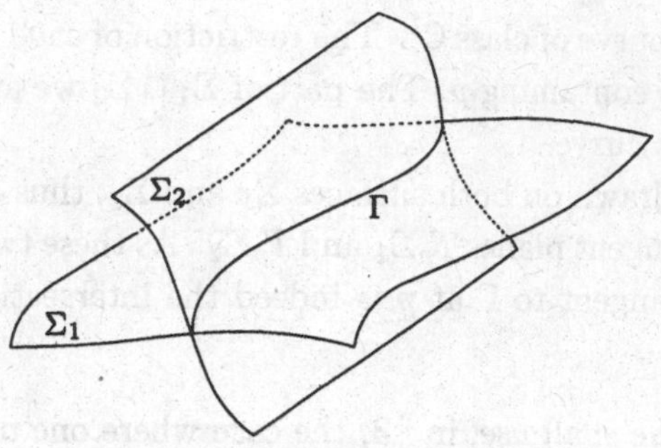

Fig. 9

Proposition 2.15. *Let Σ_1 and Σ_2 be two regular parametrized surfaces of class $\mathcal{C}^1$, the parametrizations being injective, and let p be a point of the intersection $\Sigma_1 \cap \Sigma_2$. Assume that the tangent planes $T_p\Sigma_1$ and $T_p\Sigma_2$ at p are distinct. Then, in a neighborhood of p, the intersection $\Sigma_1 \cap \Sigma_2$ is a curve, the tangent at p of which is the intersection $T_p\Sigma_1 \cap T_p\Sigma_2$ of the two tangent planes.*

Proof. Choose a frame of origin p whose third vector is parallel to neither one nor the other tangent plane, so that the two surfaces admit, in a neighborhood of p, Cartesian parametrizations:

$$z = f_1(x, y) \text{ for } \Sigma_1 \text{ and } z = f_2(x, y) \text{ for } \Sigma_2,$$

for two functions f_1 and f_2 of class $\mathcal{C}^1$, defined on a neighborhood of $(0,0) \in \mathbf{R}^2$ and such that $f_i(0,0) = 0$. The intersection $\Sigma_1 \cap \Sigma_2$ is thus the image of

$$\{(x, y) \subset U \mid f_1(x, y) = f_2(x, y)\}$$

under one Cartesian parametrization or the other.

Let us consider the function $f = f_1 - f_2$ and calculate its partial derivatives:

$$\begin{cases} \dfrac{\partial f}{\partial x}(0,0) = \dfrac{\partial f_1}{\partial x}(0,0) - \dfrac{\partial f_2}{\partial x}(0,0) \\[2ex] \dfrac{\partial f}{\partial y}(0,0) = \dfrac{\partial f_1}{\partial y}(0,0) - \dfrac{\partial f_2}{\partial y}(0,0) \end{cases}$$

If the two partial derivatives were zero, we would have $T_p\Sigma_1 = T_p\Sigma_2$, but this is contradictory to our assumption. Thus at least one of them is nonzero and we can apply the implicit function theorem(4), thus getting open intervals I and J containing 0 and such that

$$\{(x, y) \in I \times J \mid f(x, y) = 0\}$$

(4) See, *e.g.*, [Car67] for this theorem and its variants.

is a simple regular curve of class $\mathcal{C}^1$. The restriction of each f_i to $I \times J$ defines a sub-surface of Σ_i containing p. The part of $\Sigma_1 \cap \Sigma_2$ we are investigating is the image Γ of this curve.

The curve Γ is drawn on both surfaces Σ_1 and Σ_2, thus its tangent at p is contained in the tangent planes $T_p\Sigma_1$ and $T_p\Sigma_2$. As these two planes intersect along a line, the tangent to Γ at p is indeed the intersection of the tangent planes. □

Remark 2.16. We shall use, in § 3, the case where one of the surfaces is a plane.

Position of a surface with respect to its tangent plane. We have first a contact property, as for the tangent to a curve, that approximates the curve at order 1.

Proposition 2.17. *Let Σ be a regular surface defined by an injective parametrization $f : U \to \mathbf{R}^3$ and let $p_0 = f(m_0)$ be a point of Σ. The affine plane $\mathcal{P}$ defined by the equation $\varphi(p) = 0$ is the tangent plane to Σ at p_0 if and only if, in the neighborhood of m_0, one has*

$$\varphi(f(m)) = o(\|\overrightarrow{m_0m}\|).$$

In other words, the points of Σ "almost" satisfy the equation of $\mathcal{P}$.

Proof. We write first that $\varphi \circ f$ is differentiable at m_0:

$$\varphi \circ f(m) = \varphi \circ f(m_0) + d(\varphi \circ f)_{m_0} \cdot \overrightarrow{m_0m} + o(\|\overrightarrow{m_0m}\|).$$

The relation in the statement is equivalent to

$$\begin{cases} \varphi \circ f(m_0) = 0 \\ \text{and} \\ d(\varphi \circ f)_{m_0} \cdot \overrightarrow{m_0m} = 0 \text{ for all } m. \end{cases}$$

The first relation simply says that $\varphi(p_0) = 0$, that is, the plane $\mathcal{P}$ contains p_0. The second says that $d(\varphi \circ f)_{m_0} = 0$, that is, $d\varphi_{p_0} \circ df_{m_0} = 0$. Since φ is affine, the mapping $(d\varphi)_{p_0}$ is its linear part, which we call $L(\varphi)$. Our relation says that $L(\varphi) \circ df_{m_0} = 0$, namely that the image of df_{m_0} is the kernel of $L(\varphi)$, the (linear) plane directing $\mathcal{P}$.

The relation of the statement is thus indeed equivalent to the fact that $\mathcal{P}$ is the plane parallel to $T_{p_0}\Sigma$ through p_0. □

We consider now the situation at order 2. In particular, we need to assume that all the parametrizations are of class $\mathcal{C}^2$.

In a neighborhood of p_0, it can be assumed that Σ is defined by a Cartesian parametrization $z = f(x, y)$. It can also be assumed that the origin has been chosen at p_0, that is, that $f(0,0) = 0$. Let us use now the traditional so-called "Monge notation"

$$\begin{cases} p = \dfrac{\partial f}{\partial x}(0,0) \\ q = \dfrac{\partial f}{\partial y}(0,0) \end{cases}$$

so that the equation of the tangent plane at 0 is $z = px + qy$, and

$$r = \frac{\partial^2 f}{\partial x^2}(0,0) \qquad s = \frac{\partial^2 f}{\partial x \partial y}(0,0) \qquad t = \frac{\partial^2 f}{\partial y^2}(0,0).$$

The Taylor formula at order 2 in a neighborhood of 0 is

$$f(x,y) = px + qy + \frac{1}{2}(rx^2 + 2sxy + ty^2) + o(x^2 + y^2).$$

The quadratic term represents the difference (at order 2) between a point $(x, y, f(x, y))$ of the surface and the corresponding point $(x, y, \varphi(x, y))$ of the tangent plane.

The position of the surface with respect to its tangent plane is described with the help of the second derivative of f at 0, a quadratic form:

$$Q(x,y) = rx^2 + 2sxy + ty^2,$$

and the surface lies or does not lie on one side of its tangent plane (in the neighborhood of 0) if $f(x,y) - px - qy$ has or has not a constant sign, in other words, if the form Q is or is not definite.

(1) If Q is (positive or negative) definite, that is to say, if $s^2 - rt < 0$, then the function $f(x,y) - px - qy$ admits a (strict) extremum at 0; the surface lies on one side of its tangent plane. The point p_0 is said to be *elliptic* (Figure 10).

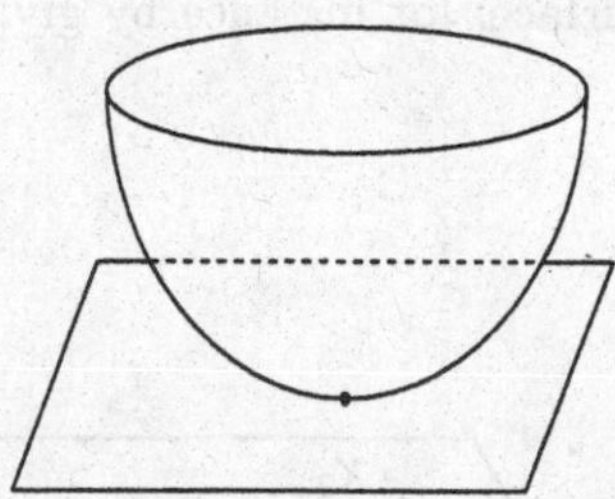

Fig. 10

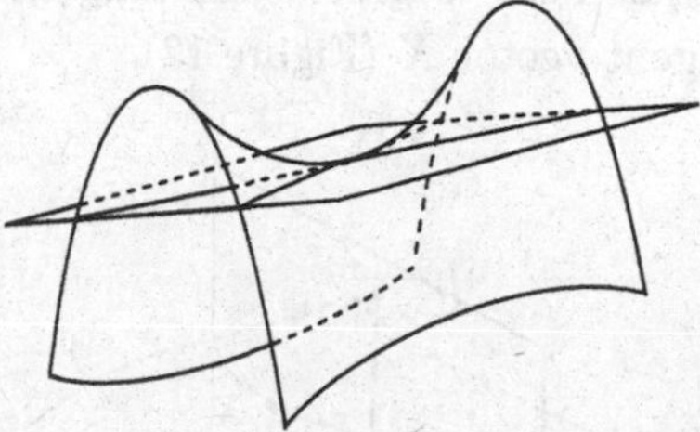

Fig. 11

(2) If Q is nondegenerate but is not definite (its sign changes), namely if $s^2 - rt > 0$, there are points on the surface that are arbitrarily close to 0 on both sides of the tangent plane; the point p_0 is said to be *hyperbolic* (Figure 11).

(3) If Q is degenerate but nonzero, namely if $s^2 - rt = 0$ but r, s and t are not all zero, it can be assumed that $Q(x, y) = \alpha x^2$; its graph is a parabolic cylinder. The surface contains points that are arbitrarily close to 0 and on the same side of the tangent plane as this graph is, but it may very well contain points that are on the other side (see Exercise VIII.11). The point p_0 is said to be *parabolic.*

(4) If Q is zero ($r = s = t = 0$), almost anything can happen (see a few examples in Exercise VIII.11). The point p_0 is said to be *planar.*

This position will be interpreted in terms of the curvature in § 3 (Corollary 3.8).

3. Metric properties of surfaces in the Euclidean space

This section is an introduction to the local metric properties of surfaces in 3-dimensional space. It was inspired by the historical presentation in Chapters 2 and 3 of the second volume of [**Spi75**]. The idea is to formalize for instance the fact that a sphere is "more curved" than a plane.

All the surfaces we will discuss in this section are of class $\mathcal{C}^2$, and even $\mathcal{C}^3$ in Theorem 3.10.

Curvature of the curves drawn on surfaces and Euler's theorem. Historically, the investigation of the metric properties of surfaces in space began with the consideration of the curves they contain. One of the results of this prehistory is due to Euler (1760). Consider a point p of Σ, the normal to the tangent plane at p, a directing vector (normal vector) $n(p)$ of this normal and an arbitrary plane P containing $n(p)$. To determine such a plane, it is equivalent to give a line tangent to the surface, for instance by giving a tangent vector X (Figure 12).

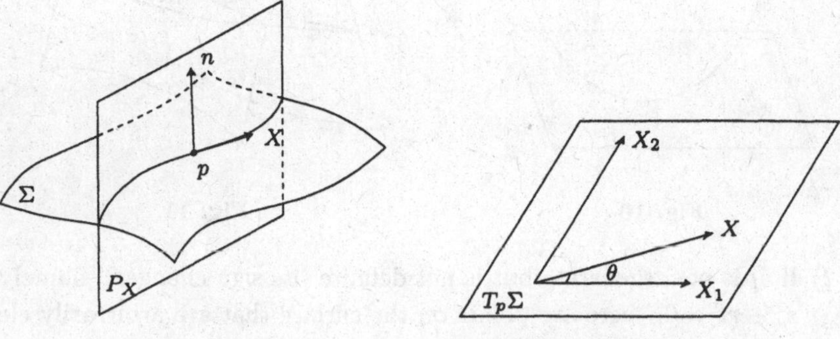

Fig. 12 **Fig. 13**

Let us fix a unit vector $X \in T_p\Sigma$. Let P_X be the plane (through p) spanned by X and $n(p)$. By virtue of Proposition 2.15, the intersection $P_X \cap \Sigma$ is a curve drawn on Σ to which the vector X is tangent. And, of course, this is a plane curve. In particular, the plane P_X being oriented by the basis $(X, n(p))$, it has an algebraic curvature[5], denoted by K_X.

We have thus a mapping $X \mapsto K_X$ which associates, to any unit vector tangent to Σ, a real number. Notice that

$$K_{-X} = K_X$$

(the parametrization of the curve has changed but the orientation of the plane P_X has changed too).

Theorem 3.1 (Euler). *If all the K_X are not equal, there exists a unique direction, represented by a vector X_1, for which K_X is minimal and a unique direction, represented by X_2, for which it is maximal. The vectors X_1 and X_2 are orthogonal and, if θ is a measure of the angle between X and X_1, one has*

$$K_X = K_{X_1} \cos^2\theta + K_{X_2} \sin^2\theta.$$

The proof will be delayed until page 289. Its statement reminds us of the existence of orthogonal axes for central conics: it has a smell of "simultaneous orthogonalization[6]"... except that there is no quadratic form to orthogonalize! We have to wait until we get a quadratic form, one of the tools that will be given to us by Gauss.

Indeed, the real history of surfaces begins with Gauss and his paper *Disquisitiones generales circa superficies curvas*[7]. The very big news is to consider the surfaces themselves, not only the curves they contain, and the great tool is the *normal vector*[8], considered as a mapping from the surface to the unit sphere, the *Gauss map*.

The Gauss mapping. We consider a regular surface Σ parametrized by a mapping $f : U \to \mathbf{R}^3$ and a point $p = f(m)$ of Σ. We want to call $n(p)$ the unit normal vector, but a choice must be made (the normal has, as any real line, two unit vectors) and we have to make it in a continuous (in p) manner. This can be achieved, *e.g.*, by choosing $n(p)$ having the same direction as the normal vector

$$\frac{\partial f}{\partial u} \wedge \frac{\partial f}{\partial v}.$$

(5) See if necessary § VII-2.

(6) That is, of Theorem VI-8.8.

(7) *General investigations on curved surfaces* (in this book, latin quotations are translated).

(8) It is said that the practices of Gauss as an astronomer are at the origin of this idea.

This definition of n is also a way to compute it. More intrinsically: the plane $\mathbf{R}^2$ is oriented by its canonical basis, the differential $(df)_m$ maps this orientation to some orientation of the tangent plane, and, as the space $\mathbf{R}^3$ is oriented, an orientation of the normal line is deduced.

Then, n can be considered as a mapping from Σ (or, rather, from U) into the unit sphere S^2 of $\mathbf{R}^3$.

For example, if Σ is the unit sphere itself, the tangent plane $T_p\Sigma$ is the plane $p^\perp$, the normal is directed by p and one can choose $n(p) = p$, namely $n = \mathrm{Id}$.

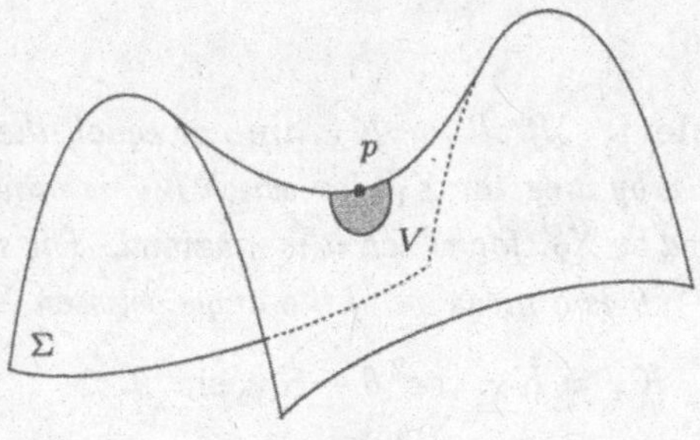

Fig. 14

The heuristic idea underlying the definition of the curvature that will be given below is rather simple. We consider a small neighborhood V of the point p in Σ (Figure 14), its signed area $\mathcal{A}(V)$ and the area of the image $\mathcal{A}(n(V))$. We try to give a meaning to the expression

$$K(p) = \lim_{V \to p} \frac{\mathcal{A}(n(V))}{\mathcal{A}(V)} .$$

This idea comes from the simple examples where:

- the surface Σ is a plane, n is constant and this formula would give $K \equiv 0$,
- the surface Σ is the unit sphere, n is the identity and this formula would give $K \equiv 1$,
- the surface Σ is the saddle depicted in Figure 14 and the area is negative.

The Gauss curvature. In the heuristic formula above, nothing is really defined. Let us thus make a detour. We need something that allows us to compute, infinitesimally, areas on a surface: we use the Euclidean metric of $\mathbf{R}^3$.

Choose an orthonormal basis (v_p, w_p) of $T_p\Sigma$. The tangent mapping to n at p maps v_p and w_p onto two vectors v'_p and w'_p of $T_{n(p)}S^2$. One *defines*

$K(p)$ as the signed area of the parallelogram constructed on the vectors v'_p and w'_p. Figure 15 represents:

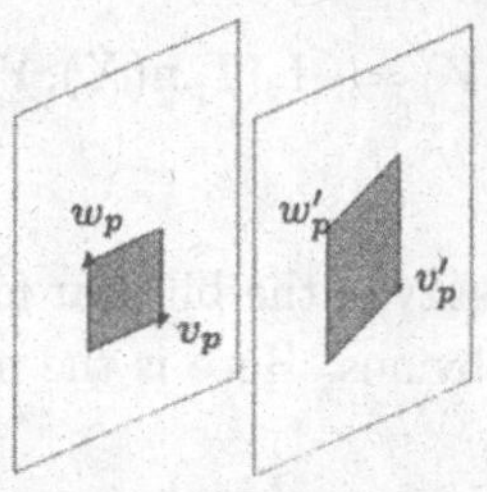

Fig. 15

- on the left, the tangent plane $T_p\Sigma$ together with its orthonormal basis,
- on the right, the oriented (parallel) plane $T_{n(p)}S^2$ and the image under the tangent mapping T_pn of the orthonormal basis.

Remarks 3.2

- We have used here the tangent mapping to $n : \Sigma \to \mathbf{R}^3$, noticing that, since n takes its values in the unit sphere (that is, $\|n(p)\|^2 = 1$ for all p), we have
$$T_pn(X) \cdot n(p) = 0$$
for any vector X of $T_p\Sigma$, thus T_pn takes its values in the tangent space to the sphere.
- The tangent planes $T_p\Sigma$ and $T_{n(p)}S^2$ are the same! Indeed, both are, in their own ways, the orthogonal of $n(p)$. The tangent mapping to n
$$T_pn : T_p\Sigma \longrightarrow T_{n(p)}S^2$$
can be considered as an endomorphism of $T_p\Sigma$. The Gauss curvature $K(p)$ is simply its determinant.

The second fundamental from. To make formulas easier to read, we let $\langle X, Y\rangle = X \cdot Y$ denote the Euclidean scalar product. We define, for any point p of Σ, and for all vectors X, Y in $T_p\Sigma$
$$\mathrm{II}_p(X, Y) = -\langle T_pn(X), Y\rangle.$$
We are going to see (Remark 3.5) that this "mysterious" formula is designed so that $\mathrm{II}_p(X, X)$ is the curvature K_X of the curve defined by the tangent vector X.

Remark 3.3. The notation II is both traditional and execrable. The symbol II is the Roman "digit" for the number 2. The bilinear form thus defined is the *second fundamental form* of the surface Σ at the point p. The reader has the right to wonder whether there is a *first* fundamental form. The answer, is, of course, yes. The first fundamental form of Σ at p is simply the restriction of the Euclidean scalar product to $T_p\Sigma$. We could have written

$$\mathrm{II}_p(X,Y) = -\mathrm{I}_p(T_pn(X),Y),$$

but we have not done it...

The most important property of the bilinear form II_p is that it is *symmetric.* This is not absolutely obvious. Here is the main ingredient in the proof of this fact.

Proposition 3.4. *Let γ be an arc length parametrized curve drawn on Σ. Let $X = \gamma'(0) \in T_p\Sigma$ be its tangent vector at p. Then*

$$\langle \gamma''(0), n(p)\rangle = \mathrm{II}_p(X,X).$$

Remark 3.5. In particular, by definition of the curvature of a plane curve, $\mathrm{II}_p(X,X)$ is the curvature K_X of the curve intersection of the plane spanned by $n(p)$ and X with the surface Σ, precisely the one that appeared in Euler's theorem (Theorem 3.1).

Proof. Differentiate $n \circ \gamma$ at 0:

$$\frac{d}{ds} n \circ \gamma|_{s=0} = T_pn(\gamma'(0)) = T_pn(X).$$

Moreover, the scalar product $\langle \gamma'(s), n(\gamma(s))\rangle$ is identically zero and its differential with respect to s as well:

$$\langle \gamma''(s), n \circ \gamma(s)\rangle + \langle \gamma'(s), \frac{d}{ds}(n\circ\gamma)(s)\rangle = 0,$$

but this gives, at $s = 0$

$$\langle \gamma''(0), n(p)\rangle = -\langle X, T_pn(X)\rangle = \mathrm{II}_p(X,X),$$

which is what we wanted to prove. □

Theorem 3.6. *The bilinear form II_p is symmetric.*

Proof. For the convenience of the reader who does not feel comfortable with derivations, I give a proof in coordinates, starting from a parametrization of the surface. Let thus $f : U \to \mathbf{R}^3$ be a parametrization of Σ. Let $N = n \circ f : U \to S^2$. In this framework, what we want to prove is the equality

$$\mathrm{II}_p\left(\frac{\partial f}{\partial u}, \frac{\partial f}{\partial v}\right) = \mathrm{II}_p\left(\frac{\partial f}{\partial v}, \frac{\partial f}{\partial u}\right).$$

We have

$$T_pn\left(\frac{\partial f}{\partial u}\right) = T_pn\left(df\left(\frac{\partial}{\partial u}\right)\right) = \frac{\partial}{\partial u}(n\circ f) = \frac{\partial N}{\partial u}.$$

But of course, as N is a normal vector, we also have

$$\left\langle N, \frac{\partial f}{\partial v} \right\rangle = 0,$$

and this, differentiating with respect to u, gives the equality

$$\left\langle \frac{\partial N}{\partial u}, \frac{\partial f}{\partial v} \right\rangle + \left\langle N, \frac{\partial^2 f}{\partial u \partial v} \right\rangle = 0.$$

The computation

$$\left\langle N, \frac{\partial^2 f}{\partial u \partial v} \right\rangle = -\left\langle \frac{\partial N}{\partial u}, \frac{\partial f}{\partial v} \right\rangle = -\left\langle T_p n \left(\frac{\partial f}{\partial u}\right), \frac{\partial f}{\partial v} \right\rangle = \mathrm{II}_p \left(\frac{\partial f}{\partial u}, \frac{\partial f}{\partial v}\right)$$

is deduced. The left-hand side is symmetric in u and v because f is a class-$\mathcal{C}^2$ function, thus the bilinear form II_p is symmetric. □

Starting from this, Euler's theorem is, as announced, an application of the simultaneous orthogonalization of quadratic forms.

Proof of Theorem 3.1. There exists an orthonormal basis (X_1, X_2) of $T_p\Sigma$ in which the second fundamental form II_p is diagonal. For each of the vectors X_i, the curvature K_{X_i} is

$$K_{X_i} = \mathrm{II}_p(X_i, X_i)$$

thanks to Proposition 3.4. These two vectors being an orthogonal basis for the second fundamental form, we have

$$\mathrm{II}_p((\cos\theta)X_1 + (\sin\theta)X_2) = K_{X_1}\cos^2\theta + K_{X_2}\sin^2\theta.$$

Notice also that, if K_{X_1} and K_{X_2} are distinct, that is, if II_p is not proportional to the Euclidean scalar product, they are indeed the extrema of the function K_X. □

Examples 3.7

- In the case of the unit sphere, choosing the orientation so that $n = \mathrm{Id}$, so that $\mathrm{II}(X, X) = -\|X\|^2$, the curvature K_X is identically equal to -1 and the Gauss curvature $K(p)$ is identically equal to 1.
- In the case of a plane, n is constant, thus $T_p n = 0$ and the form II is identically zero. The curvature K_X is identically zero.
- In the case of a cone or a cylinder, the mapping $T_p n$ is not zero but the quadratic form II_p is degenerate, therefore the curvature K_X is identically zero (Exercise VIII.15).

The directions of X_1 and X_2 are called the *principal curvature directions* at the point p under consideration. The curvatures K_{X_1} and K_{X_2} are the *principal curvatures.* As the Gauss curvature at p is the determinant of $T_p n$, it is seen by a computation in the basis (X_1, X_2) that this is the product of the principal curvatures.

Here is an application: the curvature at p is related to the local position of the surface with respect to the tangent plane at p.

Corollary 3.8. *The point p is elliptic if and only if $K(p) > 0$, hyperbolic if and only if $K(p) < 0$.*

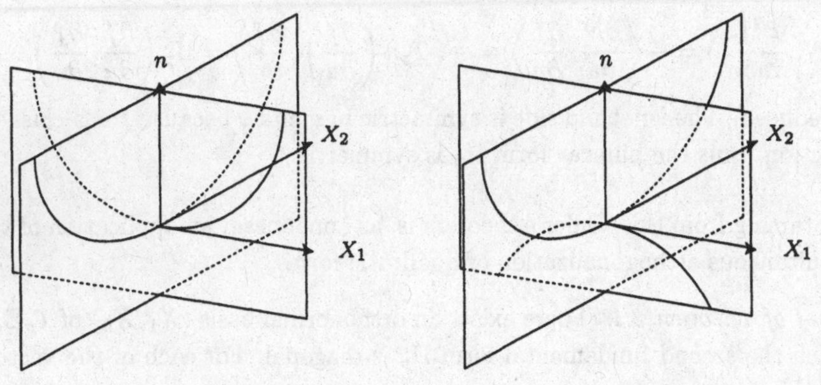

Fig. 16 **Fig. 17**

Proof. The curvature $K(p)$ is positive if and only if the second fundamental form is definite, that is, if and only if the curvature K_X has constant sign. This is equivalent to saying that the curves $P_X \cap \Sigma$ are (locally and strictly) on the same side of $T_p\Sigma$ (see Remarks VII-2.3 and Figure 16).

On the other hand, $K(p)$ is negative if and only if K_{X_1} and K_{X_2} have opposite signs, and thus if and only if the surface crosses (locally) its tangent plane (Figure 17). □

Therefore, Figure 11 did indeed show a point at which the curvature is negative. Another illustration of this fact will be found in the cartoon [**Pet80**], which I highly recommend to the reader.

Gauss' *theorema egregium*. Comparing the examples of the cone and the cylinder, one cannot refrain from thinking that a plane, a cone or a cylinder are "locally isometric" surfaces: think of a (plane) sheet of paper rolled to make a cone or a cylinder. The remarkable(9) theorem of Gauss that follows is a precise and general expression of this idea.

Definition 3.9. To say that two surfaces:

(9) This is what *egregium* means.

- on the one hand, Σ parametrized by $f : U \to \mathbf{R}^3$
- and on the other, Σ' parametrized by $g : U \to \mathbf{R}^3$

are *isometric*, is to say that, for all vectors ξ, $\eta \in \mathbf{R}^2$ and every point m of U, one has

$$\langle df_m(\xi), df_m(\eta)\rangle = \langle dg_m(\xi), dg_m(\eta)\rangle.$$

Theorem 3.10 (*Theorema egregium* of Gauss). *If two surfaces of class $\mathcal{C}^3$ are isometric, they have the same Gauss curvature.*

Remarks 3.11

- I emphasize the theorem asserts, indeed, that two isometric surfaces have the same curvature, *not* the same second fundamental form (think of the cylinder and the plane).
- One of the most remarkable aspects of this theorem is that this is not the analogue of a theorem about curves. It asserts that two surfaces with different curvatures cannot be locally isometric, while the existence of arc length parametrizations (Proposition VII-4.3) shows that all curves are locally isometric (all are isometric to $\mathbf{R}$).

Proof of Theorem 3.10. This is essentially a calculation. It is shown that the curvature can be written in terms of the first fundamental form.

Consider a parametrized surface $f : U \to \mathbf{R}^3$. At the point of parameter (u, v), the Euclidean scalar product of $\mathbf{R}^3$ defines *via* f, a scalar product on $\mathbf{R}^2$: this is the first fundamental form, in coordinates. Write the matrix of this quadratic form in the canonical basis of $\mathbf{R}^2$. This is a symmetric matrix whose entries are functions of (u, v):

$$A = \begin{pmatrix} E & F \\ F & G \end{pmatrix}, \qquad E = \left\langle \frac{\partial f}{\partial u}, \frac{\partial f}{\partial u} \right\rangle, \text{ etc.}$$

By definition, two surfaces are isometric if and only if they have the same E, F and G. We thus want to prove that the curvature K can be expressed in terms of E, F, G alone. To define the curvature, we have used the second fundamental form. Let us write its matrix in the canonical basis of $\mathbf{R}^2$ as well:

$$B = \begin{pmatrix} L & M \\ M & N \end{pmatrix}.$$

The curvature is the determinant of the second fundamental form *in a basis that is orthonormal* for the first fundamental form.

Lemma 3.12. *The Gauss curvature can be expressed, in coordinates, by the expression*

$$K(u,v)=\frac{LN-M^2}{EG-F^2}.$$

Proof. Let S be the matrix of a change of basis, from the canonical basis to an orthonormal basis, so that ${}^tSAS=\mathrm{Id}$. Notice that this equality implies

$$\det(S^2)=(\det A)^{-1}.$$

The computation of K is deduced:

$$K=\det({}^tSBS)=\det(S^2)\det B=(\det A)^{-1}\det B,$$

this giving the announced formula. □

Let us come back to our theorem. To save on notation, let ${}_u=\partial/\partial u$ (so that, *e.g.*, $f_u=\partial f/\partial u$, *etc.*). The second fundamental form is defined by the normal vector n and, as we have seen it in the proof of Theorem 3.6),

$$L=\langle f_{uu},n\rangle\,,\quad M=\langle f_{uv},n\rangle\,,\quad N=\langle f_{vv},n\rangle\,.$$

Since we are working in coordinates anyway, recall that

$$n(u,v)=\frac{f_u\wedge f_v}{\|f_u\wedge f_v\|}=\frac{f_u\wedge f_v}{\sqrt{EG-F^2}}.$$

Therefore, we have

$$L=\left\langle f_{uu},\frac{f_u\wedge f_v}{\sqrt{EG-F^2}}\right\rangle,\ M=\left\langle f_{uv},\frac{f_u\wedge f_v}{\sqrt{EG-F^2}}\right\rangle,\ N=\left\langle f_{vv},\frac{f_u\wedge f_v}{\sqrt{EG-F^2}}\right\rangle.$$

The proof consists now in the evaluation of $K(EG-F^2)^2$. Using the previous lemma, we know that

$$K(EG-F^2)^2=\langle f_{uu},f_u\wedge f_v\rangle\langle f_{vv},f_u\wedge f_v\rangle-\langle f_{uv},f_u\wedge f_v\rangle^2.$$

Recall that $\langle X,Y\wedge Z\rangle$ is the determinant of (X,Y,Z) in an orthonormal basis, here the canonical basis of $\mathbf{R}^3$, so that

$$K(EG-F^2)^2=\det(f_{uu},f_u,f_v)\det(f_{vv},f_u,f_v)-\det(f_{uv},f_u,f_v)^2.$$

The trick is to replace some of the systems of column vectors appearing in this equality by the corresponding systems of line vectors, noticing that

$$\begin{aligned}\det(X,Y,Z)\det{}^t(X',Y,Z)&=\det((X,Y,Z)\,{}^t(X',Y,Z))\\&=\begin{vmatrix}X\cdot X' & X\cdot Y & X\cdot Z\\ Y\cdot X' & Y\cdot Y & Y\cdot Z\\ Z\cdot X' & Z\cdot Y & Z\cdot Z\end{vmatrix}\end{aligned}$$

(where we have, of course, denoted by (X,Y,Z) the matrix whose columns are the vectors X, Y and Z). It is found that:

$$K(EG-F^2)^2 = \begin{vmatrix} \langle f_{uu}, f_{vv}\rangle & \langle f_{uu}, f_u\rangle & \langle f_{uu}, f_v\rangle \\ \langle f_u, f_{vv}\rangle & E & F \\ \langle f_v, f_{vv}\rangle & F & G \end{vmatrix} - \begin{vmatrix} \langle f_{uv}, f_{uv}\rangle & \langle f_{uv}, f_u\rangle & \langle f_{uv}, f_v\rangle \\ \langle f_u, f_{uv}\rangle & E & F \\ \langle f_v, f_{uv}\rangle & F & G \end{vmatrix}.$$

Let us differentiate the relations that define E, F and G, namely

$$E = \langle f_u, f_u\rangle, \quad F = \langle f_u, f_v\rangle, \quad G = \langle f_v, f_v\rangle.$$

We get

$$\langle f_{uu}, f_u\rangle = \frac{1}{2}E_u, \textit{ etc. and } \langle f_{uu}, f_v\rangle = F_u - \frac{1}{2}E_v, \textit{ etc.}$$

from which the inequality

$$\langle f_{uu}, f_{vv}\rangle - \langle f_{uv}, f_{uv}\rangle = -\frac{1}{2}G_{uu} + F_{uv} - \frac{1}{2}E_{vv}$$

follows. Putting these formulas in our computation

$$\begin{aligned} 4K(EG-F^2)^2 &= E(E_vG_v - 2F_uG_v + G_u^2) \\ &\quad + F(E_uG_v - E_vG_u - 2E_vF_v + 4F_uF_v - 2F_uG_u) \\ &\quad + G(E_uG_u - 2E_uF_v + E_v^2) \\ &\quad - 2(EG-F^2)(E_{vv} - 2F_{uv} + G_{uu}), \end{aligned}$$

an expression that allows us to calculate K in terms of E, F, G and their derivatives and in which f does not appear, this proving the theorem. □

Remark 3.13. The converse of Gauss' *theorema egregium* is wrong; a classical counter-example will be found in Exercise VIII.24.

And then[(10)]**.** We have only dealt with *local* properties and only for surfaces *contained* in $\mathbf{R}^3$. It is possible to define a surface abstractly (without needing an ambient space) or more generally a *manifold* (same thing as a surface, with more dimensions). It is also possible to speak of metric properties of such spaces and of their curvature.

For instance, the Poincaré half-plane (Exercise V.50) is an abstract surface with constant curvature equal to -1. This is related to the fact that the sum of angles in a triangle is $< \pi$, in the same way as the analogous result for the sphere (Girard's formula IV-3.1) is related to its positive curvature. Recall that we have used this result to prove the Euler formula (Theorem IV-4.4).

---

(10) After Euler and Gauss, Riemann (see volume 2 of **[Spi75]**).

Well, this was an example of a *global* result in differential geometry, a special case of the Gauss–Bonnet theorem, which relates the curvature to the global topology of the surface. See, *e.g.*, [**BG88**], [**Spi75**].

4. Appendix: a few formulas

In this appendix, I give a few calculations, in particular, that of the Gauss curvature of surfaces defined by an equation. Consider a function $F : \mathbf{R}^3 \to \mathbf{R}$ and the surface

$$\Sigma = \left\{(x_1, x_2, x_3) \in \mathbf{R}^3 \mid F(x_1, x_2, x_3) = 0\right\}$$

(for these computations in coordinates, it is better to use indices). Let ${}_i = {}_{x_i} = \partial/\partial x_i$.

Second fundamental form. The vector $\nabla_p F = (F_1(p), F_2(f), F_3(p))$ is normal to Σ, we can thus choose

$$n(p) = \frac{\nabla_p F}{\|\nabla_p F\|}.$$

The expression of the second fundamental form is deduced: we have firstly, if X is a vector in $\mathbf{R}^3$,

$$(dn)_p(X) = \frac{1}{\|\nabla_p F\|} d(\nabla_p F)(X) + \lambda(p, X)(\nabla_p F)$$

where I have not written the second term completely, because this is a vector normal to the surface and it will disappear when we will pair it with a tangent vector. The first term is computed as follows:

$$d(\nabla_p F)(X) = \left(\sum_i F_{1,i}(p)X_i, \sum_i F_{2,i}(p)X_i, \sum_i F_{3,i}(p)X_i\right).$$

Hence, if X and Y are tangent vectors to Σ at p, we have

$$\mathrm{II}_p(X, Y) = -\frac{1}{\|\nabla_p F\|} \sum_{i,j=1}^{3} F_{i,j}(p) X_i Y_j.$$

Principal curvatures. The principal curvatures at p are thus the extrema of the function (of X) $\sum F_{i,j}(p)X_iX_j$ over the set of unit tangent vectors at p, namely over

$$\left\{X \in \mathbf{R}^3 \mid \sum_{i=1}^{3} F_i(p)X_i = 0 \text{ and } \sum_{i=1}^{3} X_i^2 = 1\right\}.$$

This is a simple problem of constrained extrema (see Exercise VIII.13). The solutions are the vectors X such that there exist scalars λ and μ with

$$\frac{\partial}{\partial X_k}\left(\sum F_{i,j}(p)X_iX_j\right)=\lambda\frac{\partial}{\partial X_k}\left(\sum F_i(p)X_i\right)+\mu\frac{\partial}{\partial X_k}\left(\sum X_i^2\right),$$

that is,

$$2\sum_{i=1}^{3}F_{i,k}(p)X_i=\lambda F_k(p)+2\mu X_k.$$

At such a point X the function takes the value μ. The extrema are the reals μ such that there exists a nonzero vector X and a real λ with

$$\begin{cases}\sum_{i=1}^{3}F_{i,k}(p)X_i &= \dfrac{\lambda}{2}F_k(p)+\mu X_k\\[2ex] \sum_{i=1}^{3}F_i(p)X_i &= 0,\end{cases}$$

that is,

$$\begin{cases}(d^2F)_p\cdot X &= \dfrac{\lambda}{2}\nabla_pF+\mu X\\[2ex] \langle\nabla_pF,X\rangle &= 0.\end{cases}$$

The first equality is an equality of vectors in $\mathbf{R}^3$, the second one is an equality of scalars. We group them in an equation in $\mathbf{R}^4$:

$$\begin{pmatrix}(d^2F)_p-\mu\,\mathrm{Id} & {}^t\nabla_pF\\ \\ \nabla_pF & 0\end{pmatrix}\begin{pmatrix}{}^tX\\ \\ \dfrac{\lambda}{2}\end{pmatrix}=0.$$

Our problem amounts to the nullity of the determinant of this 4×4 matrix, an equation of degree 2 in μ. Here are thus the principal curvatures:

$$K_i(p)=-\frac{\mu_i}{\|\nabla_pF\|}$$

where μ_1 and μ_2 are the solutions of the equation

$$\begin{vmatrix}(d^2F)_p-\mu\,\mathrm{Id} & {}^t\nabla_pF\\ \\ \nabla_pF & 0\end{vmatrix}=0.$$

Gauss curvature. Eventually, we have computed the Gauss curvature

$$K(p)=K_1(p)K_2(p)=\frac{\mu_1\mu_2}{\|\nabla_pF\|^2}.$$

Exercises and problems

For all the exercises that contain examples of surfaces, I request the reader to use her preferred software to draw the surfaces in question.

Exercise VIII.1 (Equations of the surfaces of revolution). The space is endowed with an orthonormal frame in which D is the z-axis.

Assume C is described, in the plane xOz, by the parametric equations

$$\begin{cases} x = r(t) \\ z = z(t). \end{cases}$$

Prove that S is described by the parametric equations

$$\begin{cases} x = r(t)\cos\theta \\ y = r(t)\sin\theta \\ z = z(t). \end{cases}$$

Find parametric equations for the sphere, the torus of revolution, the cylinder of revolution, and the cone of revolution.

Assume now that C is described by a Cartesian equation $f(x,z) = 0$. Prove that

$$f\left(\pm\sqrt{x^2+y^2}, z\right) = 0$$

is a Cartesian equation for S.

Exercise VIII.2 (Helicoid). Draw the surface parametrized by

$$(\theta, t) \longmapsto (t\cos\theta, t\sin\theta, a\theta)$$

(a is a fixed scalar, $(\theta,t) \in \mathbf{R}^2$). Check that this is a ruled surface (spiral staircase). What can be said of the curves $t =$ constant?

Exercise VIII.3 (Plücker's conoid). For which values of (x,y) does the equation

$$z(x^2+y^2) = xy$$

define a surface? Draw this surface. Prove that this is a ruled surface.

Exercise VIII.4. Two noncoplanar lines D and D' are given. The line D' turns around D. What is the surface obtained this way?

Exercise VIII.5 (The hyperboloid of one sheet as a ruled surface). Let Γ be an ellipse in the plane xOy, parametrized by $(x(\theta), y(\theta)) = (a\cos\theta, b\sin\theta)$. With the point of parameter θ is associated the vector $w(\theta) = (dx/d\theta, dy/d\theta, 1)$. Prove that the ruled surface defined by Γ and w is a hyperboloid of one sheet (see Exercise VI.11).

Prove that a hyperboloid of one sheet is, in two ways, a ruled surface.

Exercise VIII.6. The affine space $\mathcal{E}$ is endowed with a frame. Write a Cartesian equation of the (affine) tangent plane to the surface parametrized by $(u, v) \mapsto f(u, v)$ at the point $f(u_0, v_0)$. We assume of course that the surface is regular and the parametrization injective.

Exercise VIII.7. When are the tangent vectors to the meridian and the parallel through a point of a surface of revolution independent? Can a surface of revolution be regular at a point where it intersects the axis of revolution? When this is the case, what is its tangent plane at this point?

Exercise VIII.8. Consider a ruled surface parametrized by

$$\begin{aligned} f: I \times \mathbf{R} &\longrightarrow \mathbf{R}^3 \\ (t, u) &\longmapsto \gamma(t) + uw(t) \end{aligned}$$

(where $w(t)$ does not vanish). Prove that the line spanned by $w(t)$ is contained in the tangent plane at the point $f(t, u)$. For which condition do all the points of this line have the same tangent plane? Prove that all the points of a conic or cylindric ruled surface are parabolic or planar.

Exercise VIII.9 (Swallow tail). Consider the (real, affine, 3-dimensional) space $\mathcal{E}$ consisting of all the polynomials $P \in \mathbf{R}[X]$ of the form $P(X) = X^4 + aX^2 + bX + c$. Describe by parametric equations the subset of $\mathcal{E}$ consisting of the polynomials that have a multiple root and draw it(11). Check that this is a surface and describe its singular points.

Exercise VIII.10 (Whitney's umbrella). Consider the surface parametrized by

$$(u, v) \longmapsto (uv, v, u^2).$$

Draw it (Figure 18). Prove that the half-line ($x = y = 0$, $z > 0$) is a line of

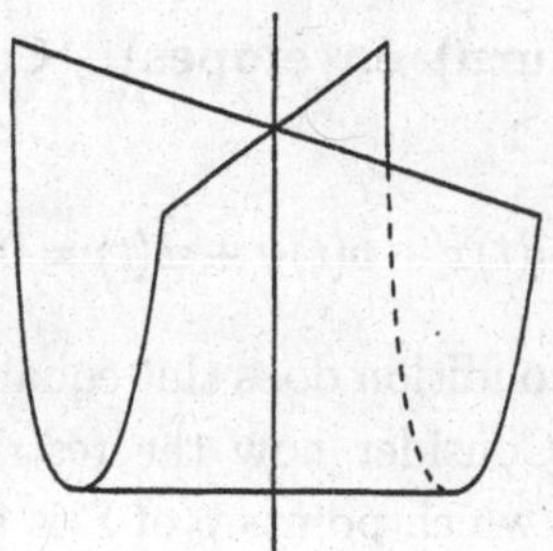

Fig. 18. The Whitney umbrella

(11) You can make *maple* or one of its colleagues do it for you.

double points and that the origin of $\mathbf{R}^3$ is a singular point. Prove that, for all $v \neq 0$, the vector

$$N(u,v) = \left(-2\frac{u}{v}, 2\frac{u^2}{v}, 1\right)$$

is normal to the surface and that it has no limit when (u,v) tends to $(0,0)$ (the tangent planes have no limit at the singular point).

Exercise VIII.11 (Parabolic and planar points). For all the following equations, draw the surface and investigate its position with respect to its tangent plane at 0: $z = x^2$, $z = x^2 + y^3$, $z = x^3$, $z = x^4$, $z = x^3 - 3xy^2$ ("monkey saddle").

Exercise VIII.12. How do the planes parallel to the tangent plane intersect a surface near an elliptic point? near a hyperbolic point?

On a large-scale map of the Vosges mountains, contemplate the level curves, then find the mountain passes and the summits.

Exercise VIII.13 (Constrained extrema). A function $F : \mathbf{R}^3 \to \mathbf{R}$ of class $\mathcal{C}^1$ is given. Prove that, if p is an extremum of h with the constraint $F(p) = 0$, then $(dh)_p$ must be a multiple of $(dF)_p$. Investigate the example of

$$F(x,y,z) = x^2 + y^2 + z^2 - 1$$

and $h(x,y,z) = z$.

Let $G : \mathbf{R}^3 \to \mathbf{R}$ be another function of class $\mathcal{C}^1$ such that $(dF)_p$ and $(dG)_p$ are independent for all p such that $F(p) = G(p) = 0$. Prove that, if p is an extremum of h with the constraint $F(p) = G(p) = 0$, then $(dh)_p$ must be a linear combination of $(dF)_p$ and $(dG)_p$.

Exercise VIII.14 ((Return) envelopes). Consider, as in § VII-1, the family of lines of equation

$$u(t)x + v(t)y + w(t) = 0$$

in the plane. Under what condition does this equality define a regular surface Σ in the (x,y,t)-space? Consider now the restriction π of the projection $(x,y,t) \mapsto (x,y)$ to Σ. For which points p of Σ is the tangent mapping

$$T_p\pi : T_p\Sigma \longrightarrow \mathbf{R}^2$$

surjective? Prove that the images in $\mathbf{R}^2$ of the points where it is not surjective (the *apparent contour* of Σ) form the envelope of the family of lines.

The cusp. Draw Σ, the projection and the apparent contour for the lines of equations

$$3tx - 2y - t^3 = 0.$$

For each point of the plane, determine of how many points of the surface it is the image. What is the relation of this question to Exercise VII.12?

Exercise VIII.15. A point O is given together with a curve Γ in a plane not containing O. Let Σ be the cone of vertex O based on Γ. Compute the curvature of Σ.

Exercise VIII.16. Let Σ be the quadric of equation

$$\frac{x_1^2}{\alpha_1} + \frac{x_2^2}{\alpha_2} + \frac{x_3^2}{\alpha_3} = 1.$$

Compute its Gauss curvature.

Exercise VIII.17 (Surfaces of revolution with constant curvature). Consider a plane curve parametrized by arc length and make it turn about a line (of the plane) it does not intersect, thus getting a surface parametrized by

$$(s, \theta) \longmapsto (g(s)\cos\theta, g(s)\sin\theta, h(s))$$

with $g'^2 + h'^2 = 1$. Prove that the curvature K is, at the point of parameter (s, θ), given by

$$K(s, \theta) = -g''(s)/g(s).$$

Assume now that K is constant.

(1) Prove that, if K is zero, the surface of revolution is a plane, a cylinder or a cone.
(2) If $K > 0$, write the general solution $(g(s), h(s))$ and check that the spheres are solutions.
(3) If $K < 0$, write the general solution $(g(s), h(s))$. Draw the obtained surface when

$$g(s) = e^s \qquad h(s) = \pm \int_0^s \sqrt{1 - e^{2t}}\, dt$$

(pseudosphere[12]).

Exercise VIII.18. Contemplate a torus of revolution and guess what are the points where the curvature is positive, *resp.* negative. Compute the curvature to check that you were right.

---

[12] The meridian curves obtained are *tractrices*. Among other properties, a tractrix is the evolute of a *catenary*, the graph of the function "hyperbolic cosine" cosh.

Exercise VIII.19. Let Σ be a *compact* surface in the Euclidean affine space. Prove that there exists a point of Σ at which the curvature is positive(13). Hint: fix a point O in the space and prove that, at the points of Σ whose distance to O is maximal, the curvature of Σ is positive.

Exercise VIII.20 (The Möbius strip). Draw the surface M parametrized by

$$(\theta, t) \longmapsto \left(\cos\theta + t\cos\left(\frac{\theta}{2}\right)\cos\theta, \sin\theta + t\cos\left(\frac{\theta}{2}\right)\sin\theta, t\sin\left(\frac{\theta}{2}\right)\right)$$

($\theta \in [0, 2\pi]$, $t \in]-\frac{1}{2}, \frac{1}{2}[$). Prove that M is homeomorphic to a Möbius strip (Exercise V.47)... but that this surface of $\mathbf{R}^3$ can*not* be obtained by gluing a strip of paper according to Figure 22 of Chapter V.

Exercise VIII.21 (Umbilics). A point p of a surface is said to be an *umbilic* if the principal curvatures coincide at p. What are the umbilics of a sphere? Prove that an ellipsoid has, in general, four umbilics, and that a torus of revolution has no umbilic at all.

Exercise VIII.22. We want to prove that a surface all the points of which are umbilics (Exercise VIII.21) is a subset of a sphere or a plane. Consider a surface parametrized by $f : U \to \mathbf{R}^3$ and assume that all its points are umbilics. Prove that there exist a constant k and a vector v such that $n = -kf + v$. If $k = 0$, prove that the image of f is contained in a plane. If $k \neq 0$, prove that the image of f is contained in a sphere centered at v/k.

Exercise VIII.23 (Geodesics). An arc length parametrized curve $s \mapsto f(s)$ drawn on a surface Σ is called a *geodesic* if $f''(s)$ is a vector normal(14) to Σ at $f(s)$ for all s. What are the geodesics of a sphere? of a plane? of a cylinder?

Exercise VIII.24. Draw the surface of revolution Σ parametrized by

$$f : (t, \theta) \longmapsto (t\sin\theta, t\cos\theta, \log t)$$

and compute its curvature.

Consider now the helicoid (Exercise VIII.2) parametrized by

$$g : (t, \theta) \longmapsto (t\cos\theta, t\sin\theta, \theta).$$

Prove that its curvature at the point of parameter (s, θ) is the same as that of Σ, but that the two surfaces are not (locally) isometric.

---

(13) This is why the examples of surfaces with everywhere negative curvature given here (*e.g.*, in Exercise VIII.17) are noncompact surfaces.

(14) That is to say, the acceleration has no tangential component. It is understood that this property is related to a local minimization of the distance.

A few Hints and Solutions to Exercises

The reader will find here a mixture of hints, solutions, comments and bibliographical references for some of the exercises. These are *not* complete solutions, even when they are rather detailed. They will exempt the reader

- neither from *writing* their solutions (some of the hints given here are really written in telegraphic language)
- nor from taking care of converses. When I write for instance "the point P lies on the circle of diameter IJ", I do not write "the locus of points P is the circle of diameter IJ".

Last remark: the reader will often need to draw a picture to understand the solution.

Chapter I

Exercises I.6 and I.7. Go back to linear algebra.

Exercise I.15. Use the fixed point theorem (Theorem 2.22).

Exercise I.23. Assume that

$$(\forall x \in E) \quad (\exists \lambda \in \mathbf{K}) \quad (f(x) = \lambda x),$$

in which, *a priori*, $\lambda = \lambda_x$ depends on x. Check that:

- if $\dim E = 1$, λ does not depend on x,
- if $\dim E \geqslant 2$, for any two independent vectors x and y, one has

  $$\lambda_{x+y}(x+y) = f(x+y) = f(x) + f(y) = \lambda_x x + \lambda_y y,$$

  thus $\lambda_x = \lambda_y$. Deduce that λ_x actually does not depend on x.

Hence we have

$$(\exists \lambda \in \mathbf{K}) \quad (\forall x \in E) \quad (f(x) = \lambda x),$$

this meaning that f is a linear dilatation.

Exercise I.24. The linear mapping $\overrightarrow{\varphi}$ maps any vector to a vector that is collinear. Using the "trick" of Exercise I.23, this is thus a linear dilatation.

Exercise I.28. Be careful that if $(\lambda_0, \ldots, \lambda_n)$ is a system of barycentric coordinates of M and if μ is any nonzero scalar, $(\mu\lambda_0, \ldots, \mu\lambda_n)$ is *also* a system of barycentric coordinates of M.

Exercise I.29. Consider for instance the barycenter M_1 of the system $((M, \alpha), (M', -1))$; prove that it is invariant and that the mapping that associates M_1 to M is a projection.

Exercise I.30. In dimensions greater than or equal to 3, you can "turn around" a line; the complement is connected. A formal proof: choose an affine frame whose origin is on the line and whose first vector directs it; we are then in the case of the line $\mathbf{R} \times \{0\}$ in $\mathbf{R} \times \mathbf{R}^{n-1}$; for $n-1 \geqslant 2$, $\mathbf{R}^{n-1} - \{0\}$ is path-connected (this is the place where you "turn around", clear?), thus

$$\mathcal{E} - D = \mathbf{R} \times (\mathbf{R}^{n-1} - \{0\})$$

is path-connected as the product of two path-connected spaces.

Similarly, the complement of a complex line in a complex plane is path-connected, as is the complement of 0 in $\mathbf{C}$.

Exercise I.31. In the case of the centroid, this is a dilatation. In the case of the orthocenter, you can exhibit three collinear points whose images are not collinear, for instance remembering that the orthocenter of a right-angled triangle is the vertex of the right angle.

Exercise I.33. If a subset has two symmetry centers, it is preserved by some translation: the composition of two central symmetries is a translation.

Exercise I.34. Use the composition of central symmetries $\sigma_{A_1} \circ \cdots \circ \sigma_{A_n}$, that is, a translation or a central symmetry according to the parity of n.

Exercise I.35. We have

$$h\left(A, \frac{1}{2}\right) \circ h\left(B, \frac{1}{2}\right) \circ h\left(C, \frac{1}{2}\right) = h\left(B', \frac{1}{8}\right).$$

If I is defined by $\overrightarrow{BI} = \overrightarrow{CB}$, the image of I by this dilatation is the midpoint J of AB. The position of B' on IJ and the construction are deduced.

Exercise I.36. You can look for a system of barycentric coordinates for the point A'' (Figure 20) in the triangle ABC; you can also use the fact that the area of the triangle $AA'C$ is one third of that of the triangle ABC.

Exercise I.37. The dilatation of center C' and ratio $\overrightarrow{C'A}/\overrightarrow{C'B}$ transforms B into A, that of center B' and ratio $\overrightarrow{B'C}/\overrightarrow{B'A}$ transforms A into C. You can use their composition.

The converse statement is a consequence of the direct one.

One has $\overrightarrow{A''C} = -\overrightarrow{A'B}$, *etc.*, thus A'', B'' and C'' satisfy the same relation as A', B' and C' do.

The points F, G and I lie on the parallel to $A'B'$ through G, *etc.*, thus I, J and K are three points on the sides of the triangle EFG. Prove that

$$\frac{\overrightarrow{IF}}{\overrightarrow{IG}} \cdot \frac{\overrightarrow{JG}}{\overrightarrow{JE}} \cdot \frac{\overrightarrow{KE}}{\overrightarrow{KF}} = 1$$

and use Menelaüs' theorem.

Exercise I.38. The simplest thing to do is to use the associativity of the barycenters. But this is also a consequence of Menelaüs' theorem.

Exercise I.39. You can use Menelaüs' theorem six times in the triangle MNP where $M = BC' \cap CA'$, $N = CA' \cap AB'$ and $P = AB' \cap BC'$.

Exercise I.40. If AA', BB' and CC' are concurrent at O, use Menelaüs in the triangles OAB, OAC, OBC and ABC.

Exercise I.41. If β and γ are the points

$$\beta = BM \cap PB', \quad \gamma = CM \cap PC',$$

the line CC' is the image of PM by a dilatation of center γ, hence γ, K and I are collinear. Similarly for β, I and J. Moreover $\beta P \gamma M$ is a parallelogram, thus β, I and γ are collinear.

For the next question, use an affinity about the line BC.

Exercise I.42. You can use Desargues' theorem to construct another point of the line, that is on the sheet of paper.

Exercise I.44. The convex hull of $A_0, \ldots, A_N$ is the image of

$$K = \left\{(x_1, \ldots, x_N) \in \mathbf{R}^N \mid 0 \leqslant x_i \leqslant 1, \sum x_i = 1\right\}$$

(a notorious compact subset of $\mathbf{R}^N$) by the continuous mapping that maps $(x_1, \ldots, x_N)$ to the point M defined by

$$\overrightarrow{A_0M} = \sum_{i=1}^{N} x_i \overrightarrow{A_0A_i}.$$

Exercise I.45. Let n be the dimension of the affine space. Extract $n+1$ independent points $A_0, \ldots, A_n$ of S (use the fact that S is not contained in a hyperplane). Let U be the set of barycenters of these $n+1$ points endowed with strictly positive masses:

$$U = \left\{ M \in \mathcal{E} \mid \overrightarrow{A_0M} = x_1\overrightarrow{A_0A_1} + \cdots + x_n\overrightarrow{A_0A_n} \text{ avec } x_i > 0 \text{ et } \sum x_i < 1 \right\}.$$

- This is an open subset.
- It is not empty since it contains the equibarycenter of the four points.
- It is contained in $\mathcal{C}(S)$, which consists of the barycenters of the points of the set S endowed with positive coefficients (Proposition I-5.6).

We have constructed a nonempty open subset U contained in $\mathcal{C}(S)$, the interior of the latter is thus nonempty.

Exercise I.49. One can use dilatations or even central symmetries and Exercise I.18.

Exercise I.50. One gets the bijection by rewriting affine transformations as compositions of mappings fixing O and translations. There is no group isomorphism as can be seen by the comparison of the centers of the two groups (and using Exercise I.49).

Exercise I.51. If necessary, there is a proof of the "fundamental" theorem, proof that has inspired the statement of this exercise, in [**Ber77**].

Chapter II

Exercise II.1. Write that $\|\lambda x + y\|^2$ is positive for all λ, that is to say that the polynomial

$$\lambda^2 \|x\|^2 + 2\lambda x \cdot y + \|y\|^2$$

keeps the same sign, or that its discriminant is negative:

$$(x \cdot y)^2 - \|x\|^2 \cdot \|y\|^2 \leqslant 0 \text{ or } |x \cdot y| \leqslant \|x\| \cdot \|y\|.$$

This is the Cauchy–Schwarz inequality, equality holds if and only if:

- the polynomial has a real double root λ (that is, $\lambda x + y = 0$ for some λ)
- and $x \cdot y \geqslant 0$ (due to the square root),

that is, if and only if x and y are collinear and have the same direction.

Exercise II.3. For all points O and G, one has

$$OA^2 = OG^2 + GA^2 + 2\overrightarrow{OG} \cdot \overrightarrow{GA} \qquad OB^2 = OG^2 + GB^2 + 2\overrightarrow{OG} \cdot \overrightarrow{GB}$$

and thus also

$$\alpha OA^2 + (1-\alpha)OB^2 = OG^2 + \alpha GA^2 + (1-\alpha)GB^2 + 2\overrightarrow{OG} \cdot (\alpha\overrightarrow{GA} + (1-\alpha)\overrightarrow{GB}).$$

The expected equality holds for some point G if only if the latter satisfies, for all point O,

$$\overrightarrow{OG} \cdot (\alpha\overrightarrow{GA} + (1-\alpha)\overrightarrow{GB}) = 0,$$

namely if G is the barycenter of $((A, \alpha), (B, 1-\alpha))$.

Exercise II.4. It can be shown that φ preserves the barycenters, for instance using Exercise II.3. Denoting the images of the points by $'$, the equality of Exercise II.3 and the preservation of distances imply that one has, for all points O, the equality

$$\alpha O'A'^2 + (1-\alpha)O'B'^2 = O'G'^2 + \alpha G'A'^2 + (1-\alpha)G'B'^2.$$

If we were sure that the point O' obtained can be any point, this would imply, still using Exercise II.3, that G' is the barycenter of $((A', \alpha), (B', 1-\alpha))$, and this would be enough to assert that φ is affine (Proposition I-2.8).

It suffices thus to prove that φ is surjective. Fix an affine frame. To simplify, let us work in a plane, so that the frame consists of three noncollinear points A, B and C. Their images A', B' and C' are not collinear either (due to the preservation of distances and the triangle inequality). Let N be a point of the plane. There is a unique point M such that

$$MA = NA', \quad MB = NB' \quad MC = NC'$$

and then $\varphi(M) = N$.

Exercise II.5. One can expand $\|f(\lambda x + \mu y) - \lambda f(x) - \mu f(y)\|^2$ or use the fact that there exist orthonormal bases, or even use Exercise II.4.

Exercise II.9. The formula for $F(M)$ is proved exactly as that of Exercise II.3 (that is a special case). The loci are a circle (maybe empty or reduced to a point), a line and, in the last case, a circle if $k \neq 1$.

Exercise II.13. Use a square exterior to the triangle, one of the sides of which is BC and a dilatation of center A.

Exercise II.14. The vector $\overrightarrow{MM'}$ is constant.

Exercise II.15. The absolute value of the ratio of the dilatation is the ratio R'/R of the radii. The center is one of the points S of the line OO' of centers that satisfy $SO'/SO = R'/R$. One can then use the last question (or answer) of Exercise II.9. Be careful with the case where the two circles have the same radius and the case where they have the same center.

Exercise II.17. If the two lines intersect at A and if the given point does not belong to one of the two lines, one can use a circle tangent to the two lines and a dilatation of center A.

Exercise II.19. Let I be the midpoint of AB. Use the dilatation of ratio $-1/2$ and center the centroid G of AMB to prove the equality $\overrightarrow{MH} = 2\overrightarrow{OI}$ (O is the center of $\mathcal{C}$). One can also use translations (see [**DC51**]) or angles (next chapter).

Exercise II.20. One can begin by determining the dilatations that transform the circumcircle to ABC into $\mathcal{C}$.

Exercise II.21. Use the "trick" of Exercise I.23.

Exercise II.23. Let J be the diagonal matrix

$$J = \begin{pmatrix} 1 & & & \\ & -1 & & \\ & & \ddots & \\ & & & -1 \end{pmatrix}.$$

Isometries have matrices A that satisfy ${}^tAJA = J$. The elements of $O_q(2)$ are the matrices of the form

$$A = \begin{pmatrix} \varepsilon \cosh t & \eta \sinh t \\ \sinh t & \varepsilon\eta \cosh t \end{pmatrix} \text{ where } t \in \mathbf{R} \text{ and } \varepsilon^2 = \eta^2 = 1.$$

Thus the group O_q is not compact when $n = 2$ (and not more compact in general): it contains a closed noncompact subgroup. The group O_q^+ has two connected components ($\eta = 1$) and the group O_q has four.

Chapter III

Exercise III.2. One can use the properties of the angles at the circumference (Proposition 1.17).

Exercise III.3. Evaluate the scalar product $\overrightarrow{AB} \cdot \overrightarrow{AC}$.

Exercise III.4. The bisectors of the angles at A and B intersect at a point I. This point is equidistant from CA and CB and thus belongs to one of the bisectors of these two lines. It also belongs to the sector defined by the angle at A and to that defined by the angle at B, it is thus inside the triangle and hence on the *internal* bisector of the angle at C. Analogous reasoning for the excircle.

Exercise III.6. This is the vector of the translation you are required to find!

Exercise III.8. The affine mappings are the

$$z \longmapsto az + b\overline{z} + c,$$

the affine transformations those for which $|a|^2 - |b|^2 \neq 0$.

Exercise III.9. Remember that $O^+(P)$ is commutative, that a nontrivial translation cannot preserve a bounded subset and that it has infinite order in the affine group.

Exercise III.10. Let $\varphi_1, \ldots, \varphi_n$ be the elements of G and let M be any point of $\mathcal{E}$. The equibarycenter of $\varphi_1(M), \ldots, \varphi_n(M)$ is fixed under all the elements of G.

Exercise III.12. The ratio of a similarity mapping $\mathcal{C}$ to $\mathcal{C}'$ is the ratio $k = \dfrac{R'}{R}$ of their radii. The center A satisfies the inequality $\dfrac{AO'}{AO} = k$. It is thus:

- on a circle (the circle of diameter SS' where S and S' are the centers of the dilatations mapping $\mathcal{C}$ to $\mathcal{C}'$, see Exercise II.15) if $R \neq R'$,
- on the perpendicular bisector of OO' if $R = R'$.

And of course, you have to study the reverse inclusion...

Exercise III.15. The composition of two inversions of the same pole is a dilatation, not an inversion.

For the conjugacy, this can be done by a computation in complex numbers. There is also a geometric way of reasoning. Assume (this is enough) that the power p of the inversion $I = I(O, p)$ is positive, and let $\mathcal{C}$ be the circle of inversion. Conjugate I by $I' = I(\Omega, k)$. Let A and B be a point of the plane and its image by I. Any circle $\mathcal{S}$ through A and B is orthogonal to $\mathcal{C}$ (Proposition 4.12). Its image $\mathcal{S}'$ by I', a circle through the images A' and B' of A and B, is orthogonal to $I'(\mathcal{C})$. Deduce that A' and B' are transformed one into the other by an inversion of circle $I'(\mathcal{C})$. The conjugate $I' \circ I \circ I'$ is an inversion of circle $I'(\mathcal{C})$.

The pole is the center of this circle, that is easily determined: let $J = I(\Omega)$, we have

$$\overrightarrow{OA} \cdot \overrightarrow{OB} = p = \overrightarrow{O\Omega} \cdot \overrightarrow{OJ}$$

hence the circle ΩAB passes through J, and therefore the line $A'B'$ (its image by I') passes through $J' = I'(J)$, that is, the pole (this being true for all points A and $B = I(A)$).

The power is determined by a direct computation using these points, it is found that

$$\frac{k^2 p}{(O\Omega^2 - p)^2}.$$

See, *e.g.*, [**DC51**].

Exercise III.16. Let O be the pole of the inversion. If A, A' and B are collinear, B belongs to the line OAA' hence B' too. Otherwise, $\overrightarrow{OA} \cdot \overrightarrow{OA'}$ is the power of O with respect to the circle $AA'B$; this is also the power of the inversion and thus we have also $\overrightarrow{OB} \cdot \overrightarrow{OB'}$, therefore B' also belongs to this circle.

Exercise III.19. You *must* draw a figure. The point C is fixed, the circle is transformed into the tangent at C, the sides AB and AD are preserved, they intersect the tangent at the images B' and D' of B and D, the lines BC and DC are transformed into the circles of diameters AB' and AD'... and all the assertions are easily justified.

Exercise III.21. It has many failings: it is not defined everywhere, it does not preserve collinearity, it is not an involution!

Exercise III.22. The theorem on the angles at the circumference and the reflection about IJ give the equality of angles of lines

$$(AB, AI) = (AI, AC).$$

Therefore, the line AI is one of the bisectors of the angle at A of the triangle ABC. The points A and I are on both sides of the line BC. The latter thus intersects the *segment* AI at a point K. Since AI is a chord of $\mathcal{C}$, the point K is inside the circle $\mathcal{C}$; it is thus on the segment BC. Therefore AI is the bisector that intersects the segment BC, namely the internal bisector.

Exercise III.23. Assume that the ABC is not right-angled (otherwise, there is nothing to prove). By symmetry, we have

$$(DB, DC) = (HC, HB)$$

and by Chasles' relation,

$$(HC, HB) = (HC, AB) + (AB, AC) + (AC, HB).$$

As HC is perpendicular to AB and HB to AC, the equality of angles of lines

$$(DB, DC) = (AB, AC)$$

follows, hence A, B, C and D are cyclic.

Assume now that the angles of the triangle are acute. The feet of the altitudes D', E' and F' are then located inside the segments BC, CA and AB. They are the images of the points D, E and F by the dilatation $h\left(H, \frac{1}{2}\right)$. One has

$$CD = CH = CE,$$

thus C belongs to the perpendicular bisector of DE. We want to apply the result of Exercise III.22. We must thus check that C and F are on both sides of DE.

The points E' and D' are on both sides of the altitude CF'. Using the dilatation $h(H, 2)$, the points E and D are thus on both sides of CF'. Therefore the line CF intersects the segment ED at a point inside the circle, hence inside the segment FC.

Exercise III.24. For the existence, one can use the compactness of the segments AB, BC and CA and the continuity of the function that, with (P, Q, R), associates $PQ + QR + RP$.

To minimize the perimeter of PQR amounts to minimizing the length of the broken line

$$R_1R + RQ + QQ_1.$$

The four points must be collinear *in the order* R_1RQQ_1.

As the angles at B and C are acute, R_1 and Q_1 are on the same side of BC as A. As the angle at A is acute, R_1Q_1 intersects the *half*-lines AB and AC of origin A. Hence R and Q belong to the segments AB and AC.

To determine P: the triangle AR_1Q_1 is isosceles at A and its angle at the vertex is the double of that of ABC. To minimize R_1Q_1 amounts to minimize $AR_1 = AP$, so that P must be the foot of the altitude.

For the end, use Exercise III.23.

Exercise III.25. Use the fact that the altitude through A has length $\leqslant R_a$ to get $\frac{1}{2}aR_a \geqslant \text{Area(ABC)}$. To prove the Erdős–Mordell inequality, use the three inequalities such as

$$aR_a \geqslant br_c + cr_b.$$

For the equality case, we must have $a^2 + b^2 = 2ab$, *etc.*, that is, $a = b = c$ (the triangle is equilateral) and $aR_a \geqslant br_b + cr_c$, *etc.*, that is, P belongs to the altitudes. Hence P is the center of the triangle.

Conversely, if ABC is equilateral and P is its center, we have $r_a = r_b = r_c$ (P is the incenter), $R_a = R_b = R_c$ (P is the circumcenter) and $R_a = 2r_a$ (P is the centroid), thus we have indeed the equality.

Exercise III.26. If AP is a bisector of the angle at A, consider the other bisector AQ and the orthogonal projections B' and C' of B and C on AQ. It is checked that

$$\frac{PB}{PC} = \frac{AB'}{AC'}$$

(since the lines PA, CC' and BB' are not parallel) and that

$$\frac{AB'}{AC'} = \frac{AB}{AC}$$

(since the right-angled triangles ABB' and ACC' are similar).

Exercise III.28. One can for instance (as this is suggested in [**Ber77**]) prove the last equality criterion by mapping A to A' and B to B' by an isometry and adjust, then prove the two others using the metric relations in the triangle (Exercises III.2 and III.3).

Exercise III.30. This is a central symmetry. To determine its center, look at the images of the contact points of the incircle with the sides of the triangle.

Exercise III.32. Write the equalities of angles of lines

$$(AB, AA') = (B'B, B'A')$$
$$(CC', CB) = (B'C', B'B)$$
$$(CD, CC') = (D'D, D'C')$$
$$(AA', AD) = (D'A', D'D)$$

and $(CB, CD) = (AB, AD)$ if we assume A, B, C and D cyclic. Adding the five equalities to get $(B'A', B'C') = (D'A', D'C')$, gives the cyclicity of A', B', C' and D'.

Exercise III.33. Because of the right angles, the circle of diameter MC contains the points P and Q and the circle of diameter MB passes through P and R. We thus have, for all points M

$$(PM, PQ) = (CM, CQ) = (CM, CA)$$

and

$$(PM, PR) = (BM, BR) = (BM, BA).$$

The points P, Q and R are collinear if and only if the equality $(PM, PQ) = (PM, PR)$ holds, that is, if and only if the points A, B, C and M are cyclic.

Exercise III.34. The dilatation $h\left(M, \dfrac{1}{2}\right)$ maps the points P', Q' and R' on the orthogonal projections P, Q and R of M on the three sides. The points P', Q' and R' are collinear if and only if P, Q and R are, that is (Exercise III.33) if and only if M belongs to the circumcircle of ABC (and the Steiner line is parallel to the Simson line $\mathcal{S}_M$).

Let D be the symmetric of the orthocenter H with respect to BC. This is a point of $\mathcal{C}$ (Exercise III.23). Assume that M is distinct from A, B, C and D. The line MP' intersects the circle at another point P'' (equal to M if MP' is tangent to $\mathcal{C}$). We have the equality of angles of lines

$$(HP', MP') = (AP'', AD)$$

from which we deduce that HP' and AP'' are parallel.

One proves also the equality

$$(AP'', MP') = (\mathcal{S}_M, MP').$$

Hence AP'' is parallel to the Simson line and HP' is parallel to AP''. We deduce that HP' is the Steiner line of M and thus that the latter passes through H.

We still have to check that the Steiner lines of the points A, B, C and D pass through H. But these are the altitudes for A, B and C and the parallel to the tangent at A through H in the case of D.

Exercise III.36. Choose A on D_1. The vertex C is deduced from B by a rotation of center A and angle $\pm\pi/3$; it is thus also on the image D_2' of D_2 by such a rotation.

Exercise III.38. Let O be the midpoint of BD. One checks that the rotation of center O and angle $\pm\pi/2$ maps P to S, thus $OP = OS$ and $OP \perp OS$. Similarly, $OQ = OR$ and $OQ \perp OR$, hence the associated linear rotation maps $\overrightarrow{PR}$ onto $\overrightarrow{SQ}$.

Therefore the diagonals of $PQRS$ are orthogonal and have the same length. The quadrilateral $PQRS$ is a square if and only if they intersect at their midpoints. Let α and β be the midpoints of the diagonals PQ and RS, ω the midpoint of AC. Notice that the quadrilateral $O\beta\omega\alpha$ is a square. Therefore $PQRS$ is a square if and only if $\beta = \alpha$, that is, if and only if $O = \omega$, that is, if and only if $ABCD$ is a parallelogram.

Exercise III.39. The line OP is the perpendicular bisector of AF but is also that of EB, hence that of $E'B'$, thus the triangle $PB'E'$ is isosceles of vertex P. The triangle OAB is equilateral and its median AB' is perpendicular to BO. Therefore P and B' are on the circle of diameter OA and we have

$$(PB', PO) = (AB', AO)$$

thus the angle at the vertex of the isosceles triangle $PB'E'$ has measure $\pi/3$ and this triangle is equilateral. Moreover, we have

$$\begin{cases} \overrightarrow{B'M} = \frac{1}{2}\overrightarrow{OC} \\ \overrightarrow{E'N} = \frac{1}{2}\overrightarrow{OD} \end{cases} \quad \text{and thus} \quad \begin{cases} B'M' = E'N \\ (\overrightarrow{B'M}, \overrightarrow{E'N}) = (\overrightarrow{OC}, \overrightarrow{OD}) \equiv \frac{\pi}{3} \mod 2\pi. \end{cases}$$

The linear rotation of angle $\pi/3$ maps $\overrightarrow{B'M}$ to $\overrightarrow{E'N}$. The rotation of center P and angle $\pi/3$ maps B' to E' and thus M to N, from which it is deduced that PMN is equilateral.

This exercise can also be solved using complex numbers and the characterization of (direct) equilateral triangles by "$a + bj + cj^2 = 0$".

Exercise III.40. With the assumption about angles, the point F is inside the triangle.

Let M be any point. Then $MA + MB + MC$ is the length of the broken line $C'M' + M'M + MC$ where M' is the image of M by a rotation of center B and angle $\pm\pi/3$. Deduce that, if there exists a solution, it is at F.

Exercise III.42. For rigid motions: all finite subgroups of the group of plane rigid motions are abelian (Exercise III.9). For isometries: $\mathfrak{A}_4$ has no index-2 subgroup.

Exercise III.45. Consider the triangle (similar to the others) Abc, where b is the orthogonal projection of A on $\mathfrak{D}$, together with its orthocenter h. The triangles ABb and ACc are similar, thus C belongs to the perpendicular to Ac at c and similarly H belongs to the perpendicular to Ah at h.

Exercise III.46. Consider the orthogonal projections m and m' of S on D and D', the center ω of the corresponding circle, the contact points t and t' of the tangents through S. If O is the midpoint of MM', it is the image of ω by a similarity of center S that maps m to M and m' to M'. Deduce that $S\omega O$ and StT are similar and eventually that T and T' belong to ωt and $\omega t'$.

Exercise III.48. One can consider the points A and A' opposite to I on $\mathcal{C}$ and $\mathcal{C}'$ respectively and M', the other intersection point of MJ and $\mathcal{C}'$. Check that $\sigma(A) = A'$ and that the triangles IMM' and IAA' are similar. As for the point P, it lies on the circle of diameter IJ.

Exercise III.50. Observe that B' and C' belong to the circle of diameter BC to obtain the relation, which also says that α belongs to the radical axis of the circumcircle of ABC and its Euler circle.

Exercise III.51. Use Menelaüs for the transversals mentioned in the text and use the cyclicity to write equalities such as $\overrightarrow{AR}\cdot\overrightarrow{BR}=\overrightarrow{CR}\cdot\overrightarrow{DR}\ldots$ then conclude.

Exercise III.52. In general, there is an inversion of pole O that maps $\mathcal{C}$ to $\mathcal{C}'$ if and only if there is a dilatation of center O that maps $\mathcal{C}$ to $\mathcal{C}'$.

Exercise III.53. The pole is one of the intersection points of the circles of diameters AC and BD.

Exercise III.54. Let O and O' be the centers of the circles, and A one of their intersection points. In the triangle $OO'A$, we have the relation

$$d^2 = R^2 + R'^2 - 2RR'\cos A$$

and we conclude by preservation of angles.

Exercise III.55. One may consider the images A', B' and C' of A, B and C by an inversion of pole D. Prove that the inequality is an equality if and only if the points A', B' and C' are collinear in this order; this is equivalent to saying that A, B, C and D are cyclic "in this order" (that is, to the fact that the quadrilateral $ABCD$ is cyclic and convex).

Exercise III.56. Consider the image of the line MM' by the inversion $I(P,\overrightarrow{PA}\cdot\overrightarrow{PM})$, which is a circle centered on Δ and passing through A and A'.

Exercise III.58. This is a circle of the pencil spanned by the two circles (the radical axis if $k=1$).

Exercise III.60. Let M be a point of the plane, not on the radical axis Δ of $\mathcal{F}$. Choose a point P on Δ. On the line PM, there is a unique point M' such that $\overrightarrow{PM}\cdot\overrightarrow{PM'}$ is the power of P with respect to the circles of the pencil $\mathcal{F}$. The center of the circle we are looking for is the intersection point of the perpendicular bisector of MM' and the line of centers of $\mathcal{F}$.

Exercise III.61. This is essentially done in the text!

Exercise III.62. One can use Exercise III.61 and the easy case where $\mathcal{C}$ and $\mathcal{C}'$ are concentric.

Exercise III.63. If O is a point of the plane and H its orthogonal projection on the line D, the image $I(O,k)(D)$ is a circle of diameter OH_1 where H_1 is the point of OH such that $\overrightarrow{OH} \cdot \overrightarrow{OH_1} = k$. If D' is another line, consider $I(O,k)(D')$, and deduce that the locus consists of the two bisectors of D and D' (minus the intersection point of D and D').

The centers of the in- and excircles are thus the poles of the inversions that transform the three sides in circles of the same radius. Let I be one of these four points. The images of the sides are the circles centered at α, β and γ; they intersect at I and at the inverses A', B' and C' of the vertices. The inverse of the circumcircle of ABC is the circumcircle of $A'B'C'$. One can then argue like this: the point I is the center of the circumcircle of $\alpha\beta\gamma$, that has the same radius as the three others, the dilatation of center I and radius 1/2 maps A', B' and C' on the midpoints of the sides of the triangle $\alpha\beta\gamma$... Hence the radius of the circumcircle of $A'B'C'$ is twice that of the Euler circle of $\alpha\beta\gamma$ and thus(1) it is equal to that of the circumcircle to $\alpha\beta\gamma$.

Up to this point, only the poles of the inversions have been used. Fix now the power. Let I be the center of the incircle of ABC, consider the inversion of pole I and power $k = OI^2 - R^2$. The circumcircle is invariant and all the circles images of the sides have the same radius R. Let K be the contact point of the incircle with the side BC and let K' be its image. one has $IK' = 2R$ and $\overrightarrow{IK} \cdot \overrightarrow{IK'} = -2Rr$ from which the expected relation is deduced. The readers who want can prove analogous relations involving the radii of the excircles (all these relations can also be proved, of course, in a simpler way).

Exercise III.66. For the end: if the function has an essential singularity at infinity, it cannot be injective. It is not hard to prove, indeed, using Weierstrass' theorem that the set of points that are reached infinitely many times is dense in **C** (see [**Car95**]). Otherwise, this is a rational function, and the injectivity assumption allows us to conclude.

Chapter IV

Exercise IV.1. Its cosine is $-1/2$.

Exercise IV.4. What is the conjugate of a translation by an affine transformation?

(1) At this point, I assume that the reader has drawn a figure.

Exercise IV.6. Use the corresponding linear result. One checks easily: if the composition is a translation, the axes are parallel, if this is a rotation, the axes have a common point and are thus coplanar... if the axes are not coplanar, the composition is a screw displacement.

Exercise IV.7. It is obvious, by definition of the reflections, that $-s_P$ is a symmetry with respect to the line $P^\perp$. Then, if D and D' are two lines, we have

$$s_{D'} \circ s_D = s_{P'} \circ s_P,$$

where P and P' are the planes orthogonal to D and D' and hence this is a rotation of axis $P \cap P'$. If the composition of the two half-turns is a half-turn, it is thus necessary (and sufficient) that P and P' are orthogonal, namely that the lines D and D' are orthogonal; the axis of the composed half-turn is then the line perpendicular to D and D'.

That the half-turns are all conjugated is a consequence of the conjugacy principle:

$$f \circ s_D \circ f^{-1} = s_{f(D)},$$

hence, to conjugate s_D and $s_{D'}$, it suffices to use a rotation that maps D to D'.

Exercise IV.10. There exists a reflection mapping u_1 to v_1 (about the perpendicular bisector plane of u_1 and v_1); the vector u_2 is mapped to another unit vector v_2', such that

$$\|v_1 - v_2'\| = \|u_1 - u_2\| = \|v_1 - v_2\|.$$

The perpendicular bisector plane of v_2 and v_2' thus contains v_1; the reflection about this plane fixes v_1 and maps v_2' to v_2. The composition of the two reflections is a rotation and has the expected properties.

Exercise IV.12. If v and w are collinear, this is easy. Otherwise, it is clear that the result of the computation is a vector of the plane spanned by v and w. The simplest thing to do is to choose a direct orthonormal basis (e_1, e_2, e_3) such that

$$e_1 = \frac{1}{\|v\|}v, \quad \langle e_1, e_2 \rangle = \langle v, w \rangle$$

and to compute in this basis.

Exercise IV.14. If the angles are equal and if A, B, C and D are four consecutive vertices, one has $AB = CD$.

Exercise IV.16. Figure 14 illustrates the property... but, to prove it, one can also look at Figure 15.

Exercise IV.17. Any isometry that preserves the cube transforms one of the two tetrahedra into one of the two tetrahedra. The group of the isometries of the cube that preserve one of the tetrahedra is an index-2 subgroup (there are indeed isometries that exchange them). Moreover, any isometry that preserves this tetrahedron preserves the cube, hence the group of isometries of the (abstract) regular tetrahedron is an index-2 subgroup in the isometry group of the cube.

Exercise IV.18. If A' and B' are the images of the two vertices A and B of the tetrahedron by an isometry, we have $A'B' = AB$. If, moreover, the isometry preserves the tetrahedron, A' and B' must be points of this tetrahedron. But the length AB is the largest possible distance between two points of the tetrahedron and it is realized by two points only if these points are vertices. Hence A' and B' are vertices of the tetrahedron.

The morphism maps an isometry to the permutation of the vertices that it determines. It is injective because an affine mapping is completely determines by the image of an affine frame. Since the reflection with respect to the perpendicular bisector plane of an edge realizes the transposition of the ends of this edge, the image of our morphism contains all the transpositions. As the latter generate the symmetric group, the morphism is surjective.

Exercise IV.20. It is easy to make a list of the rigid motions that preserve a cube, but a little less simple to prove that we have forgotten nothing. The consideration of the two tetrahedra gives the order of the group "for free" and allows us to stop once we have found twenty-four rigid motions.

The length of a main diagonal is the biggest distance between two points of the cube; this is why the isometries preserve the set of main diagonals. This way we have a homomorphism φ from the isometry group of the cube to the symmetric group $\mathfrak{S}_4$. It is clear that φ is not injective: the reflection that exchanges two parallel faces fixes each main diagonal. To prove that φ nevertheless defines an isomorphism from the rigid motion group to $\mathfrak{S}_4$ is a somewhat boring exercise. Here we know the order of the rigid motion group and it suffices to exhibit a rigid motion that induces a transposition of the main diagonals to have the surjectivity of $\varphi\ldots$ and thus also its injectivity.

Exercise IV.21. We know that the composition r is a rotation and that its axis passes through A. The point C' of the plane BCD defined by $\overrightarrow{C'B} = \overrightarrow{DC}$ is fixed (use the fact that the edges AB and CD are orthogonal). Thus the axis is AC'. We still have to determine the angle of the rotation. One can consider the image of B; it is easy to check that this is the point β defined by

$\overrightarrow{A\beta} = \overrightarrow{CB}$. If H is the orthogonal projection of B on AC', one checks that $BH\beta$ is a right-angled (at H) triangle and deduces that the angle is $\pm\pi/2$.

Exercise IV.22. The triangles GAB, GBC and GCA have the same area because the barycentric coordinates of G are $(1,1,1)$... but it is also possible to solve this exercise by transforming ABC into an equilateral triangle by an affine map of determinant 1 (that preserves the areas)!

Exercise IV.23. The existence of the Gergonne point is also an immediate consequence of Ceva's theorem.

Exercise IV.24. Let S be the south pole. I suggest that the reader draws a figure in the plane NSM. Let $\mathcal{P}'$ be the parallel plane to $\mathcal{P}$ through S and let μ be the intersection point of the line NM with $\mathcal{P}'$. We have

$$\overrightarrow{Nm} \cdot \overrightarrow{N\mu} = NS^2 = 4R^2$$

and

$$\overrightarrow{Nm} \cdot \overrightarrow{NM} = \frac{1}{2}\overrightarrow{Nm} \cdot \overrightarrow{N\mu} = 2R^2$$

(the dilatation of center N and ratio $1/2$ maps $\mathcal{P}'$ to $\mathcal{P}$; I am copying the proof of Proposition III-4.8).

Thus φ is indeed the restriction of an inversion, so that it is continuous, it is clearly invertible... and its inverse is the restriction to $\mathcal{S} - \{N\}$ of the same inversion, it is thus also continuous.

For the second part, notice that the triangles OMN and ONM' are similar and one deduces the inequality

$$\overrightarrow{OM} \cdot \overrightarrow{OM'} = ON^2 = R^2.$$

Exercise IV.26. An origin must be chosen to define the longitude! However, the *difference of two longitudes* is independent of the choice of the origin. With his watch giving Washington time, Gédéon Spilett is able to measure the difference between the longitude of the island and that of Washington.

The heroes of the "Fur Country" have certainly measured their latitude... but the height of the sun could not inform them on their longitude.

Exercise IV.27. The existence of the point M_0 and the fact that d is positive come from the continuity of the function and the compactness of its source space. Then the point A is on the side "$\overrightarrow{AM} \cdot \overrightarrow{AM_0} < \dfrac{d^2}{2}$" and it suffices to check that C is on the other side. Assume thus that there is a point M in C such that

$$\overrightarrow{M_0M} \cdot \overrightarrow{M_0A} \geqslant \frac{d^2}{2}.$$

If N is a point of the segment M_0M at the distance ε of M_0, N is in C (use the convexity assumption) and, using the metric relations in the triangle AM_0M, it is checked that, for ε small enough, one has $AN^2 < d^2$, which is contrary to the definition of d.

Exercise IV.28. It is clear that $\check{C}$ is convex, since this is an intersection of half-spaces.

Since 0 is in the interior of C, for $\varepsilon > 0$ small enough, the closed ball $B(0,\varepsilon)$ of center 0 and radius ε is contained in C. Then

$$\check{C} \subset B\left(0, \frac{1}{\varepsilon}\right).$$

Indeed, if $u \in \check{P}$, one has $u \cdot v \leqslant 1$ for all v in P and in particular for all v in $B(0,\varepsilon)$. Using $v = \varepsilon u \|u\|$, the expected inequality $\|u\| \leqslant 1/\varepsilon$ is deduced.

The inclusion $C \subset \check{\check{C}}$ holds by definition. For the inverse inclusion, consider a point A out of C; prove that it does not belong to $\check{\check{C}}$ by separating it from C by a plane as Exercise IV.27 allows us to do.

If P is the convex hull of the finite set $\{B_1, \ldots, B_N\}$, write

$$v = \overrightarrow{OM} = \sum \lambda_i \overrightarrow{OB_i}$$

with $\lambda_i \geqslant 0$ and $\sum \lambda_i = 1$, so that

$$\check{P} = \bigcap_{i=1}^{N} \left\{u \in E \mid u \cdot \overrightarrow{OB_i} \leqslant 1\right\}$$

Then $\check{P}$ is the intersection of a finite number of closed half-spaces. this is a convex polyhedron according to the characterization given in Proposition 4.3 and to the fact that it is bounded.

Exercise IV.30. Let d be the distance from the center O to the faces of P and let $M_1, \ldots, M_F$ be the centers of its faces. The faces of P are the planes of equations

$$\overrightarrow{OM} \cdot \overrightarrow{OM_i} = d^2,$$

the points of P are the points M such that

$$\overrightarrow{OM} \cdot \overrightarrow{OM_i} \leqslant d^2.$$

As for the dual of P', this is the polyhedron defined by the inequalities

$$\overrightarrow{OM} \cdot \overrightarrow{OM_i} \leqslant 1.$$

We just need to use the dilatation of center O and ratio d^2.

Exercise IV.31. This is a parallelogram. In particular, its isometry group is bigger than that of the quadrilateral we started from and hence of its dual (Exercise IV.29). Thus, they are not images one of the other by a dilatation.

Exercise IV.32. Assume the bathroom has been tiled, or the ball has been made with hexagonal pieces. Each face has six edges, and each edge belongs to two faces, thus $6F = 2E$. Therefore the Euler relation gives

$$F - E + V = \frac{E}{3} - E + V = 2, \quad \text{that is,} \quad 3V = 6 + 2E.$$

The number of edges at each vertex is a number r_s that depends on the vertex and satisfies $r_s \geqslant 3$, thus

$$3V \leqslant \sum r_s = 2E,$$

which gives the contradiction.

Exercise IV.33. The Euler formula gives twelve pentagons... but says nothing on the number of hexagons (on a real soccer ball, there are twenty hexagons). Indeed, if m is the number of hexagons and n the number of pentagons, we have $F = m + n$, $2E = 6m + 5n$, $3V = 2E$ and the Euler formula gives

$$m + n - \frac{6m + 5n}{2} + \frac{6m + 5n}{3} = 2,$$

that is, $n = 12$.

Exercise IV.34. Transform the figure by stereographic projection to see a polyhedron drawn on a sphere and evaluate its numbers of vertices, edges and faces.

Exercise IV.36. You *must* draw a figure. The great circles defined by xy, xz intersect the plane $x^\perp$ at two points u and v such that

$$\begin{aligned} y &= (\cos c)x + (\sin c)u \\ z &= (\cos b)x + (\sin b)v \end{aligned}$$

and $u \cdot v = \cos \alpha$. Evaluate $\cos a = y \cdot z$ to obtain the formula. We have thus

$$\left| \frac{\cos a - \cos b \cos c}{\sin b \sin c} \right| < 1, \text{ that is, } \cos(b + c) < \cos a < \cos(b - c)$$

from which the triangle inequality is deduced. See also [**Ber77**, Chapter 18] for these considerations on the sphere.

Exercise IV.37. Extend φ to E putting

$$\begin{cases} \widetilde{\varphi}(0) = 0 \\ \widetilde{\varphi}(v) = \|v\| \, \varphi\left(\dfrac{v}{\|v\|}\right) \text{ if } v \neq 0. \end{cases}$$

Notice that $\widetilde{\varphi}$ preserves the scalar product (because $x \cdot y = \cos d(x, y)$), then use Exercise II.5.

Exercise IV.38. The two conditions imply the inequality

$$\left|\frac{\cos a - \cos b \cos c}{\sin b \sin c}\right| < 1.$$

Let $\alpha \in]0, \pi[$ be such that

$$\cos \alpha = \frac{\cos a - \cos b \cos c}{\sin b \sin c}.$$

Fix x on S, construct two great circles through x making the angle α, put y and z on these circles at the distances b and c of x, and the spherical trigonometry formula gives $d(x, y) = a$. Check the uniqueness.

Exercise IV.39. Draw a small equilateral triangle ABC in U. Let M be its center (intersection point of the perpendicular bisectors); this point is equidistant from A, B and C. Assume that $f : U \to \mathbf{R}^2$ preserves the distances. Let A', B', C' and M' be the images of A, B, C and M. Then $A'B'C'$ is equilateral and M' is its center. We prove now that the metric relations in the two triangles are incompatible:

- in the "flat" triangle, we have $A'B' = \sqrt{3}A'M'$,
- but in the spherical triangle, we have $AB < \sqrt{3}AM$.

To prove this inequality, apply the spherical trigonometry formula (Exercise IV.36). Put $a = AB$ and $b = AM$; we have

$$\begin{aligned}\cos a &= \cos^2 b + \sin^2 b \cos \frac{2\pi}{3}\\ &= -\frac{1}{2} + \frac{3}{2}\cos^2 b.\end{aligned}$$

We want to prove that $a < \sqrt{3}b$. For b small enough, this is a consequence of the computation

$$\cos(\sqrt{3}b) = 1 - \frac{3b^2}{2} + \frac{9b^4}{24} + o(b^5) \text{ and } \cos a = 1 - \frac{3b^2}{2} + \frac{b^4}{2} + o(b^5),$$

which gives (exactly) the expected inequality. The case of a very small equilateral triangle is enough (but the inequality also holds true for the large triangles).

Exercise IV.40. The hints in the statement should be sufficient, the existence of the equilateral triangle coming from the result of Exercise IV.38.

Exercise IV.41. This is the dodecahedron constructed on the cube of vertices $(\pm 1, \pm 1, \pm 1)$ as in Exercise IV.40 and in Figure 9.

Exercise IV.43. Use the fact that φ is the restriction to $\mathcal{P}$ of an inversion (Exercise IV.24).

Exercise IV.44. See [**Cox49**] or [**Art90**] for instance.

Exercise IV.45. All the proofs, with other applications of quaternions, can be found, *e.g.*, in [**Per96**], [**MT86**], [**God71**] and [**Por69**].

Chapter V

Exercise V.4. The complement of a line in a projective plane is an affine plane, hence it is connected. The complement of $\mathbf{P}_1(\mathbf{R})$ in $\mathbf{P}_1(\mathbf{C})$ is the complement of $\mathbf{R}$ in $\mathbf{C}$ and therefore, it has two connected components (see Figure 13).

Exercise V.6. Linear algebra, that's easy!

Exercise V.7. Let $f : E \to E$ be a linear isomorphism defining the projective transformation g. A fixed point of g is the image of an eigenvector of f. Complex endomorphisms have eigenvectors (thanks to the fundamental theorem of algebra) and the same is true of real endomorphisms of odd dimensional real vector spaces (for a similar reason).

For the counter-example, start from an isomorphism of $\mathbf{R}^2$ without real eigenvalue, a rotation for instance. Here is one: $z \mapsto -1/z$ (from which rotation does it come?).

Exercise V.8. The translation in linear algebra is a special case of the (easy to prove and) very classical fact: two endomorphisms of the plane both having two independent eigenvectors (namely, two diagonalizable endomorphisms) commute if and only if they have the same eigenvectors (namely, are diagonalizable in the same basis).

Exercise V.9. The homographies of $\mathbf{P}_1$ that preserve ∞ are the affine transformations $z \mapsto az + b$ (with $a \neq 0$); those that also preserve 0 are the linear isomorphisms $z \mapsto az$ (always with $a \neq 0$). The group of homographies preserving 0 and ∞ is thus isomorphic to the multiplicative group $\mathbf{K}^\star$.

If a and b are two (distinct) points, let g_0 be a homography that maps them to 0 and ∞. Then g preserves a and b if and only if $g_0 g g_0^{-1}$ preserves 0 and ∞. The group of homographies that preserve a and b is thus conjugated to $\mathbf{K}^\star$ in $PGL(2; \mathbf{K})$, in particular, it is isomorphic to it.

Exercise V.11. Choose a basis (e_1, e_2, e_3) of the vector space E defining P in such a way that $m = p(e_1)$ and $D = P(\langle e_2, e_3\rangle)$. The line $m^\star$ is the image in $P(E^\star)$ of the plane $\langle e_2^\star, e_3^\star\rangle$ (linear forms vanishing on e_1).

The subspace of E of equation $(ae_2^\star + be_3^\star)(u) = 0$ intersects the plane $\langle e_2, e_3\rangle$ along the line spanned by $(b, -a)$. The incidence is thus the homography associated with the linear isomorphism

$$\begin{array}{ccc}\langle e_2^\star, e_3^\star\rangle & \longrightarrow & \langle e_2, e_3\rangle \\ (a, b) & \longmapsto & (b, -a).\end{array}$$

Exercise V.12. The projective hyperplanes come from two (linear) hyperplanes F and F', and the point m comes from a line D. The hypothesis is that D is contained neither in F nor in F'. Let $f : E \to E$ be the projection onto F' in the direction D. It defines, by restriction, an isomorphism

$$f|_F : F \longrightarrow F'$$

... and the projective transformation g.

The perspectivity g is also the composition of the mappings

$$\begin{array}{ccc}H & \to & m^\star \\ x & \mapsto & mx\end{array} \quad \text{and} \quad \begin{array}{ccc}m^\star & \to & H' \\ d & \mapsto & d \cap H'\end{array}$$

which are, indeed, incidences (by definition for the second one, by duality for the first one: write $H = A^\star$ for $A \in P(E^\star)$, $x \in H$ is a line of $P(E^\star)$ through A and mx its intersection point with $m^\star$).

Exercise V.13. It can immediately be checked that $f(B) = B$, $f(M) = A'$, $f(C') = N$, $f(\alpha) = \gamma$ on the one hand, and that $g(B) = B$, $g(M) = A'$, $g(C') = N$ on the other. Therefore $g = f$ and also $g(\alpha) = \gamma$, but $g(\alpha)$ is the intersection point of $\beta\alpha$ and $B'A$, thus α, β and γ are collinear.

Exercise V.14. One could choose another secant line D' and prove that there exists a homography (a perspectivity) from D to D' mapping a_i to a'_i.

Exercise V.15. The incidence $m^\star \to D$ is a homography (Exercise V.11) that maps d_i to a_i.

Exercise V.16. One has $[d_1, d_2, d_3, d_4] = [a_1, a_2, a_3, \infty]$.

Exercise V.18. Use Menelaüs' theorem.

Exercise V.19. One obtains, of course, Thales' theorem. What has been used is the fact that perspectivities are projective transformations, namely the fact that the projections are linear mappings... this is the same argument as in Chapter I (I hope that nobody had a doubt).

Exercise V.20. If BB', CC' and DD' are concurrent at m, we have the equality of cross-ratios (this is the cross-ratio of the four lines mA, mB, mC and mD). The converse is a consequence of the direct statement.

Exercise V.21. Consequence of V.16.

Exercise V.22. Let O be the center of $\mathcal{C}$. The circles $\mathcal{C}$ and $\mathcal{C}'$ are orthogonal if and only if $\overrightarrow{OA}^2 = \overrightarrow{OB}^2 = \overrightarrow{OM} \cdot \overrightarrow{ON}$, a relation which is equivalent to $[A, B, M, N] = -1$.

Exercise V.23. The construction is shown in the figure. To check that it works, one may either consider the affine situation where A is at infinity or use perspectivities.

Exercise V.24. One can send $\alpha\beta$ to infinity.

Exercise V.26. There is only one homography that fixes 1 and exchanges 0 and ∞, this is $z \mapsto 1/z$. One maps first a, b, c, d (in this order) to ∞, 0, 1, $[a, b, c, d]$, then (by $1/z$) to 0, ∞, 1, $[a, b, c, d]^{-1}$. The composition maps (in this order) b, a, c, d to ∞, 0, 1, $[a, b, c, d]^{-1}$. The latter is thus the cross-ratio $[b, a, c, d]$. To prove that the other equalities hold, one can argue similarly, with:

- the homography $z \mapsto [a, b, c, d]^{-1}z$ that fixes ∞ and 0 and maps $[a, b, c, d]$ to 1 for the first one,
- the homography $z \mapsto 1 - z$ that fixes ∞ and exchanges 0 and 1 for the second one.

Exercise V.28. Let p and p' be two distinct points exchanged by g. Let q be another point and q' be its image. We want to prove that $g(q') = q$. We compute

$$\begin{aligned}[p, p', q, q'] &= [g(p), g(p'), g(q), g(q')] \\ &= [p', p, q', g(q')], \text{ then exchanging the two points in each pair} \\ &= [p, p', g(q'), q'], \text{ hence the result.}\end{aligned}$$

As f is a homography, we have

$$[a_1, a_2, a_3, a'_j] = [a'_1, a'_2, a'_3, f(a'_j)].$$

Hence the condition $[a_1, a_2, a_3, a'_j] = [a'_1, a'_2, a'_3, a_j]$ is equivalent to $a_j = f(a'_j)$. We use the previous property.

Translation in linear algebra: an involutive endomorphism of $\mathbf{R}^2$ that is not a dilatation and that has a real eigenvalue has two distinct real eigenvalues.

Let λ and μ be these eigenvalues. Using a basis of eigenvectors and the formulas for the cross-ratio, it is checked that, for any point m,

$$[a, b, m, g(m)] = \frac{\lambda}{\mu}.$$

In particular, this cross-ratio is constant. The case where g is involutive is the special case where $\mu = -\lambda$.

Exercise V.29. The image of a circle or a line by a homography is also a circle or a line. The required images are thus determined very easily once you know the images of a few points.

Exercise V.30. Writing $E = \ell \oplus H$, the lines of E_H are identified with the graphs of the linear mappings from ℓ to H.

Exercise V.31. The composed mapping $g_2 \circ g_1^{-1}$ is a projective transformation; the images are "glued" using a projective frame of H (see Chapter 4 in [**Ber77**]).

Exercise V.32. Let D and D' be two points, a, b and c be three lines through D, a', b' and c' be three lines through D'. The lines joining the points $a \cap b'$ and $a' \cap b$, $a \cap c'$ and $a' \cap c$, $b \cap c'$ and $b' \cap c$ are concurrent.

Exercise V.33. This is a consequence of Pappus' theorem. Notice first that, however short your ruler, it is easy to draw arbitrarily long lines. Call the two points you want to join α and β. Using the ruler, you can draw two long lines through α making a very small angle and two lines through β with the same properties, so that the four lines determine a small quadrilateral. Let C and C' be two opposite vertices of this quadrilateral (see Figure 1). Draw a line D through C that intersects $C'\alpha$ at B and $C'\beta$ at A. Draw a line D' through C' that intersects βC at A' and αC at B'. Choose them so that the points A and B' are close enough and similarly the points A' and B. Thus you can draw the lines $A'B$ and AB'. They intersect at a point γ of the line $\alpha\beta$ (this is Pappus' theorem). If it is not close enough to α or β, play the same game again.

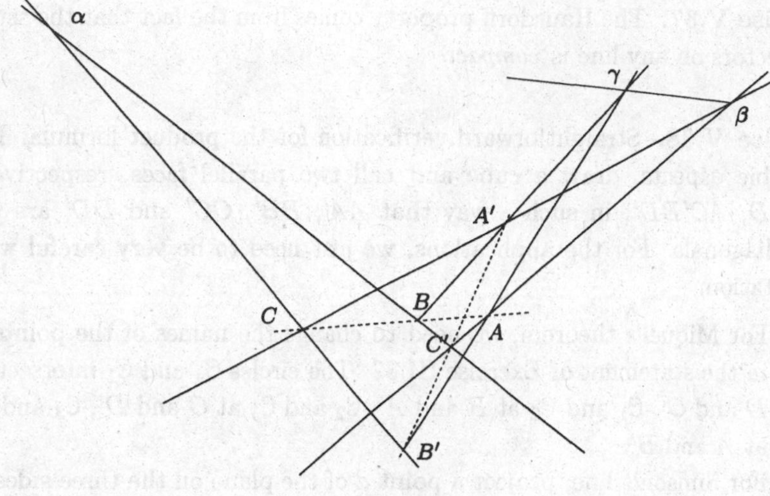

Fig. 1

Exercise V.34. Assume that AP, BQ and CR are concurrent at a point D and compute the cross-ratio of the four points of d:

$$\begin{aligned}[P', Q', R', P] &= [AP', AQ', AR', AP] \\ &= [AP', AC, AB, AD] \text{ (intersection with } BC) \\ &= [P', C, B, D'] \text{ where } D' = AP \cap BC \\ &= [DP', DC, DB, DD'] \\ &= [P', R, Q, P] \text{ (intersection with } d) \\ &= [P, Q, R, P']\end{aligned}$$

and apply the result of Exercise V.28. Incidences and perspectivities can also be used in place of intersections; this is strictly equivalent. The converse is a consequence of the direct statement.

Exercise V.35. The affine Euclidean plane is considered as an affine subspace ($z = 1$) of a Euclidean vector space of dimension 3. The line at infinity comes from the plane $z = 0$. A rotation of angle $\pm\pi/2$ in this plane is a linear isomorphism that defines the homography in question.

Exercise V.36. Apply the second theorem of Desargues (Exercise V.34) to the triangle ABC and the line at infinity. The lines AP, BQ and CR are the altitudes and the points P', Q' and R' are the points at infinity of the sides of the triangle. They correspond to the directions of the altitudes via the orthogonality relation, which is an involution, thanks to Exercise V.35. Therefore, the altitudes are concurrent.

Exercise V.37. The Hausdorff property comes from the fact that the set of unit vectors on any line is *compact.*

Exercise V.38. Straightforward verification for the product formula. For the cubic aspects, draw a cube and call two parallel faces, respectively, $A'CB'D$, $AC'BD'$, in such a way that AA', BB', CC' and DD' are the main diagonals. For the applications, we just need to be very careful with the notation.

- For Miquel's theorem, we need to change the names of the points[(2)] in the statement of Exercise III.32. The circles $\mathcal{C}_4$ and $\mathcal{C}_1$ intersect at D and C', $\mathcal{C}_1$ and $\mathcal{C}_2$ at B and A', $\mathcal{C}_2$ and $\mathcal{C}_3$ at C and D', $\mathcal{C}_3$ and $\mathcal{C}_4$ at A and B'.
- For Simson's line, project a point d of the plane on the three sides of a triangle abc at a', b', c' and let, in this order, $ABCDA'B'C'D' = abc'\infty a'b'cd$ to check that a', b' and c' are collinear if and only if a, b, c and d are cyclic.
- For the pivot, the notation of Exercise III.35 works (let D be the other intersection point of the circles $AB'C'$ and $A'BC'$ and put $D' = \infty$).

Exercise V.41. Consider Cartesian equations of the lines.

Exercise V.44. One could use the affine result (see Exercise I.51), or use a projective but analogous reasoning.

Exercise V.50. To learn more on the Poincaré half-plane or for proofs of the properties presented here, or to learn even more on *hyperbolic* geometry, see, *e.g.*, Chapter 19 of [**Ber77**] and [**Cox42**].

Chapter VI

Exercise VI.2. This is the dimension of the vector space of symmetric matrices.

Exercise VI.3. It can be written as the difference of the squares of *two* linear forms. Therefore, its rank is $\leqslant 2$.

(2) To every problem its notation.

Exercise VI.5. The bilinear form defined on $\mathbf{R}^n$ by

$$\varphi(x,y) = (Ax)\cdot y = {}^t yAx$$

is symmetric. There exists thus, thanks to Theorem 8.8, an *orthonormal* basis that is orthogonal for φ. Letting P be the matrix of the change of basis, it follows that ${}^tPAP$ is diagonal. Since P is orthogonal, ${}^tP = P^{-1}$; we have thus found a change of basis P such that $P^{-1}AP$ is diagonal.

Exercise VI.6. Use a basis of eigenvectors for the endomorphism $\widetilde{\varphi}^{-1}\circ\widetilde{\varphi}'$.

Exercise VI.9. See Figure 5 of Chapter VIII.

Exercise VI.12. No difference between ellipses and hyperbolas...

Exercise VI.17. No difference between an ellipse and a circle from the point of view of affine mappings.

Exercise VI.19. One could begin with the case of a circle and then notice that all the notions involved are affine.

Exercise VI.20. The nicest thing to do is to interpret geometrically the orthogonality condition, considering the circle of center O and radius a and the affinity of ratio b/a that fixes the major axis of the ellipse and maps the circle onto the ellipse. The vectors $\overrightarrow{OM}$ and $\overrightarrow{OM'}$ are conjugated diameters if and only if the points M and M' are the images of M_1 and M_1' with $OM_1 \perp OM_1'$. We are left with the case of the circle.

For the second theorem, write the coordinates of M and M' in terms of θ such that the affix of M_1 is $e^{i\theta}$.

Exercise VI.23. A parabola is determined by its focus and its directrix, that is to say by a point and a line not through this point. This figure is unique up to similarity (use Exercise III.13). The proper conics that are not circles are determined by focus, directrix and eccentricity, the latter being invariant by similarity...

Exercise VI.24. For a parabola, the order of the group is 2. For an ellipse or a hyperbola, this is as for a rectangle. For a circle...

Exercise VI.26. It is then the sign of q that determines the type of the conic.

Exercise VI.27. The answer is

$$\rho = \frac{ed}{1 + e\cos\theta}$$

where e is the eccentricity and d the distance from F to the directrix.

Exercise VI.29. A segment, a half-line.

Exercise VI.31. An ellipse, a hyperbola (of foci F, F')... and a line if $F' \in \mathcal{C}$.

Exercise VI.32. The first question can be solved by a computation using an equation of $\mathcal{C}$. For the second, assume MP is parallel to an asymptote. Let N be a point of MP, n its projection on D and N' its projection on PF. Thanks to the first question and to Thales' theorem, the ratio NN'/NF tends to 1 when N tends to infinity on MP. But this ratio is the sine of the angle at F in the right-angled triangle NFN'. You still have to check that the limit of this angle is the angle at P of MPF and that this is an acute angle, a contradiction.

Exercise VI.33. Solve the differential equation in $1/f$ and contemplate the solution of Exercise VI.27.

Exercise VI.34. The projections of the focus on the tangents to the parabola also lie on the tangent at the vertex (Corollary 2.17), therefore this tangent is the Simson line of the focus (see, if necessary, Exercise III.33), so that the focus F is on the circumcircle of the triangle determined by the three lines. The directrix is the image of the tangent at the vertex by the dilatation of center F and ratio 2, this is thus the Steiner line of F (see Exercise III.34).

When the focus F tends to a vertex A of the triangle, the Steiner line, which, as all Steiner lines do, passes through the orthocenter, tends to the altitude of the vertex A and the parabola degenerates into a double line.

Exercise VI.35. Call the lines AFB, AEC, DEF, DCB (have you drawn the figure?). Using Exercise VI.34, we know that the focus must lie on the circumcircles of the triangles ABC, CDE, BDF and AEF and that this point must have the same Simson line for the four triangles.

The circumcircles of ABC and BDF intersect at B, hence, either they have a common point P, or they are tangent. If they were tangent, the lines DF and AC would be parallel, which is contradictory to the general position assumption. Let thus P be the second intersection point. It lies on the circumcircle of ABC, thus its projections P_1, P_2 and P_3 on BC, CA and AB are collinear. It also lies on the circumcircle of BDF, thus its projections

P_1, P_4 and P_3 on BD, DF and FB are collinear. All the expected properties are deduced.

Exercise VI.36. See [**DC51**], [**LH61**] and Chapter 17 of [**Ber77**], where these classical results are illustrated and proved.

Exercise VI.37. Use the notation of § 1. The conic passes through A and has the equation

$$q(\overrightarrow{AM}) + L_A(\overrightarrow{AM}) = 0.$$

The line through A directed by u intersects it at the unique simple point A if and only if $q(u) = 0$. The projective conic has the equation

$$q(u) + zL_A(u) = 0.$$

It intersects the line at infinity ($z = 0$) at the points $(u, 0)$ such that $q(u) = 0$.

Exercise VI.40. Use the classification of real quadrics.

Exercise VI.42. Three points can be chosen on each of the three lines. Now remember that there is always a quadric through nine points...

Exercise VI.43. The two last pictures of the top line represent conics generating the same pencil. The same is true of the two first pictures of the bottom line.

Exercise VI.44. The intersection points of the two conics are the same as those of one of the conics with a degenerate conic of the pencil. The degenerate conics are found by solving a degree-3 equation (proposition 3.11) and the intersection of these with one of the given conics by solving degree-2 equations.

Exercise VI.45. If the point m is not on the degenerate conic $\mathcal{C}$, its orthogonal is a line that passes through the intersection point of the two lines (this is proved, *e.g.*, in coordinates in which an equation of $\mathcal{C}$ is $xy = 0$). If the point m lies on the conic...

Exercise VI.46. Let $\mathcal{C}_0$ be the degenerate conic consisting of the lines xy and zt, and let, similarly, $\mathcal{C}_1$ be that consisting of xz and yt. Consider the pencil they span. The point n is on the polar of m with respect to $\mathcal{C}_0$, and on the polar of m with respect to $\mathcal{C}_1$ (an application of Exercise VI.45); it is thus orthogonal to m for all the conics of the pencil (all the conics through x, y, z and t) and in particular to $\mathcal{C}$, the one we are interested in.

Exercise VI.47. Let u be a vector spanning the vectorial line defining the point A of the projective plane. If φ and φ' are the polar forms of two quadratic forms defining two conics spanning the pencil $\mathcal{F}$, the polars in question are the images of the vectorial planes $P_{\lambda,\lambda'}$ defined by the equivalence

$$v \in P_{\lambda,\lambda'} \iff (\lambda\varphi + \lambda'\varphi')(u, v) = 0.$$

If φ and φ' are nondegenerate, all these planes intersect along the intersection line $u^{\perp_\varphi} \cap u^{\perp_{\varphi'}}$.

Exercise VI.48. Degenerate Pascal, this gives Pappus.

Exercise VI.50. Consider an arbitrary line D through e and construct its second intersection point f with $\mathcal{C}$ using Pascal's theorem. The notations used here are consistent with those of Figure 18 and of the proof of Pascal's theorem (Theorem 4.4). The point $z = ab \cap de$ is well defined, as is the point $u = bc \cap D$; the line uz intersects cd at t, and f is found at the intersection of D and at.

To construct the tangent at f, apply Pascal's theorem to the hexagon $fbcdef$ (with $a = f$, check that the theorem remains true when replacing af by the tangent at f). Let $z' = fb \cap ed$, $u' = bc \cap fe$, $t' = u'z' \cap cd$, then the tangent is the line ft'.

Exercise VI.51. This is a conic through A, B, C and D.

Exercise VI.52. In the projective plane, a pencil of circles is a pencil of conics through the circular points. One of the degenerate conics of the pencil consists of two intersecting lines, one of which is the line at infinity and the other the radical axis. The two other degenerate conics are imaginary.

Exercise VI.53. This is a hyperbola, as it intersects the line at infinity. The asymptotes are orthogonal, so that this is a rectangular hyperbola.

Exercise VI.59. The group homomorphism $x \mapsto x^2$ gives the exact sequence

$$1 \longrightarrow \{\pm 1\} \longrightarrow \mathbf{F}_q^\star \longrightarrow (\mathbf{F}_q^\star)^2 \longrightarrow 1,$$

thus there are $\dfrac{q-1}{2}$ squares in $\mathbf{F}_q^\star$ and $\dfrac{q+1}{2}$ in $\mathbf{F}_q$. When y varies, the quantity $\dfrac{1-by^2}{a^2}$ takes $\dfrac{q+1}{2}\dots$ among which there is a square, since $2\dfrac{q+1}{2} > q$.

The proof of the "normal form" for the quadratic forms proceeds by induction on n. The $n = 1$ case is clear. The main point is the $n = 2$ case: the preliminary question gives a vector e_1 such that $Q(e_1) = 1$. Let f be an orthogonal vector. If $Q(f)$ is a square α^2, put $e_2 = f/\alpha$; the form is $x_1^2 + x_2^2$. Otherwise, $Q(f) = b$ is not a square, then a/b is a square ($\mathbf{F}_q^\star/(\mathbf{F}_q^\star)^2$ has only two elements), $a = b\alpha^2$, put $e_2 = \alpha f$, the form is $x_1^2 + ax_2^2$. For $n > 2$, start with an orthogonal basis $\varepsilon_1, \ldots, \varepsilon_n$. In the plane $\langle \varepsilon_1, \varepsilon_2 \rangle$ there is an e_1 such that $Q(e_1) = 1$, then apply the induction hypothesis to $H = e_1^\perp$.

The number a (or better to say, its class in $\mathbf{F}_q^\star/(\mathbf{F}_q^\star)^2$) is the discriminant of the form (Exercise VI.58), thus the two given forms are not equivalent. Moreover, a quadratic form belongs to one or the other type according to whether it has a trivial or nontrivial discriminant.

Exercise VI.61. As M is invertible, the symmetric matrix ${}^tMM$ is positive definite; one can take its square root for S (Exercise VI.60) and put $\Omega = MS^{-1}$.

To prove uniqueness, use the polynomial P. If $M = \Omega S = \Omega_1 S_1$, then ${}^tMM = S_1^2$. The matrix S_1 commutes with S_1^2, thus also with ${}^tMM$, with $P({}^tMM)$ and with S. The two symmetric matrices S and S_1 are diagonalizable and commute, therefore they are diagonalizable in the same basis and eventually they are equal.

Exercise VI.62. Use the polar decomposition (Exercise VI.61) $M = \Omega S$ and diagonalize S as $S = {}^t\Omega_2 D \Omega_2$.

Exercise VI.64. Two points m and m' of $\mathbf{P}_2(\mathbf{R})$ are in the same orbit if there exists a linear isomorphism $\widetilde{f}$ of $\mathbf{R}^3$ such that $\widetilde{D} = D'$ and $q \circ \widetilde{f} = q$ (D and D' are the two lines in $\mathbf{R}^3$ that define m and m'). According to Witt's theorem (Theorem 8.10), this is equivalent to the existence of an isomorphism $f : D \to D'$ such that

$$(q\,|_{D'}) \circ f = q\,|_D \,.$$

Now, on a 1-dimensional real space, there are exactly three types of quadratic forms (zero, positive or negative definite). Thus there are three orbits corresponding to:

- lines (strictly) inside the isotropy cone, where the form is negative definite,
- generatrices of the cone, where the form is zero,
- lines (strictly) outside the cone, where the form is positive definite.

Exercise VI.66. The proofs of these two theorems can be made very simple by the use of duality. See, *e.g.*, Chapter 2 of [**Tab95**].

Exercise VI.69. The only thing you have to be careful about: once you have chosen two vectors e_1 and f_1 such that $\varphi(e_1, f_1) = 1$, you have to check that the restriction of φ to the orthogonal of the plane they span is still nondegenerate; then you can use induction to construct the basis.

The dimension of E is even, $2n$ say; the dimension of F is less than or equal to n.

Chapter VII

Exercise VII.1. Take the origin at A by an affine change of coordinates—this will not affect the rationality of the solutions. The intersection points of $\mathcal{D}_t$ and $\mathcal{C}$ are the (x, tx) such that

$$q(x, tx) + L_A(x, tx) = 0$$

(with the notation of Chapter VI). This is a degree-2 equation in x; it has the root $x = 0$ (corresponding to the point A). Thus, it has the form

$$x(a(t)x + b(t)) = 0$$

where a and b are polynomial of degrees $\leqslant 2$ and $a(t) \neq 0$. The coordinates of the second intersection point are the rational functions

$$x = -\frac{b(t)}{a(t)} \qquad y = -\frac{tb(t)}{a(t)}.$$

The case of the unit circle ($x^2 + y^2 = 1$) and the point $A = (-1, 0)$ gives the classical parametrization by "$t = \tan\theta/2$" (you are requested to draw a figure!)

$$x = \frac{1-t^2}{1+t^2} \qquad y = \frac{2t}{1+t^2}.$$

Exercise VII.2. This is a "nodal cubic", with a node at the origin. A line through the origin intersects the cubic at the origin (double root) and at another point. If the slope of the line is t, one finds

$$x = t^2 - 1 \qquad y = t(t^2 - 1).$$

For the "cuspidal cubic", the parametrization

$$x = t^2 \qquad y = t^3$$

is found in the very same way.

Exercise VII.3. This is a "figure eight" or rather a "figure infinity". A Cartesian equation is

$$xy = (x^2 + y^2)^2$$

and (hence) an equation in polar coordinates $2\rho^4 = \sin(2\theta)$.

Exercise VII.4. This is a lemniscate of Bernoulli. The asymptotes (tangents at ∞) are transformed into the tangents at the pole (image of ∞), *etc.*

Exercise VII.5. The two parabolas have equations $y = x^2$ (when t tends to infinity, $x \sim t$, $y \sim t^2$) and $y^2 = 8x$ (when t tends to 0, $x \sim 1/t^2$, $y \sim 2/t$). There is a singular point (obtained for $t = 1$), which is a cusp point.

Exercise VII.6. Fix two points M and N of $\mathcal{D}$ and the two corresponding points M' and N' of $\mathcal{D}'$. The assumption is that, for P in $\mathcal{D}$, the corresponding point P' satisfies

$$\frac{\overrightarrow{M'P'}}{\overrightarrow{M'N'}} = \frac{\overrightarrow{MP}}{\overrightarrow{MN}}.$$

Let A be the center of the direct similarity that maps M to M' and N to N' (and thus also P to P'). The triangles AMN and $AM'N'$ are similar, therefore AMM' and ANN' are similar too. If H_M is the orthogonal projection of A to MM', the triangles AH_MM and AH_NN are similar, thus there exists a similarity of center A that maps the point M of $\mathcal{D}$ to H_M. Let Δ be the image of $\mathcal{D}$ under this similarity. The desired envelope is the parabola whose focus is A and tangent at the vertex is Δ.

Exercise VII.7. Let B and B' be two points of D and D' respectively such that $OB = a$ and $OB' = a$ and let F be the intersection of the perpendicular bisectors of OB and OB'. The perpendicular bisector of AA' passes through F. Let I be the midpoint of AA'. The triangle FAI is isosceles and right-angled; the locus of I is thus the image D'' of D by a similarity. The line AA' envelops a parabola of focus F whose tangent at the vertex is the line D''.

Exercise VII.8. Using Thales' theorem, it is proved that

$$\frac{\overrightarrow{FB'}}{\overrightarrow{FM}} = \frac{\overrightarrow{FB}}{\overrightarrow{FP}} \quad \text{and} \quad \frac{\overrightarrow{FB'}}{\overrightarrow{FM}} = \frac{\overrightarrow{FP}}{\overrightarrow{FC}}$$

and therefore we also have $FP^2 = \overrightarrow{FB} \cdot \overrightarrow{FC}$. Hence the point F is fixed. The envelope is a parabola with focus at F and whose tangent at the vertex is the line AP.

Exercise VII.11. The same group as for an equilateral triangle.

Exercise VII.12. Parametrize the "discriminant" curve by the multiple root t, writing

$$P(X) = (X-t)^2(X+2t),$$

which gives a and b in terms of t... and once again the cuspidal cubic. The singular point corresponds to the polynomial X^3, which has a triple root. The two components of the complement correspond to the polynomials that have:

– a unique real root (the one that contains the points $a = 0$, $b \neq 0$),
– three real roots (the other one). See also Exercise VIII.9.

Exercise VII.14. A line that is parallel to the x-axis, when reflected at the point $(\cos\theta, \sin\theta)$ of the unit circle becomes a line of slope $\tan(2\theta)$. Differentiate with respect to θ the equation

$$\sin(2\theta)x - \cos(2\theta)y = \sin\theta$$

of this line, to find the parametrization

$$x + iy = \frac{1}{4}\left(3e^{i\theta} - e^{3i\theta}\right)$$

which represents a nephroid.

Exercise VII.16. The radius of the circle $\mathcal{C}'$ is R and that of the circle $\mathcal{C}$ is r. Fix the origin at the center of the fixed circle $\mathcal{C}'$ and parametrize $\mathcal{C}'$ by $Re^{i\theta}$. Assume that, at the beginning, the point M is at R ($\theta = 0$, this is a choice of origin for θ). The hypothesis "to roll without slipping" means that

- the circles are tangent;
- when the point of tangency is at $Re^{i\theta}$, the point M that would have run the distance $R\theta$ along $\mathcal{C}'$ has run the same length along $\mathcal{C}$ (draw a figure).

A parametrization of the desired curve is easily deduced; this is

$$(R+r)e^{i\theta} \pm re^{i\frac{R+r}{r}\theta},$$

hence the result.

Exercise VII.17. To find the equations of the cycloid is very similar to what has been done in Exercise VII.16. Then, the normal has the equation

$$x\sin\frac{t}{2} + y\cos\frac{t}{2} - Rt\sin\frac{t}{2} = 0.$$

The envelope of the normals is parametrized by

$$x = R(t + \sin t) \qquad y = R(\cos t - 1).$$

This is the image of the cycloid by the translation $(\pi R, -2R)$.

Exercise VII.18. Let O be the center of the circle. With any point M (distinct from A) is associated the second intersection point M' of $\mathcal{D}_M$ with the circle. If θ is a measure of the angle $(\overrightarrow{MA}, \overrightarrow{MO})$, 4θ is a measure of the angle $(\overrightarrow{OA}, \overrightarrow{OM'})$.

Let α be a measure of the *fixed* angle (OA, D). Evaluate the angle $(\overrightarrow{M'A}, \overrightarrow{M'M})$ (you should find $\alpha - \theta - \frac{\pi}{2}$) then $(\overrightarrow{OA}, \overrightarrow{OM})$, a measure of which is $2\alpha - 2\theta - \pi$. Therefore M' turns twice faster than M in the opposite sense. The envelope is indeed a three-cusped hypocycloid.

Exercise VII.20. The matrix $A(s)$ is orthogonal, thus satisfies, for all s, the relation

$$^{t}A(s)A(s) = \mathrm{Id}.$$

Differentiate this relation with respect to s to obtain

$$^{t}\left(\frac{dA}{ds}(s)\right)A(s) + {}^{t}A(s)\frac{dA}{ds}(s) = 0.$$

Let $B(s) = {}^{t}A(s)\dfrac{dA}{ds}(s)$. The previous relation asserts that $B(s)$ is skew symmetric.

If $A(s)$ is the (orthogonal) matrix that describes the orthonormal basis $(\tau(s), n(s), b(s))$ in the canonical basis, the matrix $B(s)$ expresses the vectors $(\tau'(s), n'(s), b'(s))$ in the basis $(\tau(s), n(s), b(s))$. It is, indeed, skew symmetric.

Exercise VII.22. The parametrization of the astroid by $-3e^{i\theta} + e^{-3i\theta}$ becomes, after the change of variables $\theta = \varphi - \dfrac{\pi}{4}$ (and up to a rotation)

$$3e^{i\varphi} + e^{-3i\varphi} = 4(\cos^3\varphi + i\sin^3\varphi).$$

The ellipse is parametrized by $(x, y) = (a\cos\varphi, b\sin\varphi)$. The normal at the point of parameter φ has the equation

$$ax\sin\varphi - by\cos\varphi - (a^2 - b^2)\sin\varphi\cos\varphi = 0.$$

A parametrization of the evolute is deduced:

$$x = \frac{a^2 - b^2}{a}\cos^3\varphi \qquad y = -\frac{a^2 - b^2}{b}\sin^3\varphi.$$

Exercise VII.23. Differentiate the relation

$$C(s) = g(s) + \rho(s)n(s)$$

with respect to s to obtain the equality

$$C'(s) = \rho'(s)n(s)$$

and to conclude.

Exercise VII.25. The Steiner lines pass through the orthocenter H (Exercise III.34), their envelope is thus the point H.

The dilatation $h\left(H, \dfrac{1}{2}\right)$ transforms the circumcircle into the Euler circle $\mathcal{E}$ (Exercise II.20) thus the point μ lies on $\mathcal{E}$. Moreover

$$H \in \mathcal{D}_M = h(M, 2)(\mathcal{S}_M)$$

hence $h(M, 2)^{-1}(H) = \mu \in \mathcal{S}_M$. The point μ is the midpoint of HM; it is thus projected on BC at the midpoint of $A'M_A$ where A', the end of the altitude from A, is a (fixed) point of the Euler circle. Moreover, the triangle $A'\mu M_A$ is isosceles of vertex μ; thus the bisector of its angle at μ is its altitude, hence it is parallel to the altitude from A (a fixed direction).

In conclusion, the Simson line μM_A of M:

- passes through the point μ of circle $\mathcal{E}$,
- is such that the bisector of the angle $(A'\mu, \mathcal{S}_M)$ is parallel to the altitude from A.

Just apply now the result of Exercise VII.18.

Chapter VIII

Exercise VIII.2. The curves $\theta =$ constant are straight lines passing through the points of the z-axis, and the curves $t =$ constant are circular helices.

Exercise VIII.3. It must be assumed that x and y are not both zero. We look for the points $(0, 0, c)$ of the z-axis through which the lines contained in the surface pass. In fact, the Plücker conoid consists of two Whitney umbrellas (see Exercise VIII.10).

Exercise VIII.4. A hyperboloid of one sheet.

Exercise VIII.7. If the meridian curve is regular, the two vectors are colinear only at the points where it intersects the revolution axis. The parametrizations obtained in Exercise VIII.1 are never regular at these points. However, if the meridians have a tangent vector that is orthogonal to the axis, the surface is regular and the tangent plane is the plane orthogonal to the axis (one could use a Cartesian parametrization). Think of the example of the sphere.

Exercise VIII.8. The tangent vector $\partial f/\partial u$ is w.

The tangent plane at the point of parameter (t, u) is spanned by $w(t)$ and $\gamma'(t) + uw'(t)$. It does not depend on u if and only if $w'(t)$ is a linear combination of $\gamma'(t)$ and $w(t)$.

If the surface is a cylinder, w is constant hence w' is zero. If this is a cone, $w(t) = \gamma(t)$ (taking the origin at the vertex of the cone) and $w' = \gamma'$. In both cases, the tangent plane at a point intersects the surface along the generatrix through this point. In general, the points of these surfaces are parabolic.

Exercise VIII.9. As in Exercise VII.12, to say that P has a multiple root is to say that it can be written

$$P(X) = (X-u)^2(X-v)(X+2u+v).$$

Expressing a, b and c in terms of u and v, a parametrization of the "discriminant" surface is deduced. The surface has:

- a line of double points, when the polynomial has two double roots ($v=-u$),
- two curves of singular points corresponding to the polynomials with a triple root ($v=u$ and $v=-3u$),
- these three curves intersecting at the point $u=v=0$ corresponding to the polynomial X^4, which has a quadruple root.

Exercise VIII.10. Why "umbrella"? The points of the surface satisfy the Cartesian equation $x^2-zy^2=0$, which, conversely, describes the union of the surface and the z-axis. The part $z<0$ of this axis is considered as the stick of the umbrella.

Exercise VIII.11. All these surfaces are tangent to the plane $z=0$ at 0. Therefore, no computation is needed. The second derivative is (up to a factor 2) the quadratic part of the equation. Nevertheless, you must draw a picture.

- For $z=x^2$, we have a parabolic point, and the surface intersects the tangent plane along a line, staying on the same side of this plane; for $z=x^2+y^3$, still a parabolic point (the second derivative is the same) but the surface has points on both sides of the tangent plane.
- For the other ones, 0 is a planar point. For $z=x^3$, the surface crosses the tangent plane, which it intersects along a straight line; for $z=x^4$, it intersects it along a line, but stays on the same side; for $z=x^3-3xy^2$, the surface intersects the tangent plane along three concurrent lines, and lies alternatively above and below the plane sectors limited by these lines (there is a hollow for each of the legs of the monkey and another one for its tail).

Exercise VIII.13. One shows firstly that, if p is an extremum of h on Σ, then $(dh)_p(X)=0$ for all X in $T_p\Sigma$. Indeed, use a curve γ drawn on Σ and such that $\gamma(0)=p$ and $\gamma'(0)=X$. If p is an extremum of h, 0 is an extremum of $h\circ\gamma$, thus

$$(dh)_p(X) = (h\circ\gamma)'(0) = 0.$$

Then, $(dh)_p$ is zero on $T_p\Sigma = \operatorname{Ker}(dF)_p$ (Proposition 2.10), hence

$$(dh)_p = \lambda(dF)_p.$$

On the sphere $F^{-1}(0)$, it is clear that the height function has a maximum at the north pole and a minimum at the south pole. This is confirmed by the study above:

$$(dF)_p(X) = 2p \cdot X \qquad (dz)_p(X) = e_3 \cdot X$$

(this is the third coordinate of X) and $(dz)_p = \lambda(dF)_p$ if and only if $p = \pm e_3$.

The condition of independency of $(dF)_p$ and $(dG)_p$ ensures that the tangent planes $\operatorname{Ker}(dF)_p$ and $\operatorname{Ker}(dG)_p$ to the two surfaces are distinct. It is thus possible to apply Proposition 2.15: the intersection is a curve Γ. The same argument as above shows that, if $p \in \Gamma$ is an extremum of h, $(dh)_p$ is zero on $T_p\Gamma = \operatorname{Ker}(dF)_p \cap \operatorname{Ker}(dG)_p$. An easy linear algebra lemma allows us to conclude.

Exercise VIII.14. The tangent plane $T_p\Sigma$ is the kernel of the differential of

$$F(x, y, t) = u(t)x + v(t)y + w(t)$$

at p. This kernel contains the t-axis if and only if $\partial F/\partial t$ is zero. This way, we get again the linear system of § VII-1.

In the case of the cusp, the curve obtained is also the discriminant of the degree-3 polynomials (in t).

Exercise VIII.15. One can simply say that all the points are planar or parabolic, and therefore that the Gauss curvature is zero. This can also be checked, noticing that the normal vector is constant along the generatrices. Parametrize Γ by $t \mapsto \gamma(t)$ and the surface by $(t, u) \mapsto (1 + tu)\gamma(t) = f(t, u)$. One gets

$$\frac{\partial f}{\partial t} = (1 + u)\gamma'(t), \qquad \frac{\partial f}{\partial u} = \gamma(t)$$

so that one can choose

$$n(t, u) = \frac{\gamma'(t) \wedge \gamma(t)}{\|\gamma'(t) \wedge \gamma(t)\|}$$

which does not depend on u. Hence $\partial n/\partial u = 0$ and the Gauss curvature is zero.

Exercise VIII.16. The formulas of Appendix (§ 4) easily give the result in this case:

$$K(p) = \frac{1}{\alpha_1\alpha_2\alpha_3}\left(\frac{x_1^2}{\alpha_1^2} + \frac{x_2^2}{\alpha_2^2} + \frac{x_3^2}{\alpha_3^2}\right)^{-2}.$$

As can be expected (?), the curvature of the ellipsoids (all the α_i's positive) and that of the hyperboloids of two sheets (two among the α_i's negative) is positive, while that of the hyperboloids of one sheet, ruled surfaces (one of the α_i's positive) is negative.

Exercise VIII.17. One has

$$n(s,\theta)=\begin{pmatrix}-h'(s)\cos\theta\\ -h'(s)\sin\theta\\ g'(s)\end{pmatrix},$$

hence

$$\frac{\partial n}{\partial s}(s,\theta)=-\frac{h''(s)}{g'(s)}\frac{\partial f}{\partial s}(s,\theta),\quad \frac{\partial n}{\partial \theta}=-\frac{h'(s)}{g(s)}\frac{\partial f}{\partial \theta}(s,\theta).$$

The Gauss curvature is thus

$$K(s,\theta)=\frac{h'(s)h''(s)}{g(s)g'(s)}=-\frac{g''(s)}{g(s)}.$$

(1) If $K(s,\theta)=0$, $g(s)=as+b$, up to a change of origin for s, one thus has

$$\text{either}\ \begin{cases} g(s) &= as\\ h(s) &= \pm(\sqrt{1-a^2})s\end{cases}\ \text{or}\ \begin{cases} g(s) &= b\\ h(s) &= s.\end{cases}$$

One finds a cone (for $|a|<1$) or a plane (for $|a|=1$) in the first case, a cylinder in the second case.

(2) If $K(s,\theta)=K>0$, $g(s)=a\cos(\sqrt{K}s+b)$, one can assume that

$$\begin{cases} g(s) &= a\cos(\sqrt{K}s)\\ h(s) &= \displaystyle\int_0^s \pm\sqrt{1-a^2K\sin^2(\sqrt{K}u)}du.\end{cases}$$

This is a sphere if $a^2K=1$, otherwise the meridians are a little bit more intricate, h being given by an "elliptic integral".

(3) If $K(s,\theta)<0$, one has $g(s)=a\exp(\sqrt{-K}s)+b\exp(-\sqrt{-K}s)$.

Exercise VIII.18. The curvature of the inner tube of the tire is positive on the "tire side" and negative on the "rim side". By computation, parametrize the circle by arc length

$$g(s)=R+r\cos\frac{s}{r},\quad h(s)=r\sin\frac{s}{r}\qquad R>r.$$

Use the formula at the beginning of Exercise VIII.17 and find

$$K(s,\theta)=\frac{\cos\frac{s}{r}}{r(R+r\cos\frac{s}{r})}$$

which is positive for s/r in $[-\frac{\pi}{2},\frac{\pi}{2}]$ as we expected.

Exercise VIII.19. Fix a point O out of Σ. As the surface is compact, the function $F(p) = \|p\|^2$ reaches its maximum at some point of Σ. We show that, at a point p for which a local maximum of F is reached, we have (for a suitable choice of the normal vector n)

$$\mathrm{II}_p(X, X) \geqslant \frac{1}{\|p\|^2}$$

for any unit vector X in $T_p\Sigma$. In particular, the principal curvatures satisfy this inequality and thus the Gauss curvature is positive.

Notice firstly that $T_p\Sigma = p^\perp$, since $(dF)_p(X) = 2p \cdot X$ must vanish on $T_p\Sigma$. Choose the normal vector in a neighborhood of p in such a way that $n(p)$ has the opposite direction to p. Let X be a unit vector in $T_p\Sigma$. The plane perpendicular to $T_p\Sigma$ that contains X is the plane through O spanned by p and X. Let γ be a curve drawn on Σ, parametrized by arc length and such that $\gamma(0) = p$, $\gamma'(0) = X$. We know (Proposition 3.4) that

$$\mathrm{II}_p(X, X) = \langle \gamma''(0), n(p) \rangle.$$

Consider now the oriented plane $P_X = \langle X, p \rangle$. The point p realizes also a maximum of F on γ. Write

$$\begin{aligned}
\gamma(s) &= \gamma(0) + s\tau(0) + \frac{s^2}{2} K_X(0) n(0) + o(s^2) \\
&= p + sX + \frac{s^2}{2} K_X(0) n + o(s^2) \\
&= sX + \left(\frac{s^2}{2} K_X(0) - \|p\| \right) n + o(s^2).
\end{aligned}$$

Compute

$$\|\gamma(s)\|^2 = \|p\|^2 + s^2 \left(1 - \|p\| \, K_X(0)\right) + o(s^2).$$

If this function of s has a local maximum at 0, the coefficient of s^2 is negative, and this gives exactly the expected condition on K_X.

Exercise VIII.20. Any surface you construct with a sheet of paper is flat (its Gauss curvature is zero).

Exercise VIII.21. All the points of the sphere are umbilics. For an ellipsoid, using the formulas of the appendix (§ 4), a point is an umbilic if and only if the equation in μ given by the 4×4 determinant has a double root. One finds, if $\alpha_1 > \alpha_2 > \alpha_3$ (namely, up to a change of order, if the ellipsoid is not a surface of revolution) that the umbilics are the four points of coordinates

$$\left(\pm\sqrt{\alpha_1 \frac{\alpha_1 - \alpha_2}{\alpha_1 - \alpha_3}}, 0, \pm\sqrt{\alpha_3 \frac{\alpha_2 - \alpha_3}{\alpha_1 - \alpha_3}} \right).$$

Exercise VIII.24. Compute, for the first surface

$$n(t,\theta)=\frac{1}{\sqrt{1+t^2}}\begin{pmatrix}\sin\theta\\ \cos\theta\\ -t\end{pmatrix},$$

$$\frac{\partial n}{\partial\theta}=\frac{1}{t\sqrt{1+t^2}}\frac{\partial f}{\partial\theta},\quad \frac{\partial n}{\partial t}=-\frac{t}{(1+t^2)^{3/2}}\frac{\partial f}{\partial t}+A\frac{\partial f}{\partial\theta},$$

thus $K(t,\theta)=-\dfrac{1}{(1+t^2)^2}$. For the helicoid,

$$n(t,\theta)=\frac{1}{\sqrt{1+t^2}}\begin{pmatrix}\sin\theta\\ -\cos\theta\\ -t\end{pmatrix},$$

$$\frac{\partial n}{\partial\theta}=\frac{1}{\sqrt{1+t^2}}\frac{\partial f}{\partial t},\quad \frac{\partial n}{\partial t}=-\frac{t}{(1+t^2)^{3/2}}\frac{\partial f}{\partial\theta},$$

thus $K(t,\theta)=-\dfrac{1}{(1+t^2)^2}$. The curvatures are equal. If φ was a local isometry, it would preserve the curvature and thus be of the form

$$\varphi(t,\theta)=(\pm t,\psi(t,\theta)).$$

But, for the first surface, we have

$$\left\|\frac{\partial f}{\partial t}\right\|^2=1+\frac{1}{t^2}$$

while for the second one

$$\left\|\frac{\partial g}{\partial t}\right\|^2=1.$$

Bibliography

[Apé87] F. APÉRY – *Models of the real projective plane*, Vieweg, 1987.

[Art57] E. ARTIN – *Geometric algebra*, Interscience Publishers, Inc., New York–London, 1957.

[Art90] M. ARTIN – *Algebra*, Prentice Hall, 1990.

[Ber77] M. BERGER – *Géométrie*, CEDIC, 1977, Réédition Nathan, 1990.

[Ber94] ———, *Geometry I*, Springer-Verlag, Berlin, 1994, Translated from the 1977 French original by M. Cole and S. Levy, Corrected reprint of the 1987 translation.

[BG88] M. BERGER & B. GOSTIAUX – *Differential geometry: manifolds, curves, and surfaces*, Springer-Verlag, New York, 1988, Translated from the French by Silvio Levy.

[Bou89] N. BOURBAKI – *General topology. Chapters 1–4*, Springer-Verlag, Berlin, 1989, Translated from the French, Reprint of the 1966 edition.

[Boy87] C. BOYER – *The rainbow, from myth to mathematics*, MacMillan, 1987.

[Car67] H. CARTAN – *Calcul différentiel*, Méthodes, Hermann, 1967.

[Car95] H. CARTAN – *Elementary theory of analytic functions of one or several complex variables*, Dover Publications Inc., New York, 1995, Translated from the French, Reprint of the 1973 edition.

[CG67] H. S. M. COXETER & S. L. GREITZER – *Geometry revisited*, Mathematical Association of America, 1967.

[Cho66] G. CHOQUET – *Topology*, Academic Press, New York, 1966.

[Cox42] H. S. M. COXETER – *Non-Euclidean geometry*, University of Toronto Press, Toronto, Ont., 1942.

[Cox49] H. S. M. COXETER – *Regular polytopes*, Methuen & Co. Ltd., London, 1948; 1949.

[Cox69] ———, *Introduction to geometry*, Wiley, 1969.

[DC51] R. DELTHEIL & D. CAIRE – *Géométrie & Compléments de géométrie*, 1951, Réimpression Gabay.

[Dix76] J. DIXMIER – *Cours de mathématiques du premier cycle*, Gauthier-Villars, 1976.

[Dür95] A. DÜRER – *Underweysung des messung / mit dem zirckel und richtscheyt / in linien ebnen unnd gantzen corporen / durch Albrecht Dürer zu samen getzogen / und zu nutz aller kuntsliebhabenden mit zu gehörigen figuren : in truck gebracht : im jar M.D.X.X.V.*, in *Géométrie, présentation et traduction de Jeanne Peiffer*, Sources du savoir, Seuil, 1995, *The Painter's Manual. A Manual of Measurement of Lines, Areas, and Solids by means of Compass and Ruler assembled by Albrecht Dürer for the Use of all Lovers of Art with Appropriate Illustrations Arranged to be Printed in the Year MDXXV*, translated and with a Commentary by Walter L. Strauss, Abaris Books, New York, 1977.

[Eis41] H. EISLER – *Vierzehn Arten den Regen zu beschreiben*, Variations for flute, clarinet, viola, cello and piano, opus 70, 1941.

[Fre73] J. FRENKEL – *Géométrie pour l'élève professeur*, Actualités scientifiques et industrielles, Hermann, 1973.

[God71] C. GODBILLON – *Éléments de topologie algébrique*, Méthodes, Hermann, 1971.

[HCV52] D. HILBERT & S. COHN-VOSSEN – *Geometry and the imagination*, Chelsea, 1952.

[Her36] HERGÉ – *Le lotus bleu (The blue lotus)*, Les aventures de Tintin et Milou, Casterman, 1936.

[Her45] ———, *Le trésor de Rackham le Rouge (Red Rackham's treasure)*, Les aventures de Tintin et Milou, Casterman, 1945.

[Kom97] V. KOMORNIK – A short proof of the Erdős-Mordell theorem, *Amer. Math. Month.* (1997), p. 57–60.

[Lak76] I. LAKATOS – *Proofs and refutations*, Cambridge University Press, Cambridge, 1976, The logic of mathematical discovery, Edited by John Worrall and Elie Zahar.

[LH61] C. LEBOSSÉ & C. HÉMERY – *Géométrie, classe de mathématiques*, Nathan, 1961, Réimpression Gabay 1990.

[Lib74] P. LIBERMANN – Géométrie différentielle classique, *Encyclopædia Universalis* (1974).

[LP67] V. LESPINARD & R. PERNET – *Géométrie, Terminale C*, André Desvignes, 1967.

[MT86] R. MNEIMNÉ & F. TESTARD – *Introduction à la théorie des groupes de Lie classiques*, Méthodes, Hermann, 1986.

[Per78] G. PEREC – *Je me souviens*, P.O.L., Hachette, 1978.

[Per95] D. PERRIN – *Géométrie algébrique*, Savoirs actuels, InterÉditions, 1995.

[Per96] ———, *Cours d'algèbre*, Ellipses, 1996.

[Pet80] J.-P. PETIT – *Le géométricon*, Les aventures d'Anselme Lanturlu, Belin, 1980.

[Por69] I. PORTEOUS – *Geometric topology*, Van Nostrand, 1969.

[Ran84] Rand McNally & Company – *The new international atlas*, 1984.

[Rud87] W. RUDIN – *Real and complex analysis*, third ed., McGraw-Hill Book Co., New York, 1987.

[Sam88] P. SAMUEL – *Projective geometry*, Springer-Verlag, New York, 1988, Translated from the French by Silvio Levy, Readings in Mathematics.

[Sau86] P. SAUSER – *Algèbre et géométrie*, Ellipses, 1986.

[Ser73] J.-P. SERRE – *A course in arithmetic*, Springer-Verlag, New York, 1973, Translated from the French, Graduate Texts in Mathematics, No. 7.

[Sid93] J.-C. SIDLER – *Géométrie projective*, InterÉditions, 1993.

[Sil72] R. SILVERMAN – *Introductory complex analysis*, Dover, 1972.

[Sil01] J. SILVESTER – *Geometry, ancient and modern*, Oxford University Press, 2001.

[Spi75] M. SPIVAK – *Differential geometry*, Publish or perish, 1975.

[Ste82] I. STEWART – *Oh! Catastrophe!*, Les chroniques de Rose Polymath, Belin, 1982.

[Tab95] S. TABACHNIKOV – *Billiards*, Panoramas et Synthèses, 1, Société Mathématique de France, 1995.

[Ver73] J. VERNE – *Le pays des fourrures (Fur country)*, Hetzel, 1873.

[Ver74] ———, *L'île mystérieuse (The mysterious island)*, Hetzel, 1874.

[Wey52] H. WEYL – *Symmetry*, Princeton University Press, 1952.

Index

In this index, the *items* are put in alphabetic order of all the words (*e.g.*, "of a projective space" before "of an affine space"). The numbers are page numbers. There are two kinds of *items*, something I explain by two examples:

- *affine mapping:* this is an elementary and basic notion; there are affine mappings everywhere in this text, and the number refers to the page where the notion is defined;
- *simultaneous orthogonalization* of two quadratic forms: this is a result that is useful in many domains of mathematics (a "transversal" notion, in technocratic language), and the numbers refer to all the pages where this notion is used or mentioned.

On the other hand, the theorems that are only mentioned in the text (in general for cultural reasons) show up in this index if they appear in the text with a bibliographical reference.

Universitext

Aksoy, A.; Khamsi, M. A.: Methods in Fixed Point Theory

Alevras, D.; Padberg M. W.: Linear Optimization and Extensions

Andersson, M.: Topics in Complex Analysis

Aoki, M.: State Space Modeling of Time Series

Aupetit, B.: A Primer on Spectral Theory

Bachem, A.; Kern, W.: Linear Programming Duality

Bachmann, G.; Narici, L.; Beckenstein, E.: Fourier and Wavelet Analysis

Badescu, L.: Algebraic Surfaces

Balakrishnan, R.; Ranganathan, K.: A Textbook of Graph Theory

Balser, W.: Formal Power Series and Linear Systems of Meromorphic Ordinary Differential Equations

Bapat, R.B.: Linear Algebra and Linear Models

Benedetti, R.; Petronio, C.: Lectures on Hyperbolic Geometry

Berberian, S. K.: Fundamentals of Real Analysis

Berger, M.: Geometry I, and II

Bliedtner, J.; Hansen, W.: Potential Theory

Blowey, J. F.; Coleman, J. P.; Craig, A. W. (Eds.): Theory and Numerics of Differential Equations

Börger, E.; Grädel, E.; Gurevich, Y.: The Classical Decision Problem

Böttcher, A; Silbermann, B.: Introduction to Large Truncated Toeplitz Matrices

Boltyanski, V.; Martini, H.; Soltan, P. S.: Excursions into Combinatorial Geometry

Boltyanskii, V. G.; Efremovich, V. A.: Intuitive Combinatorial Topology

Booss, B.; Bleecker, D. D.: Topology and Analysis

Borkar, V. S.: Probability Theory

Carleson, L.; Gamelin, T. W.: Complex Dynamics

Cecil, T. E.: Lie Sphere Geometry: With Applications of Submanifolds

Chae, S. B.: Lebesgue Integration

Chandrasekharan, K.: Classical Fourier Transform

Charlap, L. S.: Bieberbach Groups and Flat Manifolds

Chern, S.: Complex Manifolds without Potential Theory

Chorin, A. J.; Marsden, J. E.: Mathematical Introduction to Fluid Mechanics

Cohn, H.: A Classical Invitation to Algebraic Numbers and Class Fields

Curtis, M. L.: Abstract Linear Algebra

Curtis, M. L.: Matrix Groups

Cyganowski, S.; Kloeden, P.; Ombach, J.: From Elementary Probability to Stochastic Differential Equations with MAPLE

Dalen, D. van: Logic and Structure

Das, A.: The Special Theory of Relativity: A Mathematical Exposition

Debarre, O.: Higher-Dimensional Algebraic Geometry

Deitmar, A.: A First Course in Harmonic Analysis

Demazure, M.: Bifurcations and Catastrophes

Devlin, K. J.: Fundamentals of Contemporary Set Theory

DiBenedetto, E.: Degenerate Parabolic Equations

Diener, F.; Diener, M.(Eds.): Nonstandard Analysis in Practice

Dimca, A.: Singularities and Topology of Hypersurfaces

DoCarmo, M. P.: Differential Forms and Applications

Duistermaat, J. J.; Kolk, J. A. C.: Lie Groups

Edwards, R. E.: A Formal Background to Higher Mathematics Ia, and Ib

Edwards, R. E.: A Formal Background to Higher Mathematics IIa, and IIb

Emery, M.: Stochastic Calculus in Manifolds

Endler, O.: Valuation Theory

Erez, B.: Galois Modules in Arithmetic

Everest, G.; Ward, T.: Heights of Polynomials and Entropy in Algebraic Dynamics

Farenick, D. R.: Algebras of Linear Transformations

Foulds, L. R.: Graph Theory Applications

Frauenthal, J. C.: Mathematical Modeling in Epidemiology

Friedman, R.: Algebraic Surfaces and Holomorphic Vector Bundles

Fuks, D. B.; Rokhlin, V. A.: Beginner's Course in Topology

Fuhrmann, P. A.: A Polynomial Approach to Linear Algebra

Gallot, S.; Hulin, D.; Lafontaine, J.: Riemannian Geometry

Gardiner, C. F.: A First Course in Group Theory

Gårding, L.; Tambour, T.: Algebra for Computer Science

Godbillon, C.: Dynamical Systems on Surfaces

Goldblatt, R.: Orthogonality and Spacetime Geometry

Gouvêa, F. Q.: p-Adic Numbers

Gustafson, K. E.; Rao, D. K. M.: Numerical Range. The Field of Values of Linear Operators and Matrices

Hahn, A. J.: Quadratic Algebras, Clifford Algebras, and Arithmetic Witt Groups

Hájek, P.; Havránek, T.: Mechanizing Hypothesis Formation

Heinonen, J.: Lectures on Analysis on Metric Spaces

Hlawka, E.; Schoißengeier, J.; Taschner, R.: Geometric and Analytic Number Theory

Holmgren, R. A.: A First Course in Discrete Dynamical Systems

Howe, R., Tan, E. Ch.: Non-Abelian Harmonic Analysis

Howes, N. R.: Modern Analysis and Topology

Hsieh, P.-F.; Sibuya, Y. (Eds.): Basic Theory of Ordinary Differential Equations

Humi, M., Miller, W.: Second Course in Ordinary Differential Equations for Scientists and Engineers

Hurwitz, A.; Kritikos, N.: Lectures on Number Theory

Iversen, B.: Cohomology of Sheaves

Jacod, J.; Protter, P.: Probability Essentials

Jennings, G. A.: Modern Geometry with Applications

Jones, A.; Morris, S. A.; Pearson, K. R.: Abstract Algebra and Famous Inpossibilities

Jost, J.: Compact Riemann Surfaces

Jost, J.: Postmodern Analysis

Jost, J.: Riemannian Geometry and Geometric Analysis

Kac, V.; Cheung, P.: Quantum Calculus

Kannan, R.; Krueger, C. K.: Advanced Analysis on the Real Line

Kelly, P.; Matthews, G.: The Non-Euclidean Hyperbolic Plane

Kempf, G.: Complex Abelian Varieties and Theta Functions

Kitchens, B. P.: Symbolic Dynamics

Kloeden, P.; Ombach, J.; Cyganowski, S.: From Elementary Probability to Stochastic Differential Equations with MAPLE

Kloeden, P. E.; Platen; E.; Schurz, H.: Numerical Solution of SDE Through Computer Experiments

Kostrikin, A. I.: Introduction to Algebra

Krasnoselskii, M. A.; Pokrovskii, A. V.: Systems with Hysteresis

Luecking, D. H., Rubel, L. A.: Complex Analysis. A Functional Analysis Approach

Ma, Zhi-Ming; Roeckner, M.: Introduction to the Theory of (non-symmetric) Dirichlet Forms

Mac Lane, S.; Moerdijk, I.: Sheaves in Geometry and Logic

Marcus, D. A.: Number Fields

Martinez, A.: An Introduction to Semiclassical and Microlocal Analysis

Matsuki, K.: Introduction to the Mori Program

Mc Carthy, P. J.: Introduction to Arithmetical Functions

Meyer, R. M.: Essential Mathematics for Applied Field

Meyer-Nieberg, P.: Banach Lattices

Mines, R.; Richman, F.; Ruitenburg, W.: A Course in Constructive Algebra

Moise, E. E.: Introductory Problem Courses in Analysis and Topology

Montesinos-Amilibia, J. M.: Classical Tessellations and Three Manifolds

Morris, P.: Introduction to Game Theory

Nikulin, V. V.; Shafarevich, I. R.: Geometries and Groups

Oden, J. J.; Reddy, J. N.: Variational Methods in Theoretical Mechanics

Øksendal, B.: Stochastic Differential Equations

Poizat, B.: A Course in Model Theory

Polster, B.: A Geometrical Picture Book

Porter, J. R.; Woods, R. G.: Extensions and Absolutes of Hausdorff Spaces

Radjavi, H.; Rosenthal, P.: Simultaneous Triangularization

Ramsay, A.; Richtmeyer, R. D.: Introduction to Hyperbolic Geometry

Rees, E. G.: Notes on Geometry

Reisel, R. B.: Elementary Theory of Metric Spaces

Rey, W. J. J.: Introduction to Robust and Quasi-Robust Statistical Methods

Ribenboim, P.: Classical Theory of Algebraic Numbers

Rickart, C. E.: Natural Function Algebras

Rotman, J. J.: Galois Theory

Rubel, L. A.: Entire and Meromorphic Functions

Rybakowski, K. P.: The Homotopy Index and Partial Differential Equations

Sagan, H.: Space-Filling Curves

Samelson, H.: Notes on Lie Algebras

Schiff, J. L.: Normal Families

Sengupta, J. K.: Optimal Decisions under Uncertainty

Séroul, R.: Programming for Mathematicians

Seydel, R.: Tools for Computational Finance

Shapiro, J. H.: Composition Operators and Classical Function Theory

Simonnet, M.: Measures and Probabilities

Smith, K. E.; Kahanpää, L.; Kekäläinen, P.; Traves, W.: An Invitation to Algebraic Geometry

Smith, K. T.: Power Series from a Computational Point of View

Smoryński, C.: Logical Number Theory I. An Introduction

Stichtenoth, H.: Algebraic Function Fields and Codes

Stillwell, J.: Geometry of Surfaces

Stroock, D. W.: An Introduction to the Theory of Large Deviations

Sunder, V. S.: An Invitation to von Neumann Algebras

Tamme, G.: Introduction to Étale Cohomology

Tondeur, P.: Foliations on Riemannian Manifolds

Verhulst, F.: Nonlinear Differential Equations and Dynamical Systems

Wong, M. W.: Weyl Transforms

Zaanen, A.C.: Continuity, Integration and Fourier Theory

Zhang, F.: Matrix Theory

Zong, C.: Sphere Packings

Zong, C.: Strange Phenomena in Convex and Discrete Geometry

Printed in Italy by Legoprint S.p.A., Lavis (Trento)